COTTON DISEASES

Cotton Diseases

Edited by

R.J. HILLOCKS
Department of Agriculture
University of Reading
Earley Gate
PO Box 236
Reading RG6 2AT, UK

C·A·B *International*

C·A·B International   Tel: Wallingford (0491) 32111
Wallingford   Telex: 847964 (COMAGG G)
Oxon OX10 8DE   Telecom Gold/Dialcom: 84: CAU001
UK   Fax: (0491) 33508

A catalogue record for this book is available from the British Library.

ISBN 0 85198 749 4✓

Typeset by Redwood Press Ltd
Printed and bound in the UK by Redwood Press Ltd, Melksham

Contents

Contributors

ALOIS A. BELL: *Research Plant Pathologist, Cotton Pathology Research Unit, USDA ARS, Southern Plains Area, Southern Crops Research Laboratory, Route 5, Box 805, College Station, TX 77840, USA.*

JOHN BRIDGE: *International Institute of Parasitology, 395a Hatfield Road, St Albans, Herts, AL4 0XU, UK.*

JUDITH K. BROWN: *Department of Plant Pathology, College of Agriculture, The University of Arizona, Forbes, Building 36, Tucson, AZ 85721, USA.*

R. J. HILLOCKS: *University of Reading, Department of Agriculture, Earley Gate, PO Box 236, Reading, RG6 2AT, UK.*

S. C. HODGES: *Extension Agronomist, Cooperative Extension Service, University of Georgia, College of Agriculture, PO Box 1209, Tifton, Georgia 31793, USA.*

M. J. JEGER: *Head, Food Crop Commodity Department, Natural Resource Institute, Chatham Maritime, Chatham, Kent, ME4 4TB, UK.*

C. M. KENERLEY: *Department of Plant Pathology & Microbiology, Texas A & M University, College Station, TX 77843-2132, USA.*

Foreword

Cotton is one of the most important and widely-grown cash crops in the world and its role as an income generator, for both sophisticated, high input farmers and low income farmers in developing countries, is well recognized. Its versatility is reflected by its products, which include spinnable fibres for the manufacture of thread, cloth and fabric; packing for bedding products and furniture; short fibres (linters) for the production of rayon, explosives, film and plastic; raw material used in producing paper and cardboard; oil; cake and meal for livestock feed; and fuel. There has been a great deal of fundamental, strategic and applied research performed on cotton, partly as a consequence of the large and profitable market that it has created for manufacturers of chemicals for weed, disease and pest control and for fertilizers. Yet, as emphasized by the editor in his preface to this book, in recent years there has been a dearth of composite publications on cotton diseases; many of these diseases are capable of causing severe crop losses and adversely affecting lint quality.

Dr Hillock's own experience as a cotton pathologist, first in Tanzania then in Zimbabwe, provided him with the knowledge necessary to edit and write a considerable portion of the book. His work on the mechanisms of resistance to Fusarium wilt and the effect of root-knot nematodes on the expression of this disease sharpened his skills as a research plant pathologist and his continuing links with developing countries in Africa have ensured that he has encountered new problems as they arise.

His fellow authors are recognized authorities in specialized areas of research and well qualified to deal with the wide range of diseases covered. At a time when sustainable and environmentally sound crop, pest and nutrient management practices are crucial to the well-being of global agriculture, the inclusion of chapters on nematodes (which are often associated with soil pathogens) and nutrient deficiency disorders is appropriate.

The chapter on viruses should appeal to a wide audience as serological and molecular tools are being used increasingly to unravel some of the mysteries that have for so long bedevilled virology research in cotton. Perhaps some of the technology developed for cotton, a woody perennial that is now usually cultivated as an annual crop, will be applicable to some of the woody species that are currently of such interest in agroforestry.

While chemicals have played, and are likely to continue to play, an important role in disease control, growing public sensitivity to the excessive use of potentially dangerous chemicals to control diseases and the ephemeral nature of effective control by some chemicals, as well as the escalating costs to develop and bring to the market so-called 'safe' chemicals, have served to highlight the attributes of durable, genetic host resistance. The advantages of such resistance, especially to the low income farmer, include no extra seed cost and a reduction in operating costs. Understandably, the increasingly important role of commercial cultivars that are resistant to diseases is emphasized throughout this book. The information generated by research into the biological control of a number of causal agents, and the results that provide a better understanding of pathogen epidemiology and the complex interactions between host, pathogens and the environment, are crucial to the development of efficient, integrated disease control strategies.

N. L. Innes
Scottish Crop Research Institute
Dundee
March 1992

Preface

Descriptive lists of cotton diseases have been published in a number of general textbooks on diseases of field crops and in review articles. Although the *Compendium of Cotton Diseases* published by the American Phytopathological Society is a useful field guide, the aim of this book is to provide a comprehensive review of published work on cotton diseases which reflects the worldwide importance of the crop and many of its diseases. The text also includes practical details, based on the experience of the individual contributors, which it is hoped will be of value to plant pathologists involved in cotton disease research.

Some of the most important cotton diseases are widely distributed. Bacterial blight and Fusarium wilt occur in all the main areas of cotton production throughout the world including China, the USA and the Indian subcontinent. Considerable success has been achieved in breeding for resistance to these diseases. As a result, losses caused by bacterial blight in the USA in recent years have been negligible. However, the appearance in Africa of a new race of the bacterial blight pathogen capable of attacking previously immune lines, reminds us of the dynamic interaction between the genotypes of host and pathogen and that the task of breeding for resistance is never complete.

Changes in agronomic practice and types of cotton grown have, over the years, altered the relative importance of some pathogens in the cotton disease complex. The use of seed treatment chemicals in the USA, for instance, has been widely adopted and has greatly reduced the importance of both seedling disease and boll rot caused by *Colletotrichum* spp. On the other hand, the increased interest in growing Pima and other cultivars derived from *Gossypium barbadense* in the USA, Israel and India has increased the importance of leaf spot disease caused by *Alternaria macrospora*.

I would like to express my appreciation for their support during the conceptual stages of the book to Professor D.L. Hawksworth and Dr J.M. Waller of the International Mycological Institute and to Dr K.R. Anthony, Agricultural Consultant, formerly Senior Agricultural Adviser with the Overseas Development Administration. It would not have been possible to complete the book without access to library facilities at the University of Reading and the Natural Resources Institute. I would like to thank the Overseas Development Administration (UK), and the International Cotton Advisory Committee (Washington, DC) for contributions towards the cost of including the colour plates. Finally, I am grateful to Dr Sarah Simons of Reading University's Department of Agriculture for her proofreading, encouragement and patience in improving my word processing skills.

R.J. Hillocks
Malawi
February 1992

Introduction

Cotton belongs to the genus *Gossypium* of the family Bombacaceae in the order Malvales. Eight sections are recognized within the genus *Gossypium* (Hutchinson *et al.*, 1947), consisting of 30 species of xerophytic annual sub-shrubs, perennial shrubs or small trees. Six of the sections contain the wild lintless diploid cottons, occurring in the arid regions of tropical and sub-tropical Africa, Asia, Australia and America. The cultivated species are contained in the two remaining sections, Herbacea and Hirsuta. Section Herbacea contains two species, *G. herbaceum* and *G. arboreum*, the so-called Old World linted diploids, each having a number of cultivated races. Mutation to the linted type is believed to have occurred in *G. herbaceum* race *africanum*, the wild form of which is found in the dry bush regions of southern Africa. Although races of *G. herbaceum* and *G. arboreum* are to be found growing as 'backyard cottons' in Africa and Asia, the main areas of commercial production are the Indian subcontinent and China.

The vast majority of cotton produced throughout the world is derived from cultivars of two species in the section Hirsuta. *G. hirsutum* (or Upland cotton) and *G. barbadense* (Sea Island or Egyptian cotton), the New World linted cottons, are allopolyploids derived from hybridization between two diploid species. *G. hirsutum* may have arisen from a cross between an Old World linted species similar to *G. herbaceum* and a New World diploid species such as *G. thurburi*. *G. barbadense* could have arisen from a similar cross between *G. herbaceum* and *G. raimondii*. How the two diploid species came together is a matter of some controversy which it would be inappropriate to enter into in the present work (see Purseglove, 1968).

G. barbadense can be distinguished from *G. hirsutum* in the field by having a more open canopy and its leaves are palmate and more deeply divided. The flowers tend to be larger, are yellow in colour compared with the white or cream of *G. hirsutum* and contain a red spot on the inside base of the corolla tube. *G. barbadense* also has a longer and more pointed boll shape, requires a longer period to reach maturity, and therefore usually requires irrigation and warmer growing conditions.

The cultivated cottons have two types of cellulose hairs growing from the epidermis of the seed. The short hairs or seedcoat fuzz is not removed from the seed during the ginning process and cannot be spun. It may be removed from the seed before planting and is sometimes used to produce a high-quality paper. The long hairs or lint is removed from the seed by the cotton gin and this can be spun. The content of the cotton boll which is harvested and delivered to the cotton ginnery consists of seed and fibres, collectively re-ferred to as seed cotton.

The length of the lint fibres is a heritable characteristic which varies both between and within *Gossypium* species. Commercial cottons are classified on the basis of length of the lint fibre, or 'staple length'. Most of the cultivars of *G. herbaceum* and *G. arboreum* produce lint which is under $^{13}/_{16}$ in. and there-fore classified as short staple. Medium staple cotton is derived from cultivars of *G. hirsutum*. Having a staple length of $^7/_8$–$^{31}/_{38}$ in., it accounts for about 20% of world production. Medium-long staple cotton has a fibre length of 1–$1^3/_{32}$ in., is derived from *G. hirsutum* and accounts for about 60% of world production. Long staple cotton is $1^1/_8$–$1^5/_{16}$ in. in length, is derived mainly from *G. barbadense*, but also from some long staple cultivars of *G. hirsutum*, and accounts for about 10% of world production. Extra long staple, having a fibre length of over $1^3/_8$ in., is derived exclusively from *G. barbadense* and is grown mainly in Egypt and Sudan.

Although the centre of origin of the main cultivated cottons is tropical America, the crop is now grown in numerous countries scattered over five continents. The top five producers in 1990/91 were China, USA, the former Soviet Union, India and Pakistan. Production in Australia has increased greatly over the last 10–15 years and it is now the tenth largest producer in the world. Egypt is the only country on the African continent among the top 10 producers. The other main producers in Africa are Sudan, Mali, Ivory Coast and Zimbabwe, but the crop is an important source of income to smallholders in a number of other African countries, notably Tanzania and Mozambique.

Cotton is cultivated under a wide range of farming systems. The crop may be grown on large plantations with a highly mechanized production system, often requiring heavy use of fertilizer and pesticides. However, cotton can also be an excellent cash crop in small-scale farming systems, where its tolerance of poor soil conditions and drought, together with its indeterminate growth habit, allows it to produce some yield under conditions when most other crops would fail.

References

HUTCHINSON, J.B., SILOW, R.A. AND STEPHENS, S.G. (1947) *The Evolution of Gossy-pium*. Oxford University Press, London.

PURSEGLOVE, J.W. (1968) *Tropical Crops Dicotyledons*, Vol. 2. Longman, London.

1 Seedling Diseases

R.J. Hillocks

Introduction

Seedling disease in cotton is a worldwide problem, often causing serious stand loss where it is not controlled. In the USA, it is a major source of production loss, estimated at 2.8% per year over the 30-year period from 1952 to 1981 (Halloin, 1983) and 3.7% for 1989/90 (Blasingame, 1990).

Atkinson (1892) was the first to describe cotton seedling disease; then Hansford (1929) in Uganda and Miller and Weindling (1939) in the USA began the work of identifying the pathogens involved.

A number of soil- and seed-borne micro-organisms can infect cotton seedlings individually or in association as a disease complex. More than 40 fungi have been isolated from diseased cotton seedlings, although only some of these have been shown to be primary pathogens (Table 1.1). Most of the primary pathogens are fungi with wide host ranges but bacteria and viruses are sometimes involved. Parasitic soil-inhabiting nematodes may also be associated with these fungi and seedling damage is often increased when the crop is attacked by both groups of organisms (e.g. Brodie and Cooper, 1964; Cauquil and Shepherd, 1970).

The disease syndrome encompassed by the term seedling disease includes any host–pathogen interaction which debilitates or kills the plant between planting and about 1 month after emergence. The symptoms may be divided for convenience into four groups depending on the stage of development when damage occurs and which part of the plant is affected.

1. The first stage where damage can occur is immediately after planting, resulting in seed decay.
2. Under sub-optimal conditions for plant growth, the newly emerging roots and shoot are vulnerable to attack, resulting in pre-emergence damping-off.

1

Table 1.1. Micro-organisms identified as primary seedling pathogens.

Organism	References	Optimal temp. for disease (°C)
FUNGI		
Alternaria alternata	Maier, 1965	25–30
Alternaria macrospora	Bashan and Levanony, 1987	25–30
Alternaria solani	Maier, 1965	25–30
Ascochyta gossypii	Smith, 1950	25–30
Colletotrichum gossypii	Bollenbacher and Fulton, 1967	22–25
Fusarium moniliforme	Fulton and Bollenbacher, 1959	20–25
Fusarium oxysporum	Colyer, 1988	20–25
Fusarium solani	Batson and Trevathan, 1988	21
Macrophomina phaseolina	Dwivedi and Chaube, 1985	30–35
Pythium aphanidermatum	Devey et al., 1982	20–25
Pythium ultimum	Ayers and Lumsden, 1975	18–20
Rhizoctonia solani	Brown and McCarter, 1976	24–32
Sclerotium rolfsii	Gottlieb and Brown, 1941	25–30
Thielaviopsis basicola	Mauk and Hine, 1988	16–20
BACTERIA		
Xanthomonas campestris	Logan, 1960	26–28
VIRUS		
Leaf crumple virus	Brown et al., 1987	
NEMATODES		
Meloidogyne incognita	Carter, 1981	30
Rotylenchulus reniformis	Brodie and Cooper, 1964	30

3. The seedling remains vulnerable to attack by soil-borne fungi after emergence, especially when growth is slow, and infection of the hypocotyl at this stage is referred to as post-emergence damping-off.
4. Finally, symptoms may become apparent on the cotyledons and first leaves, as a distinct spotting or a more generalized blight.

Seed Decay

CONDITIONS FAVOURING INFECTION

Seed deterioration occurs due to exposure to adverse weather conditions before harvest or to poor storage conditions, both of which might lead to contamination of the seed by micro-organisms. Provided the seed has been stored under dry conditions, moist conditions which favour the growth of seed rot pathogens will not occur until planting. The micro-organisms involved contaminate the seed prior to harvest, or invade the seed from the surrounding soil, after planting. If the seed is not heavily contaminated before planting, then significant seed decay normally occurs in wet, cool conditions when germination is slow.

CAUSAL ORGANISMS

The fungi most often associated with deteriorated seed are *Fusarium* spp. (Pratt, 1926; Chester, 1938; Klitch, 1986), *Colletotrichum gossypii* (Arndt, 1956), *Rhizopus* spp. (Davis *et al.*, 1981) and *Pythium* spp. (Devay *et al.*, 1982).

PREVENTATIVE MEASURES

Field deterioration can be prevented by ensuring that seed cotton does not remain on the plant for long periods after the boll has split. It may, in some cases, be possible to adjust the planting date so that there is less risk of the bolls opening during wet weather. Contamination of the seed after harvest can be minimized by correct storage procedures. Cotton seed can be stored for several years without loss of viability in a well-aerated, dry atmosphere at 20–25 °C. Acid delinting of seed prior to storage will remove seedcoat fuzz and adhering trash, which might harbour microbial contaminants. Acid delinting can be conveniently combined with gravity grading; the heavier seeds have a lower incidence of internal infection and give a higher germination percentage (Chester, 1938).

Low-temperature germination is a good indicator of field performance. Seeds which germinate at 18 °C are usually free of infection and are also more able to withstand attack by soil micro-organisms (Halloin and Bourland, 1981). As part of the multi-adversity resistance (MAR) breeding programme in the USA, lines have been selected which do not support mould growth at 13.3 °C. Such lines appear to be more resistant to seed decay (Bird, 1978).

Damping-off

Pre-emergence damping-off occurs when the emerging radicle or plumule is attacked so that the seedling fails to emerge above soil level. In post-emergence damping-off, the hypocotyl is attacked, causing a cortical rot which may kill or weaken the seedling. Both symptoms are caused by a similar range of pathogens, but *Pythium* spp. are probably the most important pre-emergence pathogens, and *Rhizoctonia solani* is the most common cause of post-emergence damping-off throughout the world.

RHIZOCTONIA SOLANI

Symptoms and losses

Rhizoctonia solani may be responsible for seed decay and pre-emergence damping-off when the seed is planted into cool soil (Fulton *et al.*, 1956). Once the seedling emerges, the young tissue of the hypocotyl can be subject to attack, causing post-emergence damping-off. The fungus invades the plant at soil level, producing a sunken lesion, due to cortical decay, which girdles the hypocotyl, causing the seedling to collapse. In wet conditions, fungal mycelium may sometimes be seen growing from the lesion, which can extend upwards several centimetres from soil level. As the plant matures, the stem becomes increasingly resistant to infection. However, if conditions remain favourable for disease development, the period of susceptibility may be prolonged. Older plants are rarely killed by the disease but often survive in a weakened condition, bearing the mark of the stem-girdling lesion at the base of the stem (Fig. 1.1). This condition has been known as 'soreshin' in the USA since the turn of the century (Atkinson, 1892; Neal, 1942).

Morphology and taxonomy

The genus *Rhizoctonia* contains nearly 100 species of non-sporulating fungi, many of which are root pathogens. The pathogen is characterized morphologically by brown-pigmented vegetative hyphae which are 5–13 μm in width and a cell length of over 100 μm. The mycelium grows rapidly, branching almost at right angles to the hyphal cell and close to the distal septum. The branch hyphae typically exhibit a constriction at the point of branching (Fig. 1.2). The fungus also produces barrel-shaped moniloid cells in chains or groups, which are more than twice as wide as the vegetative hyphae but only 20–40 μm in length. They may be loosely aggregated or compacted into sclerotia. The sclerotia are dark-pigmented with wide variations in size (Parmeter and Whitney, 1970).

The perfect state is a basidiomycete, *Thanatephorus cucumeris* (Frank) Donk., which has occasionally been found on infected cotton seedlings (Ullstrup, 1939). The basidia are relatively short and barrel-shaped with smooth thin-walled, prominently apiculate spores, which are oblong in shape and unilaterally flattened (Talbot, 1971).

Variation

Variations in cultural, physiological or pathogenic characteristics among isolates conforming to the general morphological description of *R. solani* have been observed by a number of workers, who have attempted to use these differences as a basis for grouping isolates (e.g. Exner, 1953). Richter and Schnider (1953) showed that the capacity for hyphal anastomosis provided an indication of relationships within groups of isolates. Parmeter *et al.* (1969) examined 200 isolates, of which 138 conformed to *R. solani* on the basis of the then current species concept. Among these were isolates which fell into one of four groups previously described by Richter and Schnider (1953). Anastomosis occurred between isolates from the same group, which therefore appeared to be genetically related. Some pathogenic specialization was associated with these groups. Anastomosis groups (AGs) 1–4 had a wide host range but AG 2 showed some specialization towards the crucifers, and isolates in AG 3 came predominantly from potato.

Nutritional requirements

The fungus does not have any specialized nutrient requirements, growing well on a wide range of media. The nature of the media does, however, have a considerable effect on the form of growth. For instance, on water agar hyphal cells tend to be long and narrow but they are shorter and wider on potato dextrose agar (PDA). On most media, vegetative hyphae grow embedded in the medium with little aerial growth. However, more aerial growth is produced on PDA than on less nutrient-rich media such as carrot agar (Maier and Staffeldt, 1960).

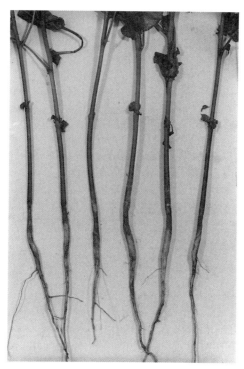

Fig. 1.1. Soreshin of cotton caused by *Rhizoctonia solani*.

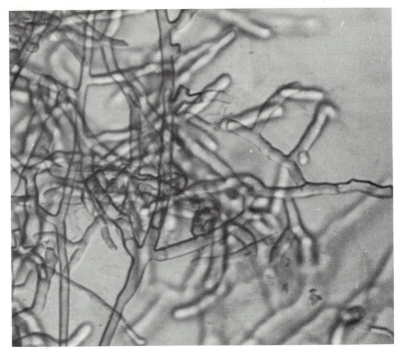

Fig. 1.2. Hyphae of *Rhizoctonia solani* growing in culture on PDA showing branching structure (× 400).

Generalizations about the ability of *R. solani* to utilize particular sources of nitrogen and carbon are difficult to make as there is considerable variation between isolates. Most sugars, and, to a lesser extent, alcohols, with the exception of ribose and rhamnose, support good growth. Many isolates decompose cellulose, particularly in the absence of soluble carbohydrate (Kohlmeyer, 1956; Bateman, 1964). Common nitrogen compounds can be utilized, although compounds, such as ammonium nitrate, which promote acid production may limit growth. Some isolates are able to tolerate pH as low as 2.6 and as high as 11.0 but growth is most rapid in the range 5.0 to 7.0. A more detailed review of the physiology of the fungus is given by Sherwood (1970).

Survival and growth in the soil

R. solani is a common component of the soil microflora in many agricultural soils. The inoculum consists of sclerotia and basidiospores, in addition to hyphae, which can grow rapidly through the soil under favourable conditions. The fungus may survive and remain pathogenic in soil for extended periods as a saprophyte on organic debris, or on the roots of numerous crop and weed species (Daniels, 1963; Oshima *et al.*, 1963).

Using the Rossi–Cholodney soil-plate method, Blair (1943) was able to show that *R. solani* could grow in unsterilized, unamended soil. However, growth through non-sterile soil appears to be dependent on the presence of a food base in the form of culture medium, associated with the initial inoculum in Blair's experiments, or infested organic debris under natural conditions (Boyle, 1956). In sterile soil, growth is less dependent on the initial food base as antagonism by other soil micro-organisms is eliminated.

Saprophytic colonization of crop residues is important in the epidemiology of the fungus. Boyle (1956) was unable to obtain infection of groundnut (*Arachis hypogaea*) in a sandy soil from which the organic matter had been removed. The fungus remained pathogenic if stems and leaves of groundnut, maize (*Zea mays*) or cotton were incorporated into the upper layer of soil.

Results from comparative studies on the competitive saprophytic ability (CSA) of the fungus depend greatly on the conditions under which the studies were carried out. Of particular importance is the degree of competition in the soil and the carbon : nitrogen ratio of the substrates. The fungus may exhibit a high degree of saprophytic ability when the population of competitors is low, and a high or low CSA depending on the C : N ratio (Davey and Papavizas, 1963). CSA tends to decrease with increasing carbon levels and to increase with increasing nitrogen (Martinson, 1963).

Isolation and maintenance in culture

Water agar is suitable for initial isolation of *R. solani* from infected host tissue. The addition of lactic acid and/or antibiotics is often helpful in reducing the growth of bacterial contaminants. Streptomycin sulphate at approximately 100 µg ml^{-1} is the antibiotic most often used. Isolation from infested soil can be done by the dilution-plate method, in which a soil suspension in sterile water is spread thinly over a selective medium. Various methods have been described for direct isolation from soil. For instance, the immersion-plate method was first described by Chesters (1940) and modified by Wood and Wilcoxson (1960). A Petri dish of a suitable medium covered by a perforated sterile disc is pressed against the soil profile and held in place. Two days later the dish is recovered from the soil, the disc is removed and fungal colonies are subcultured on to fresh plates. The immersion-tube method is similar but a test-tube with small holes in the sides replaces the Petri dish (Chesters, 1940). Another technique involves the use of baits to encourage colonization by the pathogen – an organic substrate which can be placed in the soil and removed after a suitable period, when colonizing fungi can be isolated on an agar medium. Papavizas and Davey (1959) used segments of buckwheat stem as bait, although better results have subsequently been obtained with stem segments of bean and cotton plants (Sneh *et al.*, 1966).

PDA is the preferred medium for growing *Rhizoctonia* after initial isolation, particularly with cotton isolates (Maier and Staffeldt, 1960). The optimum temperature range for mycelial growth *in vitro* is 25–30 °C.

Many storage methods are described in the literature and reports vary considerably concerning the length of time isolates remain viable and pathogenic on PDA plates or slants. Pathogenicity may be retained for 6–12 months on sealed agar plates stored in the refrigerator. The length of time for which agar cultures remain viable can be increased by storage under a layer of mineral oil. Cotton isolates can be recovered after 12 months or more if the fungus is cultured on sterile grain which is then dried in a desiccator and stored at 5 °C (R.J. Hillocks, unpublished). Kreitlow (1950) recommends grinding the grain to a powder after drying and was able to recover the fungus after 16 months of storage.

Preparation of inoculum

For experiments conducted in pots, relatively small quantities of inoculum are required and macerated agar cultures may simply be mixed with the upper layers of soil before planting (e.g. Lehman, 1940; Sinclair, 1957). For field experiments, larger quantities of inoculum are required. Owen and Gay (1964) obtained high levels of infection after mixing infected cotton seed and seedlings into the soil of a trial area, three weeks before planting. A mixture of sand and maize meal provides a good medium for bulk culturing. This method was used in Zimbabwe for field evaluation of fungicides for control of seedling disease caused by *R. solani* (Hillocks and Chinodya, 1988). The cultures were mixed with field soil in drums which were carried to the field and the soil was used to cover the seed at planting. As an alternative to moving large amounts of soil to the trial area, Cole and Cavill (1977) used a method in which cultures on PDA were dried and macerated and the powder was added to field plots.

Host–parasite relations

Carbohydrate and amino acids, exuded into the rhizosphere by roots and seeds of most plants, stimulate spore germination and hyphal growth of many soil-borne fungal parasites (Shroth and Hildebrand, 1964; Rovira, 1969). A number of factors often associated with increased incidence of seedling disease are also known to increase root exudation, for example, low temperature and low light intensity (Rovira, 1956). Exudation of amino acids from cotton roots is increased by a period of exposure to low temperature, which has been linked to increased susceptibility to root rot caused by *R. solani* (Hayman, 1969; Shao and Christiansen, 1982).

Although there are differences between isolates with respect to the mode of penetration into the host (e.g. Dodman *et al.*, 1968a, b), direct penetration through the cuticle and epidermis is common in a number of crops, including cotton (Khadga *et al.*, 1963). Hyphae grow over the surface of the root or hypocotyl following the junction-lines of the cell walls. A dome-shaped infection cushion is then formed, either by hyphae curling back on themselves and repeatedly branching, becoming short and irregularly swollen, or, as seems to be the case with cotton, by repeated terminal branching of a single hypha. This tangle of tightly compacted hyphae forming the infection cushion varies in size from 300 to 500 μm in diameter, rising 120–170 μm above the host surface (Khadga *et al.*, 1963). Little is known about the stimulus to the formation of infection cushions, or exactly how they are attached to the host surface. There appears to be some host specificity in cushion formation. Flentje (1957) has reported that an isolate from crucifers formed cushions on epidermal tissues from radish stems but not on similar tissue from non-cruciferous plants.

Infection pegs penetrating through the cuticle from infection cushions have been observed in cotton plants, where only a single peg arose from each cushion (Nakayama, 1940). The cotton plant is also readily invaded by *R. solani* through ruptures in the epidermis, where lateral roots emerge (Khadga *et al.*, 1963).

Penetration by an infection peg may be effected either by mechanical pressure or with the aid of enzymes. Studies of infection by *R. solani* on Egyptian cotton suggested that penetration is by mechanical pressure (Risk *et al.*, 1984b). Once the epidermis is penetrated, the fungus grows intracellularly, with little evidence of host injury in advance of the pathogen (Khadga *et al.*, 1963). Enzymatic degradation following penetration leads to tissue discoloration and collapse. The resulting shrinkage of invaded cells is reflected in the sunken nature of the brown-coloured lesion at the base of the hypocotyl, typical of cotton seedlings suffering post-emergence damping-off.

Cotton plants become more resistant to penetration by the fungus as they become older (see Fig. 1.3) due to increased cuticle thickness, lignification and reduction in root exudation (Hunter and Guinn, 1968). If conditions become favourable to rapid seedling growth soon after infection, a suberized cell layer

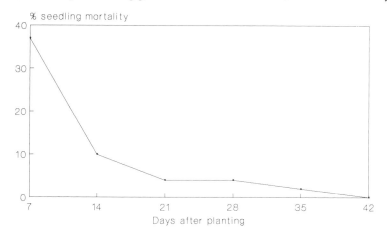

Fig. 1.3. Weekly counts of seedlings with post-emergence damping-off caused by *Rhizoctonia solani*, expressed as a percentage of stand counts in non-inoculated control plots. Cotton Research Institute, Kadoma, Zimbabwe.

is produced below the invaded outer tissues which limits pathogen growth, allowing the cotton plant to recover (Fahmy, 1931).

Factors affecting infection

The optimum temperature for infection of cotton seedlings cannot be stated independently of other environmental factors, inoculum concentration and variation between isolates. However, Jones *et al.* (1926) found that the incidence of soreshin was greatest at temperatures considerably below the optimum for growth of the pathogen. Leach (1947) reported that disease incidence with several damping-off pathogens was inversely related to the ratio between the coefficient of velocity of seedling emergence and the growth rate of the fungus. The view that, at high inoculum concentrations, severe damping-off occurs over a wide temperature range has been confirmed by work conducted in Zimbabwe (Hillocks and Chinodya, 1988). Indeed, soreshin has been reported as both a low-temperature disease and a high-temperature disease of cotton, depending on which strain of the pathogen is involved. Hunter and Staffeldt (1960) demonstrated the presence of strains causing maximum disease at different temperatures. A mildly virulent strain caused most damage at 24 °C, a moderately virulent strain was most damaging at 32 °C and a virulent strain was highly pathogenic over a wide temperature range from 24 °C to 32 °C.

R. solani grows optimally at 100% relative humidity, but infection can take place over a wide range of moisture levels. The optimum for development of damping-off in a number of crops has been variously quoted at 20–80% of field capacity (Baker and Martinson, 1970). Saturation levels reduce pathogenicity due to poor aeration, which reduces the growth rate (Blair, 1943) and CSA (Papavizas and Davey, 1961) of the fungus. However, excessive soil moisture often predisposes cotton seedlings to infection by reducing their rate of growth.

Several herbicides, mostly in the dinitroaniline group, have an effect on the incidence of infection by *Rhizoctonia*. Growth of most soil-borne fungi is inhibited in culture by these herbicides, and yet there are reports that trifluralin increases infection by *R. solani* (Neubauer and Avizohar-Hershenson, 1973; Khalifa *et al.*, 1987). In contrast to these results, experiments reported by Grinstein *et al.* (1976) showed that the dinitroanilines reduced infection by *R. solani, Verticillium dahliae* and *Fusarium* spp. in a number of crops. In the case of cotton, host resistance increased at $1 \mu g \, g^{-1}$ of soil, but, in some experiments, there was a decrease in resistance at higher application rates. El-Khadem *et al.* (1979) also found that at the recommended rates trifluralin, dinitramine and fluometuron greatly reduced the incidence of damping-off due to increased host resistance and reduced saprophytic activity of the fungus. The herbicides EPTC and linuron are also reported to reduce the incidence of damping-off caused by *R. solani*, by suppressing saprophytic activity of the pathogen (El-Khadem and Papavizas, 1984).

Control

R. solani is found at varying inoculum levels in most agricultural soils, but steps can be taken to avoid conditions leading to an increase in inoculum or to increased host susceptibility. In parts of the USA, low temperature early in the growing season is the main reason for stand loss due to seedling disease, but losses can be greatly reduced by delayed planting (Fulton *et al.*, 1956).

Inoculum increase can be minimized by rotation with crops which are poor hosts for the pathogen. In California, infection of cotton by *Rhizoctonia* is more severe in crops which follow lucerne or sugarbeet than with continuous cotton. Leach and Garber (1970) found that disease incidence was lowest in cotton planted after a crop of wheat, but very high in cotton following a cowpea green manure.

Conflicting results have been obtained concerning whether the addition of organic matter to the soil increases or decreases losses to damping-off. The critical factor is the effect of the residues on the carbon : nitrogen ratio of the soil (Davey and Papavizas, 1963). The degree of parasitism shown by *Rhizoctonia* is directly related to the degree of saprophytic activity (Martinson, 1963), and CSA decreases as the carbon content is increased. Addition of organic matter to the soil can be used, therefore, to reduce the incidence of damping-off, provided the C : N ratio is high.

Any measure which encourages rapid emergence and growth of the seedling will help to reduce disease incidence – for example, taking care not to plant the seed too deep and giving attention to land levelling to remove areas prone to waterlogging. If a pre-emergence herbicide is used, it is important to apply the correct dose as some compounds can predispose cotton to infection (El-Khadem and Papavizas, 1984).

Seedling diseases caused by *R. solani* are readily controlled by fungicides and this is the main control method used. PCNB (pentachloronitrobenzene) was found to be highly effective against *Rhizoctonia* infection (Brinkerhoff *et al.*, 1954) and was widely adopted for use on cotton in the USA. The activity of PCNB is, however, highly specific to *Rhizoctonia* and, to a certain extent, *Sclerotium*, and its use may lead to an increase in the incidence of infection by *Pythium* spp. (Gibson *et al.*, 1961) unless combined with etridiazole. Another disadvantage of the compound is that it is less toxic to some isolates of the pathogen than to others and prolonged use may therefore lead to a build-up of PCNB-tolerant strains (Shatla and Sinclair, 1963). PCNB has recently become unavailable in some countries because it contains small quantities of BHC, considered to be a hazardous compound.

One of the earliest systemic fungicides used for seedling disease control in cotton was chloroneb, which was effective against *Rhizoctonia* but had to be combined with mercurial compounds to be effective in the presence of phycomycete pathogens. Where it is still used, chloroneb may be combined with TCMTB or captafol to increase its spectrum of control. Since the early 1970s, mercurial seed-dressings have been withdrawn because of their toxicity to

Table 1.2. Some fungicides used against cotton seedling pathogens.

Common name	Activity[a]	Target[b]	Reference
Benodanil	(S)	R,F,C	Cole and Cavill, 1977
Captafol	(P)	B	Minton, 1984
Captan	(P)	B	Minton and Green, 1980
Carbendazim	(S)	B	Chauhan, 1987
Carboxin	(S)	R,C,S	Borum and Sinclair, 1968
Chloroneb	(S)	R,Py	Darag and Sinclair, 1969
Etridiazole	(P)	Ph,Py	Papavizas et al., 1977
Femaminosulf	(P)	Py	Garber et al., 1979
Fenfuram	(P)	R	Hillocks et al., 1988
Furmecyclox	(S)	R	Papavizas et al., 1979
Hexachlorophene	(P)	B	Minton, 1984
Imazalil	(S)	Al,As,F,T	Minton, 1984
Iprodione	(P)	Al,C,F,R,S	Minton, 1984
Mancozeb	(P)	B	Minton, 1984
Metalaxyl	(S)	Py	Nelson, 1988
PCNB	(P)	R,S	Brinkerhoff et al., 1954
Pencycuron	(P)	R	Yamada, 1986
TCMTB	(P)	B,R	Minton, 1984
Thiabendazole	(S)	B	Hillocks, unpublished
Thiram	(P)	B	Minton, 1984
Tolclofos-methyl	(P)	R	Hillocks et al., 1988

[a] Activity: (P) protective; (S) systemic.
[b] Target organisms: Al, *Alternaria*; As, *Aspergillus*; B, broad spectrum;
C, *Colletotrichum*; F, *Fusarium*; Ph, *Phytophthora*; Py, *Pythium*; R, *Rhizoctonia*;
S, *Sclerotium*; T, *Thielaviopsis*.

mammals, and a range of new compounds have become available which are active against *Rhizoctonia* and other members of the seedling disease complex (Table 1.2). Among those which show some systemic activity, Vitavax is probably the most widely used as a cotton seed treatment, especially in the USA (Davis *et al.*, 1981). More recently, compounds such as tolclofos-methyl, pencycuron and the systemic compound furmecyclox have become available which are highly active against the fungus, the last at very low dosage rates. Carbendazim was found to be effective in controlling a seedling disease complex involving *Rhizoctonia*, *Fusarium* and *Macrophomina*.

Chemicals for seedling disease control may be applied to the seed before planting or to the furrow at planting. In the drier areas of the USA, where the crop is grown under irrigation, fungicide applied to the seed gives good disease control, but, in higher-rainfall areas, it is necessary to apply the chemical to all soil in contact with the emerging seedling for effective control (Bell and Owen, 1963). Seed-dressing has the advantage of low cost, only 200 to 600 g of product being required to treat 100 kg of delinted seed. Also, the compound can be applied using simple methods such as shaking it with the seed in a drum. Sometimes the fungicide is added to the seed as a slurry, in which case, the seed must be dried to facilitate planting. Where mechanical planting is used, the fungicide can be added to the seed as it leaves the hopper. In areas which normally experience a high incidence of seedling disease, seed-dressing may not give adequate control, particularly of post-emergence damping-off, and an in-furrow treatment must be used.

For in-furrow treatment, the compound is usually applied at the rate of 2–4 kg ha^{-1}, sprayed on to the seed and covering soil as the seed leaves the hopper, or applied by knapsack sprayer in the case of hand-planting. In most cases where hand-planting is the norm, seed-dressing should give adequate control of seedling disease.

Weindling (1932) was the first to report that the pathogenicity of *Rhizoctonia* in soil could be reduced by *Trichoderma lignorum* and that this was related to the production of a toxin. Despite the half-century which has passed and the considerable research work conducted since that first report, it has proved difficult to convert experimental results into a practical method of control in the field. However, a recent study has shown that a preparation of *T. harzianum* on wheatbran, applied to the soil, can reduce disease incidence in cotton caused by *R. solani* (Elad *et al.*, 1980). Effective disease control was later obtained with a seed-dressing preparation of the same organism (Elad *et al.*, 1982). Pelleting with one isolate of *T. harzianum* and three of *T. viride* increased seed germination rate and reduced the incidence of damping-off. *T. harzianum* was as effective in controlling *R. solani* as was quintozene at 5 g kg^{-1} seed (Alagarsamy *et al.*, 1987). A reduction in the incidence of damping-off in cotton has also been obtained with preparations of *Gliocladium virens* (Howell, 1982) and *Pseudomonas fluorescens* (Howell and Stipanovic, 1980).

Because of the non-specialized nature of the pathogenicity of *R. solani*, it is not surprising that little success has been achieved in producing resistant varieties. There have been few reports in the literature since Luthra and Vasudeva (1941) reported their failure to select cotton for resistance to the fungus. Poswal *et al.* (1986) reported work on the inheritance of resistance to *R. solani* in Upland cotton, concluding that resistance was controlled by a complex of minor genes. In Egypt, some success has been achieved in selecting resistant individuals from an inoculated population, but this work was conducted with *G. barbadense* (Risk *et al.*, 1984a,b).

PYTHIUM SPP.

Symptoms and Pythium *species*

Juvenile or succulent host tissue is very susceptible to infection by *Pythium* spp., which commonly infect the seed and radical, causing seed rot and pre-emergence damping-off. The seedling hypocotyl can also be affected at

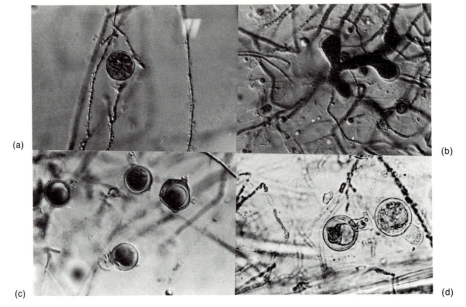

Fig. 1.4. (a) *Pythium ultimum* oogonium containing oospores with sessile antheridium. (b) *P. ultimum* sporangium. (c) *P. aphanidermatum* oogonium containing oospore with intercalary antheridium. (d) *P. aphanidermatum* sporangium. (× 650.) (Courtesy Dr G. Hall, IMI.)

soil level, causing post-emergence damping-off, which is very similar in appearance to that caused by *R. solani*. At a later stage, when the stem and main roots have developed secondary wall thickening, infection is restricted to feeder roots, which may cause stunting and chlorosis (Hendrix and Campbell, 1976).

Several species of *Pythium* have been isolated from diseased cotton seedlings. *P. ultimum* is the one most often isolated in the USA (e.g. Arndt, 1943; Ayers and Lumsden, 1975; Howell and Stipanovic, 1980). In the San Joachim Valley of California, *P. aphanidermatum* is also a frequent isolate (Devay *et al.*, 1982). *P. irregulare* and *P. sylvaticum* are also found, but much less frequently and they are less pathogenic than the other two species (Johnson and Chambers, 1973). In Egypt, the species associated with seedling damage were *P. ultimum*, *P. butleri*, *P. heterothallium* and *P. intermedium* (Moubasher *et al.*, 1984).

Morphology

Most *Pythium* species grow rapidly on culture media. *P. aphanidermatum* can cover a 9 cm agar plate in 24 h at 26 °C (Waterhouse and Waterson, 1964). The mycelium is white, well developed and woolly with hyphae 4–7 μm in diameter. In addition to sporangia, oogonia and antheridia are also produced in culture and within host tissues. There is a wide variation between the species in optimal temperature for growth in culture but the main feature used to distinguish them is size and shape of the sporangium (Fig. 1.4).

Losses

In areas where the growing season is short and cotton must be planted at soil temperatures below 23 °C, stand loss to *P. ultimum* can be great enough to reduce yields. Devay *et al.* (1982) reported that survival of seedlings in California ranged from 90% down to 22% with inoculum concentrations in the range 0–217 propagules g^{-1} of soil, 50% mortality occurring at 100 propagules g^{-1} of soil.

Aetiology

Pythium spp. survive in the soil in the absence of a host by saprophytic growth and the production of resistant resting structures. They are not vigorous competitors and have a restricted saprophytic ability, competing poorly for food bases which are already colonized (Barton, 1958, 1961). However, *P. ultimum* is well adapted as a primary colonizer of fresh organic residues, such as recently fallen cotton leaves (Hancock, 1981). Soil moisture is important to saprophytic growth of the fungus, which is tolerant of high moisture content (Griffin, 1963).

The pathogen can survive for short periods as zoospores and sporangia but also produces oospores which survive much longer periods in the absence of a host. Germination of resting structures is stimulated by root-exudates (Barton, 1957; Chang-Ho, 1970). Rapid germ-tube elongation allows the fungus to avoid antagonism by other soil organisms in the early stages of pathogenesis (Hancock, 1977). *Pythium* spp. can infect cotton radicles by a germ-tube from a single zoospore within two hours (Spencer and Cooper, 1967). The fungus is able to reproduce in plant tissues, producing both sporangia and oospores. Towards the end of the growing season, there is an initial increase in the population of *P. ultimum* on leaf litter accumulated on the soil surface, followed two to three months later by a rapid decline which may be assisted by parasitic micro-organisms (Hancock, 1981).

Factors affecting infection

In the USA, *Pythium* is most damaging to cotton seedlings at low temperatures and high soil moisture content. Working with *P. ultimum*, Arndt (1943) found that, at 18 and 20 °C and 60% soil moisture content, all inoculated plants were killed. However, disease incidence was low at 30 °C. High temperatures also increase the rate of population decline in the absence of fresh host material. The population was found to increase most rapidly on fresh cotton leaves at 15–20 °C but increase was slow at 9–11 °C and declined at temperatures above 32 °C (Hancock, 1977).

The effect of temperature on disease incidence may vary according to which of the species are involved. Klisiwicz (1968), for instance, found that, for damping-off of sunflower, more damage was caused at low temperature by *P. ultimum* and *P. irregulare*, but that *P. aphanidermatum* was more damaging at higher temperatures. Generally it is the latter species which is the more important pathogen in warmer areas and it is associated with damping-off of cotton in West Africa (Ebbels, 1976).

A similar relationship exists between disease incidence and soil moisture levels. In those species which produce zoospores from germinating resting structures, high soil moisture is required for infection. However, soil moisture level is less critical in species which produce germ-tubes (Biesbrock and Hendrix, 1970).

Environmental variables affect the relative importance of members of the seedling disease complex. In general, *Pythium* spp. are more important in areas experiencing cool, wet weather at planting and *Rhizoctonia* is more important in warmer areas. Sometimes either pathogen may become a problem in the same area but which predominates depends on the temperature at planting (Johnson and Chambers, 1973).

Other soil factors, such as its physical and chemical properties or the nature of its microflora, also exert an influence on the disease complex. In Tennessee, for instance, *Pythium* is the most important member of the seedling disease complex on clay soils but is less important on sandy soils (Johnson and Doyle, 1986).

Control

Once established in the soil, resting structures of *Pythium* spp. are almost impossible to eliminate, but the cotton seedling can be protected from infection by seed dressings and in-furrow application of suitable fungicides (see Table 1.2 above), such as metalaxyl.

Biological control with soil micro-organisms may have some potential for the control of *Pythium* diseases in cotton. Howell and Stipanovic (1980) reported that the treatment of cotton seed with the bacterium *Pseudomonas fluorescens* or with its antibiotic, pyoluteorin, increased seedling survival when seed was planted into pots containing soil infested with *P. ultimum*. The decrease in seed colonization and improvement in seedling emergence obtained with the bacterium can be as good as that achieved with metalaxyl (Loper, 1988).

Gliocladium virens also controls damping-off caused by *P. ultimum* by antibiosis (Howell, 1982; Howell and Stipanovic, 1983). Control of seed rot and damping-off by *Pythium* has been obtained with seed treatment preparations of *Enterobacter cloacae* and *Erwinia herbicola*. These bacteria were able to suppress seed colonization during the first 24 hours of germination. The level of control was equal to that of metalaxyl at 25 °C but less effective than the fungicide at 15 °C (Nelson, 1988).

Although it has proved difficult to detect variability in resistance to *Pythium* damping-off among cotton cultivars (Mathre and Otts, 1967), certain cold-tolerant Yugoslavian lines have shown a degree of resistance to *P. ultimum* (Fulton *et al.*, 1962). A small but significant difference in susceptibility between cultivars was also apparent in tests conducted by Johnson (1979) and was detected in the succeeding selfed generations, derived from the more resistant individuals (Johnson and Palmer, 1985). Poswal and El-Zik (1986) reported that narrow-sense heritability for resistance in cotton was low but resistance could be detected and that it was polygenically inherited and controlled by a complex of minor genes.

COLLETOTRICHUM SPP.

History and distribution

Before the use of mercurial seed dressings in the USA, *Colletotrichum* was the main pathogen of the seedling disease complex. Results from a survey of seedling disease reported by Miller and Weindling (1939), in which diseased plants from the main cotton-growing states were examined, showed *Colletotrichum* to be the most frequently isolated pathogen. It was isolated from 56.8% of seedlings examined compared with only 1.4% for *R. solani* and this predominance was considered to be due to the seed-borne nature of the pathogen. Similar results were obtained from the following year's survey, although on this occasion, the fungus was isolated much less frequently from diseased seedlings collected in Texas and Oklahoma than from the more easterly states (Miller and Weindling, 1940), where the conditions were more conducive to boll rot caused by the same pathogen. *Colletotrichum* continued to attract considerable interest from researchers during the 1950s, e.g. Arndt (1953) and Fulton *et al.* (1956), found a high proportion of diseased seedlings in Arkansas infected with the fungus. As late as 1966, *Colletotrichum* comprised 11% of the fungi isolated from diseased seedlings in the USA (Bollenbacher and Fulton, 1967). However, it gradually became less important as treatment of cotton seed with fungicides became more widespread, and is today a minor component of the seedling disease complex in the USA.

Colletotrichum is also reported as a seedling pathogen in Africa. It is the major cause of damping-off in the Central African Republic (Cognee, 1960), but of minor importance in Tanzania (Ebbels, 1976). It has been isolated occasionally from apparently healthy seed and from seedlings suffering from post-emergence damping-off in Zimbabwe (R. J. Hillocks, unpublished).

Symptoms

Colletotrichum affects the cotton seedling as part of the anthracnose syndrome, which causes symptoms at all stages of crop development but is most damaging to the bolls (see Chapter 7). At the seedling stage it is mainly responsible for post-emergence damping-off and soreshin (Arndt, 1944;

Cognee, 1960). Although the fungus causes cortical rot at the base of the hypocotyl, similar to that caused by *R. solani*, the lesions are reddish brown rather than dark brown and may appear along the length of the hypocotyl. In addition, the fungus can cause spotting on the cotyledons.

Taxonomy and morphology

Colletotrichum isolated from diseased cotton seedlings is often referred to as *C. gossypii* South, having a *Glomerella* perfect state (e.g. Davis *et al.*, 1981). This species has straight conidia and there is no taxonomic evidence to suggest it is distinct from *C. gloeosporioides*, the conidial state of *Glomerella cingulata* (Stonem.) Spauld. & Von Shrenck.

Cognee (1960) isolated a second species, with curved conidia, from cotton seedlings which he did not identify. However, it is likely that this species was *C. capsici* (Syd.) Butl. & Bisby, a predominantly tropical or sub-tropical species with a wide host range, for which the perfect state is unknown.

Both species produce conidia within an acervulus bearing dark setae, which, under humid conditions, are conspicuous on the surface of host tissue. The acervulus of *G. cingulata* may be as large as 500 μm in diameter, whereas that of *C. capsici* rarely exceeds 250 μm.

G. cingulata is a very variable species and the following description is given by Mordue (1971a). The host was not specified. Conidia are cylindrical with obtuse ends, hyaline, aseptate and uninucleate. They are 9–24 × 3–6 μm in size, formed on unicellular, hyaline or faintly brown, cylindrical phialidic conidiophores. Perithecia may be produced on host material and are solitary or aggregated, globose, dark brown to black in colour and 85–300 μm in diameter. The ascus is clavate to cylindrical, thickened at the apex and 35–80 × 8–14 μm in size and contains eight ascospores. Ascospores are narrowly oval to cylindrical or fusiform, sometimes slightly curved, unicellular and hyaline (see Fig. 1.5a).

C. capsici (Fig. 1.5b) is distinguished from *C. gloeosporioides* mainly by having falcate conidia which are 16–30 × 2.5–4 μm in size. Perithecia have not been observed (Mordue, 1971b).

Isolation and growth in culture

The fungus has no special nutrient requirements, both species growing well on standard media. It is readily isolated from infected tissue and, if acervuli are present, this is best done by transferring conidia to the medium. On PDA, colonies are first white, becoming grey, and mycelial growth is slow, with marked concentric zonation if exposed to light and dark periods. Acervuli are produced in abundance, with typical pink masses of conidia. Appressoria are formed on mycelium of old cultures. In the case of *G. cingulata*, perithecia may be formed in older cultures.

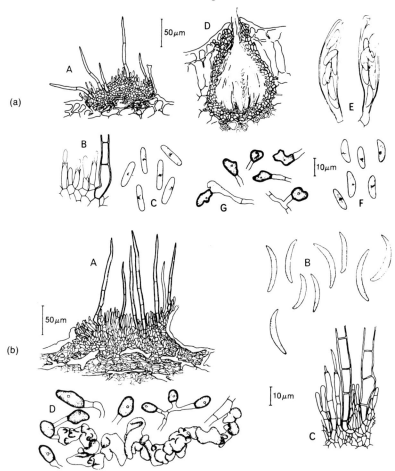

Fig. 1.5. (a) *Glomerella cingulata*: A, acervulus; B, conidiophores; C, conidia; D, perithecium; E, asci; F, ascospores; G, appressorium. (With permission of CABI.) (b) *Colletotrichum capsici*: A, acervulus; B, conidia; C, conidiophores; D, appressoria. (With permission of CABI.)

Factors affecting infection

The fungus is carried both on and within the seed (e.g. Arndt, 1953) due to its ability to infect the bolls. In the USA, seed infection rates were high when frequent rainfall occurred after boll-split (Arndt, 1956). Despite the decline in the importance of *Colletotrichum* as a seedling disease pathogen, it continues to be one of the most commonly isolated fungi from cotton seed (Klitch, 1986).

Infected seed and crop residues, therefore, provide the initial focus of infection. Damping-off occurs in the temperature range 20–26 °C (Davis *et al.*, 1981) but the optimum conditions for infection are high humidity and a temperature of 25 °C. Infection is greatly reduced below 20 °C and does not occur at 36 °C (Arndt, 1944).

Control

Control measures are aimed at eliminating sources of primary inoculum. Seed infection can be minimized by measures to reduce the incidence of boll rot, or by taking planting seed from areas which normally experience dry conditions immediately before harvest. The application of fungicide to the seed is widely practised in the USA and has greatly reduced the importance of anthracnose in cotton (Smith, 1959; Davis *et al.*, 1981).

Bollenbacher and Fulton (1967) conducted investigations into variability among *Gossypium* spp. for resistance to *Colletotrichum*. They found that all varieties of *G. hirsutum* tested were susceptible to a virulent isolate, but some Acala varieties were resistant to a less virulent isolate of the pathogen. *G. arboreum* var. Nanking was resistant to both isolates.

THIELAVIOPSIS BASICOLA

History and distribution

Thielaviopsis basicola (Berk.) & Br., causing black root rot, was first reported in Arizona (King and Barker, 1939), where it was isolated from a variety of *G. barbadense* suffering from an internal collar rot. Only later was it recognized as a seedling disease of Pima (King and Presley, 1942) and Upland cotton (Presley, 1947).

In the USA, the disease occurs in Texas (Leyendecker, 1952), Mississippi (Presley, 1947) and New Mexico (Staffeldt, 1959) and is widely distributed in the San Joachim Valley of California (Mathre *et al.*, 1966). It has also been reported from Peru (Bazan and de Sequira, 1949) and Egypt (Linsey, 1981). The fungus is distributed almost throughout the world (CMI Distribution Map No. 218) with a wide host range but is not a common pathogen of cotton outside the USA and the former Soviet Union (Nazarova *et al.*, 1987). It has been isolated as a minor component of the seedling disease complex in Spain (Melero-Vara and Jimenez-Diaz, 1990) and has recently been recorded for the first time on cotton in Australia (Allen, 1990).

Cultivars of *G. barbadense* are more susceptible than Upland (*G. hirsutum*) cultivars, and the disease has become more important in Arizona with increasing cultivation of Pima cottons (Mauk and Hine, 1988).

Symptoms and losses

Although infection normally occurs at the seedling stage, causing a cortical rot, older plants can become infected, producing a collar rot (King and Barker,

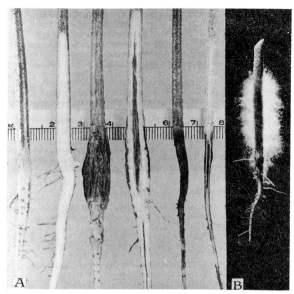

Fig. 1.6. Symptoms caused by *Thielaviopsis basicola*: A, swelling and blackening of the cortex at base of the stem; B, the fungus growing from infected cortical tissue (Blank *et al.*, 1953).

1939). This symptom is more commonly seen in cultivars of *G. barbadense*, in which the leaves and young stem tissue suddenly collapse, the stelar tissue is stained a purplish black and there is pronounced swelling of the crown.

Infection of the seedling affects the roots and the portion of the hypocotyl below soil level. Infected cortical tissue turns black (Fig. 1.6), but the vascular tissue, which is often not attacked, remains white. Diseased seedlings are stunted and easily pulled from the soil. Upland cultivars are rarely killed by the disease if soil temperatures are above 20 °C, but affected plants may still be stunted, with a pronounced swelling in the crown area just below soil level (Blank *et al.*, 1953).

If soil temperatures are below 20 °C at the beginning of the season, substantial stand loss to the disease can occur in Upland cultivars, especially on clay soils (Presley, 1947). Seedling death tends to occur at a later stage than is common with post-emergence damping-off. In experiments conducted with inoculated soils, Blank *et al.* (1953) found that symptoms first became visible above ground four to five weeks after planting. Infected seedlings were stunted compared with non-inoculated controls and seedling death began five weeks after planting. In experiments conducted by King and Presley (1942), about a third of seedlings died in soil inoculated with the pathogen. Similar results were reported by Mauk and Hine (1988) with stand losses of 28% and 32% respectively for *G. hirsutum* and *G. barbadense*, when soil temperature was in the range 16–20 °C. In California, however, although the pathogen is widely distributed, inoculum levels are too low for the disease to have a significant effect on crop yields (Mathre *et al.*, 1966).

Causal organism

T. basicola is a dematiaceous hyphomycete fungus with a wide host range, affecting 15 plant families, most of which are members of the Leguminosae and Solanaceae. Two types of single-celled conidia are produced by the fungus. The macroconidia (chlamydospores) are brown and subrectangular and are produced in short chains consisting of 5–8 spores, each 5–8 × 10–16 µm in size. The microconidia (endoconidia) are cylindrical, subhyaline and truncate at the ends and are 4–6 × 8–20 µm in size (Subramanian, 1968) (Fig. 1.7).

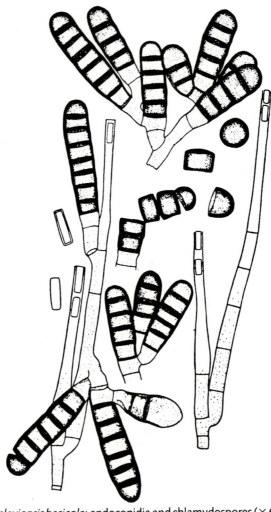

Fig. 1.7. *Thielaviopsis basicola*: endoconidia and chlamydospores (× 650) (Ellis, 1971).

Aetiology

As the fungus has such a wide host range, it is likely to exist at low population levels in non-agricultural soils, infecting the roots of wild species. Soil inoculum would then build up with the cultivation of susceptible crop species.

The pathogen survives in the absence of a host in the form of chlamydospores (Tsao and Bricker, 1966) and the endoconidia may also survive in wet soil for up to eight months (Papavizas and Lewis, 1971). Following the addition of macerated cultures of the fungus to soil in which seedlings were growing, Linderman and Toussoun (1968b) observed germination, close to the roots, of chlamydospores, conidia and mycelial fragments within 20 hours.

Penetration of epidermal cells and root hairs took place within 24 hours and new conidia were observed 12 hours later. Germ-tubes penetrated the host tissue directly without the formation of specialized structures. After penetration, the hyphae filled the invaded cell before penetrating the next wall. By five days after infection, the endodermis was reached. The endodermis presents a barrier to the advancing fungus but can be breached at high inoculum levels, when spores are produced abundantly in the vessels (Mathre *et al.*, 1966). Symptoms become apparent ten days after inoculation as stunting of the seedling, accompanied by root decay. With cultivars of *G. hirsutum* there is often little external evidence of infection by the time the plants are three months old, especially if temperatures are close to optimum for plant growth. However, the pathogen may still be isolated from plants which have recovered from infection (Mauk and Hine, 1988).

Factors affecting infection

The disease is more severe in crops grown on clay soils than on sandy soils. This is probably related to the high soil moisture requirement for chlamydospore germination. The endoconidia are less dependent on soil moisture but require a supply of exogenous nutrient for germination, whereas wetting of the soil is sufficient to stimulate chlamydospore germination (Linderman and Toussoun, 1967a). The disease often appears after irrigation, especially if the soils are slow to drain (King and Presley, 1942). Linderman and Toussoun (1967b) were unable to obtain infection of a number of hosts, including cotton, in dry soil.

Survival of chlamydospores and endoconidia is enhanced by low temperature (10 °C) (Papavizas and Lewis, 1971) and germination is greatest at temperatures between 20 and 33 °C. Some germination occurs even at 9 °C, but not at 3 °C. While spore germination occurs over a wide temperature range, germ-tube elongation is inhibited at temperatures above 30 °C (Mathre and Ravenscroft, 1966). Disease symptoms are most severe under cool, wet conditions. Blank *et al.* (1953) conducted inoculation experiments at temperatures in the range 20–26 °C and found that black root rot was most severe at the lowest temperature. In Arizona, the disease is more severe in early plantings, when soil temperatures are lower. Mauk and Hine (1988) found in their experiments that damage was greatest at soil temperatures of 16–20 °C, when up to 30% of seedlings were lost.

Naturally occurring inoculum levels vary widely and have a major effect on disease severity. The fungus is widespread in the San Joachim Valley of California but, in most infested fields, population levels are too low to cause severe disease. Mathre *et al.* (1966) estimated that 100 propagules g^{-1} of soil was about the threshold level to cause significant root rot in seedlings and that inoculum levels of 10^3–10^4 propagules g^{-1} were required to produce stunting in older plants. In Arizona, inoculum levels can reach 600 propagules g^{-1} in fields sown to the susceptible Pima cultivars (Mauk and Hine, 1988).

Isolation and growth in culture

T. basicola can be readily isolated from infected plants and grows well on standard media such as PDA. Cultures are initially ash-grey on PDA, producing endoconidia abundantly within two days at 28 °C and chlamydospores two days later. With the appearance of chlamydospores, the culture becomes blackish brown in colour (King and Presley, 1942). For the production of large numbers of spores required in the preparation of inoculum, Papavizas and Lewis (1971) recommended the use of Czapek dox agar with 0.25% yeast extract.

The fungus may be isolated from soil using a selective medium containing PCNB, rose bengal and antibiotics (Tsao and Bricker, 1966). Quantitative estimates of soil populations have been made by dilution plating on a selective medium, using carrot discs as baits, or by placing diluted soil on to carrot discs and counting the number of discs on which the fungus grows. Chlamydospore formation allows the pathogen to be easily identified in mixed cultures (Tsao and Canetta, 1964).

Control

As a disease of cool conditions, black root rot can be controlled by delaying planting until soil temperatures rise above 20 °C. Also, the disease becomes progressively more severe with increasing soil moisture levels and care should be taken to avoid over-irrigation on heavy soils infested with the fungus.

Fungicidal control is not economically viable in cotton (Linsey, 1981), although imazalil is used to control the disease in other crops.

Rotation with monocotyledonous crops can be worthwhile in reducing the rate of inoculum build-up. Linsey (1974) found that the fungus sporulated well on the roots of cotton and alfalfa but not on barley and sorghum. Ten weeks after planting sorghum, soil populations of the fungus had increased by a factor of 10, compared with factors of 20 and 100 respectively under crops of cotton and alfalfa.

FUSARIUM SPP.

Pathogenicity of Fusarium species

Species of *Fusarium* are frequently isolated from diseased cotton seedlings but are, with the exception of *F. oxysporum* f.sp. *vasinfectum* (see Chapter 4), usually considered to be of secondary importance (Bollenbacher and Fulton, 1959; Johnson *et al.*, 1978; Johnson and Doyle, 1986). However, very early in the history of the crop in Australia, Pratt (1926) reported that *F. moniliforme* caused poor germination of cotton seed. He was able to cause root rot in seedlings inoculated with the fungus and inoculation of the boll led to internal infection of the seed. Around the same time in the USA, Woodroof (1927) also found that *F. moniliforme* could cause root rot.

In surveys conducted between 1938 and 1940 in the USA, *F. moniliforme* was isolated from 29% of diseased seedlings examined and it was the second most frequently isolated fungus after *Colletotrichum* in all the cotton-growing states except Texas and Oklahoma, where it was the most commonly isolated fungus (Miller and Weindling, 1939).

Over a four-year period, 45% of fungi isolated from diseased seedlings in Arkansas were *Fusarium* spp., mainly *F. oxysporum*, *F. solani* and *F. moniliforme*, but they exhibited only mild pathogenicity (Fulton and Bollenbacher, 1959). In another study, Johnson *et al.* (1978), found that in Tennessee, *Fusarium* spp. were second to *Pythium* spp. in frequency of isolation from seedlings suffering from post-emergence disease, but they were the least pathogenic group of fungi isolated. Some of the *Fusarium* isolates caused necrotic lesions on the hypocotyls but did not cause seedling mortality. Similarly, Johnson and Doyle (1986) found that *Fusarium* spp. were frequently isolated from infected seedling hypocotyls, but their frequency of isolation was negatively correlated with the field disease index.

In contrast to these results, other workers have isolated *Fusarium* spp. which are highly virulent on cotton seedlings (Roy and Bourland, 1982; Batson and Borazjani, 1984). Colyer (1988) attempted to resolve these conflicting views. Initially, fungi were isolated from diseased seedlings collected in Louisiana. *Fusarium* spp. represented 42% of all fungi isolated, compared with 40% for *Rhizoctonia*. The main species were *F. oxysporum* and *F. solani*, with low numbers of *F. equisiti* and *F. graminearum*. Pathogenicity tests were conducted at a low temperature (21 °C), showing that most isolates were at least moderately pathogenic. Some isolates of *F. solani* were the most pathogenic, causing mortality in 8% of inoculated plants. Many of the isolates caused root and hypocotyl necrosis. Batson and Trevathan (1988) also reported the pathogenicity of *F. solani* towards cotton seedlings. When maize cobs inoculated with the fungus were added to soil into which cotton seed was planted, the pathogen caused significant reductions in emergence and many of the emerging seedlings exhibited root discoloration.

Isolation

Fusarium species are widely distributed in soils and a number of selective media have been designed for their isolation. The selective medium most often used contains PCNB (Nash and Snyder, 1962). After initial isolation, all species grow well on standard media, such as PDA, although Booth (1971) based his descriptions on cultures growing on potato sucrose agar at pH 6.5–7.

Morphology

The species can be distinguished on the basis of morphological differences in the structure of the conidiophore, the shape and size of the conidia and the presence or absence of chlamydospores. The microconidiophore structure provides the simplest means of separating *F. oxysporum* from *F. solani*, and the polyphialides of *F. moniliforme* var. *subglutinans* and absence of chlamydospores separate it from *F. oxysporum*. For further details of *Fusarium* taxonomy see Booth (1971, 1977) and Nelson *et al.* (1983). *Fusarium oxysporum* is described more fully in Chapter 4.

Control

Colyer (1988) noted that the high incidence of *Fusarium* spp. now associated with cotton seedling disease in the USA, may be due to the fact that control strategies are aimed primarily at *Rhizoctonia* and *Pythium*. He suggested that future management practices should include control of *Fusarium* spp.

Among the fungicides available for seedling disease control, thiabendazole might be expected to give some control of *Fusarium* spp. and is also effective against *R. solani*.

OTHER FUNGI CAUSING DAMPING-OFF

Sclerotium rolfsii Sacc. is more often associated with collar rot in older plants, but has been recorded as a seedling pathogen in Arizona (Gottlieb and Brown, 1941) and Ebbels (1976) considers it to be the second most important cotton seedling disease pathogen after *Alternaria* in Africa.

Macrophomina phaseolina (Tassi) Goid. is considered in more detail as the causal organism of charcoal rot in older plants (Chapter 5) but can cause seedling disease in India (Dwivedi and Chaube, 1985; Chauhan, 1988).

In Spain, *Phytophthora palmivora* has been identified as a minor component of the seedling disease complex (Melero-Vara and Jimenez-Diaz, 1990).

In addition to their role in causing spots on the leaves and cotyledons, *Alternaria* spp. in New Mexico are considered by Maier (1965) to be invaders of the root/stem transition zone. *Alternaria* spp. were isolated from 14% of diseased seedlings, the main species being *A. solani*, *A. gossypina*, *A. tenuis* and *A. humicola*. Forty-eight of the isolates were capable of primary parasitism and most were capable of secondary invasion of lesions caused by *T. basicola*, *R. solani* and *Fusarium* spp.

Seedling Blight

Spotting of the cotyledons and/or lesions on the hypocotyl can be caused by *Colletotrichum* spp., *Cercospora gossypina*, *Alternaria* spp. and *Ascochyta gossypii* Woron. Although usually not very damaging to the crop, under certain conditions, the latter two fungi are capable of causing a more generalized blight. This affects the growing point of the plant, sometimes causing seedling mortality (both these pathogens are described in the section on leaf spots, Chapter 6).

Smith (1950) reported that Ascochyta blight could cause considerable stand loss in Alabama. The disease was sporadic or locally epidemic in 1947–1949, becoming widespread in 1950, with stand losses in some areas of up to 50%. The initial focus of infection was infested seed and crop residues, with epidemics being induced by prolonged periods of cool, wet weather. Smith (1950) coined the term 'wet weather blight' for the disease. Often plants recover from infection when warm conditions return. Plants weakened by nutrient deficiency or pest attack are more susceptible to the disease.

Alternaria spp. are among the most commonly isolated fungi from cotton seed (e.g. Klich, 1986) and the most ubiquitous fungi in cotton fields throughout the season, usually causing only minor spots on the leaves and bolls. However, if conditions are warm and humid during germination and emergence, the cotyledons may become infected from contaminated seed and crop residues. The spots are light brown in colour, bounded by a purple margin, and 2–8 mm in diameter (Fig. 1.8). If initial inoculum levels are high and weather conditions suitable, spotting may be severe enough to cause shedding of the cotyledons. Cotyledons are more susceptible than the leaves and the plants usually recover with the appearance of the first leaves. Cultivars of *G. barbadense* are more susceptible than most cultivars of *G. hirsutum* at all stages of crop development. Ebbels (1976) has described a condition in Tanzania caused by *A. macrospora* in which the cotyledons and growing point take on a charred appearance, resulting in seedling mortality and stand loss. This condition was reported to be more severe in unthrifty plants, especially when seed rates were too high, giving crowded stands before thinning, and in fields close to woodland.

Some virus diseases can be damaging to the seedling but most are insect-transmitted and symptoms appear late in the season as insect populations

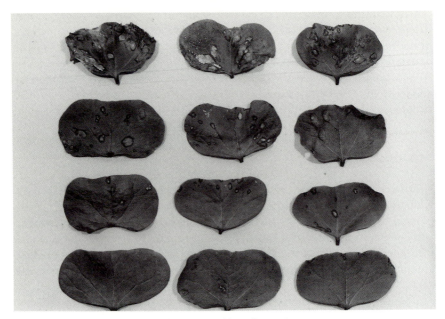

Fig. 1.8. Symptoms on cotyledons caused by *Alternaria macrospora*.

build up. Cotton leaf crumple virus, which is transmitted by the whitefly (*Bemisia tabaci*) (Laird and Dickson, 1959), occurs mainly in California and Arizona. The virus causes stunting of seedlings but plants usually recover by the end of the season (Brown *et al.*, 1987). Virus diseases are considered in Chapter 9.

Bacterial blight caused by *Xanthomonas campestris* pv. *malvacearum* is the only damaging bacterial disease of cotton (see Chapter 2) and often begins at the seedling stage. The bacterium is seed-borne and also overwinters in crop residues. Both these sources provide inoculum for infection of the cotyledons during germination and emergence.

Associations between Seedling Pathogens and Nematodes

A number of plant parasitic, soil-inhabiting nematodes can attack the cotton plant from germination onwards. Some of these cause significant damage to the crop on their own (see Chapter 10), but may also have an indirect effect by increasing root infection by fungal pathogens.

One of the best known interactions between a fungus and a plant parasitic nematode is that between *Fusarium oxysporum* f.sp. *vasinfectum* and *Meloidogyne incognita*, which is described in Chapter 4.

Soil fumigation with nematicides has often resulted in reduced incidence of seedling disease (Reynolds and Hansen, 1957; White, 1962). However, this provides only indirect evidence for the involvement of nematodes in increased seedling disease and could be due, at least partly, to the effect of the chemical on other members of the soil microflora. Further evidence for an interaction between *R. solani* and *Meloidogyne* spp. is provided by results from inoculation experiments in pots. Reynolds and Hansen (1957) showed that infection with *R. solani* was greater when *M. incognita* was present than in its absence. Similar results have been obtained by Carter (1975a,b) in the USA and by Rizk *et al.* (1986) in Egypt. There is also some evidence that increased disease incidence may be due to nematode-mediated changes in host physiology, rather than simply root damage caused by the feeding nematode (Carter, 1981). *M. incognita* is also reported to increase the incidence of black root rot when it occurs in sandy soils (Leyendecker *et al.*, 1953; White, 1962).

Cauquil and Shepherd (1970) investigated the effect of root-knot nematode on infection of cotton seedlings by several soil-borne fungi. Synergistic effects were detected between the nematode and *Alternaria alternata, F. oxysporum* f.sp. *vasinfectum, Colletotrichum gossypii* and *R. solani*. No disease occurred in plants inoculated with *F. moniliforme* with or without the nematode. The full effects of the nematode were masked in the case of interactions with *Colletotrichum* and *R. solani* because disease symptoms were severe in the absence of the nematode. With both *A. alternata* and *F. oxysporum*, severe disease occurred only in the presence of the nematode.

The main effect of nematode damage may be to prolong the period for which the plant remains susceptible to seedling disease pathogens. Brodie and Cooper (1964) showed that seedlings grown in soil inoculated with the reniform nematode (*Rotylenchulus reniformis*), the root-knot nematode (*Meloidogyne* spp.) and the sting nematode (*Hoplolaimus tylenchiformis*) were susceptible to *R. solani* for longer than in the absence of nematodes. With respect to infection with *Pythium* spp., only the root-knot nematodes prolonged the period of susceptibility.

References

ALAGARSAMY, G., MOHAN, S. AND JEYARASAN, R. (1987) Effect of seed pelleting with antagonists in the management of seedling disease of cotton. *Journal of Biological Control* 1, 66–7.

ALLEN, S.J. (1990) *Thielaviopsis basicola*: a new record on cotton in Australia. *Australian Plant Pathology* 19, 24–5.

ARNDT, C.H. (1943) *Pythium ultimum* and the damping off of cotton seedlings. *Phytopathology* 33, 607–10.

ARNDT, C.H. (1944) Infection of cotton seedlings by *Colletotrichum gossypii* as affected by temperature. *Phytopathology* 34, 861–9.

ARNDT, C.H. (1953) Survival of *Colletotrichum gossypii* on cotton seeds in storage. *Phytopathology* 43, 220.

ARNDT, C.H. (1956) Cotton seed produced in South Carolina in 1954 and 1955, its viability and infestation by fungi. *Plant Disease Reporter* 40, 1001–4.

ATKINSON, G.F. (1892) Some diseases of cotton. *Alabama Agricultural Experimental Station Bulletin* 41, 1–65.

AYERS, A.W. AND LUMSDEN, R.D. (1975) Factors affecting production and germination of oospores of three *Pythium* species. *Phytopathology* 65, 1094–100.

BAKER, R. AND MARTINSON, C. A. (1970) Epidemiology of diseases caused by *Rhizoctonia solani*. In: Parmeter, J.R. (ed.), Rhizoctonia solani: *Biology and Pathology*. University of California Press, Los Angeles and London, pp. 172–88.

BARTON, R. (1957) Germination of oospores of *Pythium mamillatum* in response to exudates from living seedlings. *Nature* 180, 613–14.

BARTON, R. (1958) Occurrence and establishment of *Pythium* in soils. *Transactions of the British Mycological Society* 41, 207–22.

BARTON, R. (1961) Saprophytic activity of *Pythium mamillatum* in soils. 2. Factors restricting *Pythium mamillatum* to pioneer colonisation of substrates. *Transactions of the British Mycological Society* 44, 105–18.

BASHAN, Y. AND LEVANONY, H. (1987) Transfer of *Alternaria macrospora* from cotton seed to seedling: light and scanning electron microscopy of colonisation. *Journal of Phytopathology* 120, 60–8.

BATEMAN, D.F. (1964) Cellulase and *Rhizoctonia* disease of bean. *Phytopathology* 54, 1372–7.

BATSON, W.E. AND BORAZJANI, A. (1984) Influence of four species of *Fusarium* on emergence and early development of cotton. *Proceedings of the Beltwide Cotton Production Research Conference*, p. 20.

BATSON, W.E. AND TREVATHAN, L.E. (1988) Suitability and efficacy of ground corncobs as a carrier of *Fusarium solani* spores. *Plant Disease Reporter* 72, 222–5.

BAZAN, C. AND DE SEGURA, M. (1949) Podredumbre radicular de algodoner. *Centro Nacional de Investigacion y Experimentacion Agricola Bulletin* 37, Lima, Peru.

BELL, D.K. AND OWEN, J.H. (1963) Effect of soil temperature and fungicide placement on cotton seedling damping-off caused by *Rhizoctonia solani*. *Plant Disease Reporter* 47, 1016–21.

BIESBROCK, J.A. AND HENDRIX, F. F. (1970) Influence of soil water and temperature on root necrosis of peach caused by *Pythium* species. *Phytopathology* 60, 880–2.

BIRD, L.S. (1978) Tamcot CAMD-EA multiadversity resistant cotton variety. *Texas Agricultural Experimental Station Bulletin* L-1720.

BLAIR, I.D. (1943) Behaviour of the fungus *Rhizoctonia solani* Khun in the soil. *Annals of Applied Biology* 30, 118–27.

BLANK, L.M., LEYENDECKER, P.J. AND NAKAYAMA, R.M. (1953) Observations on the black root rot symptoms on cotton seedlings at different soil temperatures. *Plant Disease Reporter* 37, 473–6.

BLASINGAME, D. (1990) Disease loss estimate committee report. *Proceedings of the Beltwide Cotton Production Research Conference*, National Cotton Council, Memphis, Tennessee, p. 4.

BOLLENBACHER, K. AND FULTON, N.D. (1959) Disease susceptibility of cotton seedlings from artificially deteriorated seeds. *Plant Disease Reporter* Supplement 259, 222–7.

BOLLENBACHER, K. AND FULTON, N.D. (1967) Susceptibility of *G. arboreum* and *G. hirsutum* to seedling anthracnose. *Plant Disease Reporter* 51, 632–6.

BOOTH, C. (1971) *The Genus* Fusarium. Commonwealth Mycological Institute, Kew, Surrey, England, 237pp.

BOOTH, C. (1977) *Fusarium: A Laboratory Guide to the Identification of the Major Species*. Commonwealth Mycological Institute, Kew, Surrey, 55pp.

BORUM, D.E. AND SINCLAIR, J.B. (1968) Evidence for systemic protection against *Rhizoctonia solani* with vitavax in cotton seedlings. *Phytopathology* 58, 976–7.

BOYLE, L.W. (1956) The role of saprophytic media in the development of southern blight and root rot on peanuts. *Phytopathology* 46, 7–8 (Abstr.).

BRINKERHOFF, L.A., OSWALT, E.S. AND TOMLINSON, J.T. (1954) Field tests with chemicals for the control of *Rhizoctonia* and other pathogens of cotton seedlings. *Plant Disease Reporter* 38, 467–75.

BRODIE, B.B. AND COOPER, W.E. (1964) Relation of parasitic nematodes to post-emergence damping-off in cotton. *Phytopathology* 54, 1023–7.

BROWN, E.A. AND McCARTER, M.C. (1976) Effect of seedling disease caused by *Rhizoctonia solani* on subsequent growth and yield of cotton. *Phytopathology* 66, 111–15.

BROWN, J.K., MIHAIL, J.D. AND NELSON, M.R. (1987) Effects of cotton leaf crumple virus on cotton inoculated at different growth stages. *Plant Disease Reporter* 71, 699–703.

CARTER, W.W. (1975a) Effects of soil temperature and inoculum levels of *Meloidogyne incognita* and *Rhizoctonia solani* on seedling disease of cotton. *Journal of Nematology* 7, 229–33.

CARTER, W.W. (1975b) Effects of soil texture on the interactions between *Rhizoctonia solani* and *Meloidogyne incognita* on cotton. *Journal of Nematology* 7, 234–6.

CARTER, W.W. (1981) The effect of *Meloidogyne incognita* and tissue wounding on severity of seedling disease of cotton caused by *Rhizoctonia solani*. *Journal of Nematology* 13, 374–6.

CAUQUIL, J. AND SHEPHERD, R.L. (1970) Effect of root-knot nematode–fungi combinations on cotton seedling disease. *Phytopathology* 60, 448–51.

CHANG-HO, Y. (1970) The effect of pea root exudate on the germination of *Pythium aphanidermatum* zoospore cysts. *Canadian Journal of Botany* 48, 1501–14.

CHAUHAN, M.S. (1987) Comparative efficacy of fungicides for the control of seedling disease of cotton due to *Rhizoctonia* spp. *Indian Journal of Mycology and Plant Pathology* 16, 335–7.

CHAUHAN, M.S. (1988) Relative efficiency of different methods for control of seedling disease of cotton caused by *Rhizoctonia bataticola*. *Indian Journal of Mycology and Plant Pathology* 18, 25–30.

CHESTER, K.S. (1938) Gravity grading, a method for reducing seed-borne disease in cotton. *Phytopathology* 28, 755.

CHESTERS, C.G.C. (1940) A method of isolating soil fungi. *Transactions of the British Mycological Society* 24, 352–4.

COGNEE, N. (1960) Premières observations sur les fontes de semis de cottoniers à la Station de Bambari. *Phytiatrie–Phytopharm* 9, 207–22.

COLE, D.L. AND CAVILL, M.E. (1977) Use of selected fungicides as seed dressings for the control of *Rhizoctonia solani* in cotton. *Rhodesian Journal of Agricultural Research* 15, 45–50.

COLYER, P.D. (1988) Frequency and pathogenicity of *Fusarium* spp. associated with seedling diseases of cotton in Louisiana. *Plant Disease Reporter* 72, 400–2.

DANIELS, J. (1963) Saprophytic and parasitic activities of some isolates of *Corticium solani*. *Transactions of the British Mycological Society* 46, 485–502.

DARAG, I.E.M. AND SINCLAIR, J.B. (1969) Evidence of systemic protection against *Rhizoctonia solani* with chloroneb in cotton seedlings. *Phytopathology* 59, 1102.

DAVEY, C.B. AND PAPAVIZAS, G.C. (1963) Saprophytic activity of *Rhizoctonia* as affected by the carbon–nitrogen balance of certain organic soil amendments. *Proceedings of the American Soil Science Society* 27, 164–7.

DAVIS, R.G., BIRD, L.S., CHAMBERS, A.Y., GARBER, R.H., HOWELL, C.R., MINTON, E.B., STERNE, R. AND JOHNSON, L.F. (1981) Seedling disease complex. In: Watkins, G.M. (ed.), *Compendium of Cotton Diseases*. American Phytopathological Society, St Paul, Minnesota, pp. 13–20.

DEVAY, J.E., GARBER, R.H. AND MATHERSON, D. (1982) Role of *Pythium* spp. in the seedling disease complex of cotton in California. *Plant Disease Reporter* 66, 151–4.

DODMAN, R.L., BARKER, K.R. AND WALKER, J.C. (1968a) Modes of penetration by different isolates of *Rhizoctonia solani*. *Phytopathology* 58, 31–3.

DODMAN, R.L., BARKER, K.R. AND WALKER, J.C. (1968b) A detailed study of the different modes of penetration by *Rhizoctonia solani*. *Phytopathology*, 58, 1271–6.

DWIVEDI, T.S. AND CHAUBE, H.S. (1985) Effect of fungicides on the emergence and infection of cotton seedlings by *Macrophomina phaseolina* (Tassi) Goid. *Indian Journal of Mycology and Plant Pathology* 15, 295–6.

EBBELS, D.L. (1976) Diseases of Upland cotton in Africa. *Review of Plant Pathology* 55, 747–63.

ELAD, Y., CHET, I. AND KATAN, J. (1980) *Trichoderma harzianum*: a biocontrol agent effective against *Sclerotium rolfsii* and *Rhizoctonia solani*. *Phytopathology* 70, 119–21.

ELAD, Y., KALFON, A. AND CHET, I. (1982) Control of *Rhizoctonia solani* in cotton by seed coating with *Trichoderma* spp. spores. *Plant and Soil* 66, 279–81.

ELLIS, M.B. (1971) *Dematiaceous Hyphomycetes*. Commonwealth Mycological Institute, Kew, Surrey.

EL-KHADEM, M. AND PAPAVIZAS, G.C. (1984) Effect of the herbicides EPTC and linuron on cotton diseases caused by *Rhizoctonia solani* and *Fusarium oxysporum* f.sp. *vasinfectum*. *Plant Pathology* 33, 411–16.

EL-KHADEM, M. ZAHRAN, M. AND EL-KAZZAZ, M. (1979) Effect of the herbicides trifluaradin, dinitramine and fluomethuron on *Rhizoctonia* diseases in cotton. *Plant and Soil* 51, 463–70.

EXNER, B. (1953) Comparative studies of four rhizoctonias occurring in Louisiana. *Mycologia* 45, 698–719.

FAHMY, T. (1931) The sore-shin disease and its control. *Ministry of Agriculture Egypt Technical and Scientific Service Bulletin* 108.

FLENTJEN, N.T. (1957) Studies on *Pellicularia filimentosa* (Pat.) Rogers. 3. Host penetration and resistance and strain specialisation. *Transactions of the British Mycological Society* 40, 322–36.

FULTON, N.D. AND BOLLENBACHER, K. (1959) Pathogenicity of fungi isolated from diseased cotton seedlings. *Phytopathology* 49, 684–9.

FULTON, N.D., WADDLE, B.A. AND THOMAS, J.A. (1956) Influence of planting date on fungi isolated from diseased cotton seedlings. *Plant Disease Reporter* 40, 556–8.

FULTON, N.D., WADDLE, B.A. AND BOLLENBACHER, K. (1962) Varietal resistance to seedling disease in cotton. *Phytopathology* 2, 10 (Abstr.).

GARBER, R.H., DEVAY, J.E., WEINHOLD, A.R. AND MATHERSON, D. (1979) Relationship of pathogen inoculum to cotton seedling disease control with fungicides. *Plant Disease Reporter* 63, 246–50.

GIBSON, I.A.S., LEDGER, M. AND BOEHM, E. (1961) An anomalous effect of PCNB on the incidence of damping-off caused by *Pythium* sp. *Phytopathology* 51, 531–3.

GOTTLIEB, M. AND BROWN, J.G. (1941) *Sclerotium rolfsii* on cotton in Arizona. *Phytopathology* 31, 944–6.

GRIFFIN, D.M. (1963) Soil factors and the ecology of fungi. 2. Behaviour of *Pythium ultimum* at small soil-water suctions. *Transactions of the British Mycological Society* 46, 368–72.

GRINSTEIN, A., KATAN, J. AND ESHEL, Y. (1976) Effect of dinitroaniline herbicides on plant resistance to soilborne pathogens. *Phytopathology* 66, 517–22.

HALLOIN, J.M. (1983) Thirty-year summary of cotton disease loss estimates: crop years 1952–1981. *Proceedings of the Beltwide Cotton Production Research Conference*, pp. 3–4.

HALLOIN, J.M. AND BOURLAND, F.M. (1981) Deterioration of planting seed. In: Watkins, G.M. (ed.), *Compendium of Cotton Diseases*. American Phytopathology Society, St Paul, Minnesota, pp. 11–13.

HANCOCK, J.G. (1977) Factors affecting soil populations of *Pythium ultimum* in the San Joachim valley of California. *Hilgardia* 45, 107–21.

HANCOCK, J.G. (1981) Longevity of *Pythium ultimum* in moist soils. *Phytopathology* 71, 1033–7.

HANSFORD, C.G. (1929) Cotton diseases in Uganda. *Empire Cotton Growing Review* 6, 10–26.

HAYMAN, D.S. (1969) The influence of temperature on the exudation of nutrients from cotton seeds and on preemergence damping off by *Rhizoctonia solani*. *Canadian Journal of Botany* 47, 1663–9.

HENDRIX, F.F. AND CAMPBELL, W.A. (1976) Pythiums as plant pathogens. *Annual Review of Phytopathology* 11, 77–98.

HILLOCKS, R.J., CHINODYA, R. AND GUNNER, R. (1988) Evaluation of seed dressing and in-furrow treatment with fungicides for control of seedling disease in cotton caused by *Rhizoctonia solani*. *Crop Protection* 7, 309–13.

HOWELL, C.R. (1982) Effect of *Gliocladium virens* on *Pythium ultimum*, *Rhizoctonia solani* and damping-off of cotton seedlings. *Phytopathology* 72, 496–8.

HOWELL, C.R. AND STIPANOVIC, R.D. (1980) Suppression of *Pythium ultimum*-induced damping-off of cotton seedlings by *Pseudomonas fluorescens* and its antibiotic, pyoluteorin. *Phytopathology* 70, 712–15.

HOWELL, C.R. AND STIPANOVIC, R.D. (1983) Gliovirin, a new antibiotic from *Gliocladium virens*, and its role in the biocontrol of *Pythium ultimum*. *Canadian Journal of Microbiology* 29, 321–4.

HUNTER, R.E, AND GUINN, G. (1968) Effect of root temperature on hypocotyls of cotton seedlings as a source of nutrition for *Rhizoctonia solani*. *Phytopathology* 58, 981–4.

HUNTER, R.E. AND STAFFELDT, E.E. (1960) Effect of soil temperature on the pathogenicity of *Rhizoctonia solani* isolates. *Plant Disease Reporter* 44, 793–5.

JOHNSON, L.F. (1979) Susceptibility of cotton seedlings to *Pythium ultimum* and other pathogens. *Plant Disease Reporter* 63, 59–62.

JOHNSON, L.F. AND CHAMBERS, A.Y. (1973) Isolation and identity of three species of *Pythium* that cause cotton seedling blight. *Plant Disease Reporter* 57, 848–52.

JOHNSON, L.F. AND DOYLE, J.H. (1986) Relationship of seedling disease of cotton to characteristics of loessial soils in Tennessee. *Phytopathology* 76, 286–90.

JOHNSON, L.F. AND PALMER, G.K. (1985) Symptom variability and selection for reduced severity of cotton seedling disease caused by *Pythium ultimum*. *Plant Disease* 69, 298–300.

JOHNSON, L.F., BAIRD, D.D., CHAMBERS, A.Y. AND SHAMIYEH, N.B. (1978) Fungi associated with post emergence seedling disease of cotton in tree soils. *Phytopathology* 68, 917–20.

JONES, L.R., JOHNSON, J. AND DICKSON, J.G. (1926) Wisconsin studies upon the relation of soil temperature to plant disease. *Wisconsin Agricultural Experimental Station Bulletin* 71, 144 pp.

KHADGA, B.B., SINCLAIR, J.B. AND EXNER, B.B. (1963) Infection of seedling cotton hypocotyl by an isolate of *Rhizoctonia solani*. *Phytopathology* 53, 1331–6.

KHALIFA, M.A.S., EL-DEEB, S.T., KADOUS, E.A., HASSAN, A. AND SOLIMAN, F.S. (1987) Herbicide plant disease relationships: effect of soil herbicides on *Rhizoctonia* damping-off on cotton seedling. *Mededelingen van de Faculteit Landbouwwetenschappen, Rijksuniversiteit Gent* 52, 1233–43.

KING, C.J. AND BARKER, H.D. (1934) An internal collar rot on cotton. *Phytopathology* 29, 751 (Abstr.).

KING, C.J. AND PRESLEY, J.T. (1942) A root rot of cotton caused by *Thielaviopsis basicola*. *Phytopathology* 32, 752–61.

KLICH, M.A. (1986) Mycoflora of cotton seed from southern USA: a three year study of distribution and frequency. *Mycologia* 78, 706–12.

KLISIWICZ, J.M. (1968) Relation of *Pythium* spp. to root rot and damping-off of safflower. *Phytopathology* 58, 1384–6.

KOHLMEYER, J. (1956) Uber den Cellulose-abbau durch einige phytopathologene Pilze. *Phytopathologische Zeitschrift* 27, 147–82.

KREITLOW, K.W. (1950) Longevity of inoculum of *Sclerotinia trifolionum* prepared from cultures grown on grain. *Phytopathology* 40, 16.

LAIRD, E.F. AND DICKINSON, R.C. (1959) Insect transmission of leaf crumple virus of cotton. *Phytopathology* 49, 324–7.

LEACH, L.D. (1947) Growth rates of host and pathogen as factors determining the severity of pre-emergence damping-off. *Journal of Agricultural Research* 75, 161–79.

LEACH, L.D. AND GARBER, R.H. (1970) Control of *Rhizoctonia solani*. In: Parmeter, J.R. (ed.), Rhizoctonia solani: *Biology and Pathology*. University of California Press, Los Angeles and London, pp. 189–99.

LEHMAN, S.G. (1940) Cotton seed dusting in relation to control of seedling infection by *Rhizoctonia* in the soil. *Phytopathology* 30, 847–53.

LEYENDECKER, P.J. (1952) Root rot of cotton caused by *Thielaviopsis basicola* discovered in the Upper Rio Grande Valley of Texas. *Plant Disease Reporter* 36, 53.

LEYENDECKER, P.J., BLANK, L.M. AND NAYAYAMA, R.M. (1953) Further observations on black root-rot of cotton in the Upper Rio Grande Valley of Texas and New Mexico. *Plant Disease Reporter* 37, 130–3.

LINDERMAN, R.G. AND TOUSSOUN, T.A. (1967a) Behaviour of chlamydospores and endoconidia of *Thielaviopsis basicola* in nonsterilised soil. *Phytopathology* 57, 729–31.

LINDERMAN, R.G. AND TOUSSOUN, T.A. (1967b) Pathogenesis of *Thielaviopsis basicola*. *Phytopathology* 57, 1007.

LINDERMAN, R.G. AND TOUSSOUN, T.A. (1968b) Pathogenesis of *Thielaviopsis basicola* in non-sterile soil. *Phytopathology* 58, 1578–83.

LINSEY, D.L. (1974) Influence of certain plants on populations of *Thielaviopsis basicola* in soil. *Proceedings of the Beltwide Cotton Production Research Conference*, p. 19.

LINSEY, D.L. (1981) Black root rot. In: Watkins, G.M. (ed.), *Compendium of Cotton Diseases*. American Phytopathological Society, St Paul, Minnesota, pp. 47–8.

LOGAN, C. (1960) An estimate of the effort effect of seed treatment in reducing cotton crop losses caused by *Xanthomonas malvacearum* (E.F. Smith) Dowson. *Empire Cotton Growing Review* 37, 241–3.

LOPER, J.E. (1988) Role of fluorescent siderophore production in biological control of *Pythium ultimum* by a *Pseudomonas fluorescens* strain. *Phytopathology* 78, 166–72.

LUTHRA, J.C. AND VASUDEVA, R.S. (1941) Studies on the root rot disease of cotton in the Punjab. 10. Varietal susceptibility to the disease. *Indian Journal of Agricultural Science* 11, 410–21.

MAIER, C.R. (1965) The importance of *Alternaria* spp. in the cotton seedling disease complex in New Mexico. *Plant Disease Reporter* 49, 904–9.

MAIER, C.R. AND STAFFELDT, E.E. (1960) Cultural viability of selected isolates of *Rhizoctonia solani* and *Thielaviopsis basicola* and the variability in their pathogenicity to Acala and Pima cotton respectively. *Plant Disease Reporter* 44, 956–61.

MARTINSON, C.A. (1963) Inoculum potential relationships of *Rhizoctonia solani* measured with soil microbiological sampling tubes. *Phytopathology* 53, 634–8.

MATHRE, D.E. AND OTTS, J.D. (1967) Sources of resistance in the genus *Gossypium* to several soil-borne pathogens. *Plant Disease Reporter* 51, 864–6.

MATHRE, D.E. AND RAVENSCROFT, A.V. (1966) Physiology of germination of chlamydospores and endoconidia of *Thielaviopsis basicola*. *Phytopathology* 56, 337–42.

MATHRE, D.E., RAVENSCROFT, A.V. AND GARBER, R.H. (1966) The role of *Thielaviopsis basicola* as a primary cause of yield reduction in cotton in California. *Phytopathology* 56, 1213–16.

MAUK, P.A. AND HINE, R.B. (1988) Infection, colonisation of *G. hirsutum* and *G. barbadense* and development of black root rot caused by *Thielaviopsis basicola*. *Phytopathology* 78, 1662–7.

MELERO-VARA, J.M. AND JIMENEZ-DIAZ, R.M. (1990) Etiology, incidence and distribution of cotton seedling damping-off in southern Spain. *Plant Disease Reporter* 74, 597–600.

MILLER, P.R. AND WEINDLING, R. (1939) A survey of cotton seedling diseases in 1939 and the fungi associated with them. *Plant Disease Reporter* 23, 210–14.

MILLER, P.R. AND WEINDLING, R. (1940) A survey of cotton seedling diseases in 1940 and the fungi associated with them. *Plant Disease Reporter* 24, 260.

MINTON, E.B. (1984) Report of the cottonseed treatment committee for 1983. *Proceedings of the Beltwide Cotton Production Research Conference*, p. 11.

MINTON, E.B. AND GREEN, J.A. (1980) Germination and stand with cottonseed treatment fungicides: formulations and rates. *Crop Science* 20, 5–7.

MORDUE, J.E.M. (1971a) *Glomerella cingulata*. *CMI Descriptions of Pathogenic Fungi and Bacteria*, No. 35. Commonwealth Agricultural Bureaux, Farnham Royal, Bucks, UK.

MORDUE, J.E.M. (1971b) *Colletotrichum capsici*. *CMI Descriptions of Pathogenic Fungi and Bacteria* No. 317. Commonwealth Mycological Institute, Kew, Surrey.

MOUBASHER, A.H., MALEN, M.B. AND EL-SHAROUNY, H.M. (1984) Studies on the genus *Pythium* in Egypt. 3. Survey of root infecting fungi associated with damped-off cotton seedlings. *Review of Plant Pathology* 65, 486.

NAKAYAMA, T. (1940) A study on the infection of cotton seedlings by *Rhizoctonia solani* (Japanese). *Annals of the Phytopathological Society of Japan* 10, 93–103.

NASH, S.M. AND SNYDER, W.C. (1962) Quantitative estimates by plate counts of the bean root rot *Fusarium* in field soil. *Phytopathology* 52, 567–72.

NAZAROVA, O.N., TEREKHOVA, V.A. AND D'YAKOV, Y.U.T. (1987) Vegetative hybridisation of *Thielaviopsis basicola* (Berk et BR). Ferraris. *Review of Plant Pathology* 68, 27.

NEAL, D.C. (142) *Rhizoctonia* infection of cotton and symptoms accompanying the disease beyond the seedling stage. *Phytopathology* 32, 641–2.

NELSON, E.B. (1988) Biological control of *Pythium* seed rot and preemergence damping-off of cotton with *Enterobacter cloacae* and *Erwinia herbicola* applied as seed treatments. *Plant Disease Reporter* 72, 140–2.

NELSON, P.E., TOUSSOUN, T.A. AND MARASAS, W.F.O. (eds) (1983) Fusarium *Species: an Illustrated Manual for Identification*. Pennsylvania State University Press, University Park, Pensylvania.

NEUBAUER, R. AND AVIZOHAR-HERSHENSON, Z. (1973) Effect of the herbicide, trifluralin, on *Rhizoctonia* disease of cotton. *Phytopathology* 63, 651–2.

OSHIMA, N., LIVINGSTONE, C.H. AND HARRISON, M.D. (1963) Weeds as carriers of two potato pathogens in Colorado. *Plant Disease Reporter* 47, 466–9.

OWEN, J.H. AND GAY, J.D. (1964) Tests of soil fungicides under uniform conditions for control of cotton damping-off caused by *Rhizoctonia solani*. *Plant Disease Reporter* 48, 480–3.

PAPAVIZAS, G.C. AND DAVEY, C.B. (1959) Isolation of *Rhizoctonia solani* Kuehn from naturally infested and artificially inoculated soils. *Plant Disease Reporter* 43, 404–10.

PAPAVIZAS, G.C. AND DAVEY, C.B. (1961) Saprophytic behaviour of *Rhizoctonia solani* in soil. *Phytopathology* 51, 693–9.

PAPAVIZAS, G.C. AND LEWIS, J.A. (1971) Survival of endoconidia and chlamydospores of *Thielaviopsis basicola* as affected by soil environmental factors. *Phytopathology* 61, 108–13.

PAPAVIZAS, G.C., LEWIS, J.A., LUMSDEN, R.D., AYER, W.A. AND KANTZES, J.G. (1977) Control of *Pythium* blight on bean with ethazol and prothiocarb. *Phytopathology* 67, 1293–9.

PAPAVIZAS, G.C., LEWIS, J.A. AND O'NEIL, N.R. (1979) BAS 389 a new fungicide for control of *Rhizoctonia solani* in cotton. *Plant Disease Reporter* 63, 569–73.

PARMETER, J.R. AND WHITNEY, H.S. (1970) Taxonomy and nomenclature of the imperfect state. In: Parmeter, J.R. (ed.), Rhizoctonia solani: *Biology and Pathology*. University of California Press, Los Angeles and London, pp. 7–19.

PARMETER, J.R., SHERWOOD, R.T. AND PLATT, W.D. (1969) Anastomosis grouping among isolates of *Thanatephorus cucumeris*. *Phytopathology* 59, 1270–8.

POSWAL, M.A.T., EL-ZIK, K.M. AND BIRD, L.S. (1986) Gene action and inheritance of resistance to *Rhizoctonia solani* and *Pythium ultimum* in cotton seedlings. *Phytopathology* 76, 1107 (Abstr.).

PRATT, C.A. (1926) A disease of Queensland cotton seed. *Empire Cotton Growing Review* 3, 103–11.

PRESLEY, J.T. (1947) *Thielaviopsis* root rot of cotton in Mississippi. *Plant Disease Reporter* 31, 152 (Abstr.).

REYNOLDS, H.W. AND HANSEN, R.G. (1957) *Rhizoctonia* disease of cotton in the presence and absence of the cotton root-knot nematode in Arizona. *Phytopathology* 47, 256–61.

RICHTER, H. AND SCHNIDER, R. (1953) Untersuchungen zur morphologischen und biologische Differenzierung von *Rhizoctonia solani*. *Phytopathologische Zeitschrift* 20, 167–226.

RISK, R.H., MOHAMED, H.A., AND FOUAD, M.K. (1982a) Attempts to produce lines resistant to *Rhizoctonia solani* in Egyptian cotton. *Agricultural Research Review* 60, 67–82.

RISK, R.H., MOHAMED, H.A. AND FOUAD, M.K. (1982b) Histopathological studies of cotton seedlings infected with *Rhizoctonia solani*. *Agricultural Research Review* 60, 83–92.

RISK, R.H., EL-ERAKYS, S., RIAD, F.W. AND YOUSEF, B.A. (1986) Studies on the relationship between root-knot nematode, *Meloidogyne incognita*, and the soil-inhabiting fungus *Rhizoctonia solani* on cotton. *Agricultural Research Review* 61, 185–92.

ROVIRA, A.D. (1956) Plant root exudates in relation to the rhizospere effect. 1. The nature of root exudates from oats and peas. *Plant and Soil* 7, 178–94.

ROVIRA, A. D. (1969) Plant root exudates. *Botanical Review* 35, 35–63.

ROY, K.W. AND BOURLAND, F.M. (1982) Epidemiological and mycofloral relationships in cotton seedling disease in Mississippi. *Phytopathology* 72, 868–72.

SCHROTH, M.N. AND HILDEBRAND, D.C. (1964) Influence of plant exudates on root infecting fungi. *Annual Review of Phytopathology* 2, 101–32.

SHAO, F.M. AND CHRISTIANSEN, M.N. (1982) Cotton seedling radicle exudates in relation to susceptibility to *Verticillium* wilt and *Rhizoctonia*. *Phytopathologische Zeitschrift* 105, 351–9.

SHATLA, M.M. AND SINCLAIR, J.B. (1963) Tolerance to pentacloronitrobenzene among cotton isolates of *Rhizoctonia solani*. *Phytopathology* 53, 1407–11.

SHERWOOD, R.T. (1970) The physiology of *R. solani*. In: Parmeter, J.R. (ed.), Rhizoctonia solani: *Biology and Pathology*. University of California Press, Los Angeles and London, pp. 69–92.

SINCLAIR, J.B. (1957) Laboratory and greenhouse screening of various fungicides for control of *Rhizoctonia* damping-off of cotton seedlings. *Plant Disease Reporter* 41, 1045–50.

SMITH, A.L. (1950) *Ascochyta* seedling blight of cotton in Alabama in 1950. *Plant Disease Reporter* 34, 233–5.

SMITH, A.L. (1959) Progress with problems in cotton disease control. *Plant Disease Reporter* Supplement 259, 199–204.

SNEH, B., KATAN, J., HENIS, Y. AND WAHL, I. (1966) Methods for evaluating inoculum density of *Rhizoctonia solani* in naturally infested soils. *Phytopathology* 56, 74–8.

SPENCER, J.A. AND COOPER, W.E. (1967) Pathogenesis of cotton (*G. hirsutum*) by *Pythium* species: zoospore and mycelium attraction and infectivity. *Phytopathology* 57, 1332–8.

STAFFELDT, E.E. (1959) *Thielaviopsis basicola* a part of the cotton (*Gossypium hirsutum*) seedling disease complex in New Mexico. *Plant Disease Reporter* 43, 506–8.

SUBRAMANIAN, C.V. (1968) *Thielaviopsis basicola*. *CMI Descriptions of Pathogenic Fungi and Bacteria*, No. 70. Commonwealth Agricultural Bureaux, Farnham Royal, Bucks, UK.

TALBOT, P.H.B. (1970) Taxonomy and nomenclature of the perfect stage. In: Parmeter, J.R. (ed.), Rhizoctonia solani: *Biology and Pathology.* University of California Press, Los Angeles and London, pp. 20–31.

TSAO, P.H. AND BRICKER, J.L. (1966) Chlamydospores of *Thielaviopsis basicola* as surviving propagules in natural soils. *Phytopathology* 56, 1012–14.

TSAO, P.H. AND CANATTA, A.C. (1964) Comparative study of quantitative methods used for estimating the population of *Thielaviopsis basicola* in soil. *Phytopathology* 54, 633–5.

ULLSTRUP, A.J. (1939) The occurrence of the perfect stage of *Rhizoctonia solani* in plantings of diseased cotton seedlings. *Phytopathology* 29, 373–4.

WATERHOUSE, G.M. AND WATERSON, J.M. (1964) *Pythium aphanidermatum. CMI Descriptions of Pathogenic Fungi and Bacteria*, No. 36. Commonwealth Agricultural Bureaux, Farnham Royal, Bucks, UK.

WEINDLING, R. (1932) *Trichoderma lignorum* as a parasite of other fungi. *Phytopathology* 22, 837–45.

WHITE, L.V. (1962) Root-knot and the seedling disease complex of cotton. *Plant Disease Reporter* 46, 501–4.

WOOD, F.A. AND WILCOXSON, D. (1960) Another screened immersion plate for isolating fungi from soil. *Plant Disease Reporter* 44, 594.

WOODROOF, N.C. (1927) A disease of cotton roots produced by *Fusarium moniliforme* Sheld. *Phytopathology* 17, 227–37.

YAMADA, Y. (1986) Monceren (pencycuron), a new fungicide. *Japan Pesticide Information* 48, 16–22.

2 Bacterial Blight

R.J. HILLOCKS

History and Distribution

Bacterial blight probably originated in India (Knight, 1984b), although the disease was first reported in the USA by Atkinson (1891), who gave the names angular leaf spot, blackarm and bacterial boll rot to the various stages in the syndrome of the disease. Between 1900 and 1905, E.F. Smith established Koch's postulates, successfully reproducing symptoms of the disease by inoculation of leaves and bolls and he was able to demonstrate that angular leaf spot and blackarm were caused by the same organism (Smith, 1920). Edgerton (1912) included bacterial blight in his description of cotton boll rots. The earliest epidemiological work showed that secondary spread within the crop was due to wind-blown rain (Faulwetter, 1917). Much of the early work carried out in the USA is reviewed by Faulwetter (1917) and by Smith (1920).

Following on from work conducted in the USA, bacterial blight was identified on cultivars of G. barbadense in Sudan, where Massey (1930) investigated the role of soil temperature and soil moisture in primary infection of the seedling. Epidemiological studies were also carried out in Sudan by Findlay (1928) and extensive work was done by Stoughton (1930, 1931, 1932) using controlled environment conditions.

Around the same time that Findlay was working on the disease, bacterial blight was reported to be causing crop losses in Uganda (Hansford, 1932) and, with the expansion of cotton production in Tanganyika after the Second World War, blight became a problem there also.

The first programme to breed for resistance to the disease was initiated in Sudan by research officers with the Cotton Research Corporation (e.g. Knight, 1946). Methods adopted there were soon implemented in Uganda and most of the early blight-resistant varieties grown in many African countries were derived from the Uganda programme (Innes and Jones, 1972).

The disease was first reported in India in 1918, but, in both the USA and Asia, it took another 50–60 years before the disease reached serious proportions, becoming widespread in the USA during the 1950s (Schnathorst *et al.*, 1960) and in India in the 1970s (Verma, 1986).

Bacterial blight has now been recorded in almost every country in the world which grows cotton (CMI Distribution Map No. 57). It is an important disease in India (e.g. Kotasthane and Agrawal, 1970), Pakistan (Hussain, 1984), China (Tarr, 1958), South East Asia (Dizon and Reyes, 1987), the former Soviet Union (Tarr, 1985), South America (Cedano and Delgado, 1986) and Australia (Allen, 1986) and also occurs in Europe (Ciccarone, 1959). In Africa, the disease is widespread but less important in countries which experience hot, dry weather for most of the cotton-growing season, such as Egypt (Fahmy, 1930), than in the wetter parts of West Africa or the cooler highland areas of Central and East Africa, which experience intermittent periods of wind and heavy rain during the growing season.

A number of reviews of work done on the disease have been produced over the years, the most comprehensive of which are those by Wickens (1953) and Lagière (1960) concerning the disease in Africa and by Brinkerhoff (1963) concerning the work in the USA, a general review by Innes (1983) and an excellent monograph by Verma (1986).

Symptoms

The cotton plant may be affected by bacterial blight at all stages of its development, beginning with the seedling. The disease spreads from the cotyledons to the leaves, followed by the main stem and the bolls. Symptoms at each stage have been given different descriptive names according to which organ or growth stage is affected. The disease may therefore be referred to by any one of these names; blackarm, for instance, is commonly used in the early literature. However, they all describe symptoms of the same disease syndrome, which correctly should be known as bacterial blight unless a specific symptom is discussed.

The earliest signs of the disease may be found on the cotyledons of young seedlings and this symptom is known as cotyledon or seedling blight. Small dark green 'water-soaked' spots, which are circular or irregular in shape, become visible on the under-side (see Plate 1A) and then on the upper surface of the cotyledon, usually along the margin. The lesions spread inwards and, in susceptible cultivars, the cotyledon becomes distorted. Under favourable conditions for this disease, infection spreads from the cotyledon down the petiole to the stem, resulting in stunting or death of the seedling.

On the leaves, the symptoms are known as angular leaf spot (ALS) because the lesions are delimited by the veinlets, giving them an angular outline (Fig. 2.1). As on the cotyledons, the spots are initially water-soaked and more obvious on the underside of the leaf. Individual spots are usually

Fig. 2.1. Symptoms of angular leaf spot on a cotton cultivar susceptible to bacterial blight.

2–5 mm in size. They become dark brown or black in colour as they dry out and may be numerous on a susceptible host, coalescing to affect large areas of the leaf, which then becomes distorted and necrotic (Fig. 2.2 and Plate 1B)

Fig. 2.2. Symptoms of vein blight caused by *Xanthomonas campestris* pv. *malvacearum*.

Fig. 2.3. Leaf distortion and necrosis caused by severe angular leaf spot symptom of bacterial blight.

finally being shed. Another common leaf symptom occurs when lesions extend along the sides of the main veins. This may be seen together with or in the absence of ALS and is referred to as vein blight (Fig. 2.3).

Another external symptom affecting the leaves is the appearance of roughly circular, chlorotic patches, 1–3 cm in diameter, which are surrounded by small angular spots (see Plate 1C). This symptom seems to be more common in some areas than others, having been observed in the USA early in the history of the disease (Bryan, 1932) and is common in Zimbabwe. It seems likely that this symptom is due to invasion of the vascular tissue by the bacterium. Vascular infection has been demonstrated by Bhagwat and Bhide (1962) in India and by Wickens (1956) in East Africa, who reported that vascular infection can result in vein blight and stem lesions which can be severe at the seedling stage.

In the susceptible reaction, infection spreads from the leaf lamina down the petiole to the stem. The resulting sooty black lesion gives rise to the term blackarm by which the disease is often known (see Plate 1D). Blackarm lesions may completely girdle the stem, causing it to break in windy conditions or under the weight of developing bolls, with resulting loss of fruiting branches. Blackarm is the most severe manifestation of bacterial blight, occurring only on the most susceptible cultivars of *G. hirsutum*. It is more often seen on cultivars of *G. barbadense* but is very rare on the diploid cottons, *G. arboreum* and *G. herbaceum* (Verma, 1986).

Bacterial blight symptoms can also appear on the bracts (see Plate 1E), and infection of the flower bud or young boll causes shedding. On larger bolls, infection is first seen as a small (2–5 mm), round, water-soaked spot (see Plate 1F), becoming dark brown with age and 10 mm or more in diameter on a susceptible host. In wet weather, lesions on the boll may be numerous, usually clustered close to the suture or at the base of the boll, under the epicalyx. In severe cases, the lesion penetrates the boll wall to cause an internal rot. Internal infection of the mature boll does not often occur in the absence of insect damage. Insect pests such as the stainer bug (*Dysdercus* spp.) introduce the bacteria into the boll through feeding punctures. If the boll is not completely destroyed by internal invasion, then the lint is stained yellow by bacterial growth and seed from infected bolls often carries the bacteria on its surface.

Seed Infection

In addition to surface contamination of the seed, which occurs when the pathogen grows on the lint fibres, internal infection may take place directly through the micropyle, or may occur as a result of infestation of the crop by the stainer bug. The stainer carries the bacterium passively on its stylet, introducing it into the seed, which it pierces during feeding (Brinkerhoff and Hunter, 1963).

Failure to completely remove the bacterium from the seed during acid delinting with sulphuric acid gave the first indication that seed could be internally contaminated. Up to 20% of plants grown from contaminated seed developed cotyledon blight under conditions favouring the disease (Tennyson, 1936). In India, Patel and Kulkarni (1950) estimated that about 10% of seed harvested from infected bolls carried internal inoculum. Primary infection of the seedling originating from internal seed infection was regarded as a significant factor in disease epidemiology in Oklahoma, where 6.4% of seeds from infected bolls carried the pathogen (Brinkerhoff and Hunter, 1963). In Sudan, the pathogen was isolated from 29% of seed harvested from a heavily infected susceptible variety, despite the seed having been delinted and surface sterilized in mercuric chloride (El-Nur, 1970a).

For commercial seed in the USA, Schnathorst (1964) obtained values of between 0% and 4% infection, depending on the source. He also demonstrated that seed obtained from a disease-free crop could become externally contaminated, if it was ginned after seed cotton from an infected crop. Although the amount of contamination in a commercial crop may be low, Tarr (1961) stated that a single infected plant per 6000 plants was sufficient to initiate an epidemic in a susceptible variety, under Sudanese conditions.

The pathogen may survive in seed for periods as long as two years, although viability declines with storage. Hunter and Brinkerhoff (1964) found that the percentage of infected seedlings which resulted from planting seed which had been inoculated declined from 60% after 6 months' storage to 1% after 56 months.

Losses

Bacterial blight can cause stand losses and loss of vigour at the seedling stage, but the main cause of yield loss is the blackarm symptom, due to the loss of fruiting branches (Arnold, 1965).

In Sudan, complete crop failure was not unknown when highly susceptible varieties of *G. barbadense* were grown. El-Nur (1970b) reported that losses of 20% were common and yield loss from severe infection could be as high as 77%. Knight (1948b) considered that the incorporation of blight resistance into Sudanese varieties of *G. barbadense* was responsible for yield increases of up to 64%.

In the USA, Bird (1959) obtained losses as high as 34% in susceptible varieties following inoculation. An epidemic in New Mexico in 1949 caused an estimated 35–59% reduction in yield (Leyendecker, 1950). Despite the use of resistant varieties, bacterial blight in all its manifestations was estimated to cause a total reduction of 3.6% in the North American crop for the 1977 season (Bird, 1981). Although the disease varies in severity from season to season, it has become less important in the USA over the last decade and losses were negligible during the 1989/90 season (Blasingame, 1990).

In North West India, where the crop is grown under irrigation, losses of 5–20% are often experienced (Verma and Singh, 1971b). Tarr (1958) estimated that losses from the disease in other Asian countries, including China, Pakistan and the USSR, were of the order of 20–30%.

Loss estimates have usually been made by comparing the yield from resistant and susceptible varieties, either inoculated or uninoculated with the blight bacterium (e.g. Bird, 1959), or by measuring yield increment from controlling the disease with chemicals. Varying results have been obtained with both methods because of the difficulty in restricting the spread of the pathogen into control plots and the fact that the high levels of infection required occur only under certain weather conditions. The effect on yield of seed dressing to control bacterial blight has been thoroughly investigated in East Africa. In summarizing these experiments, Wickens (1961) concluded that seed treatment could increase yields of seed cotton by as much as 30%, although part of this increase might have been due to control of other seedling diseases. Arnold (1965) reported that, in Tanzania, the disease could reduce seed cotton yields by up to 350 kg ha^{-1} in a susceptible variety, and that a gain of 340 kg ha^{-1} could be obtained from controlling the disease with a chemical seed dressing. In India much larger gains can apparently be achieved when the disease is controlled by a combination of seed treatment and foliar sprays. Mathur *et al.* (1973) reported an increase from 1009 kg ha^{-1} of seed cotton in the untreated control plots to 2238 kg ha^{-1} from plots treated with streptomycin sulphate.

Causal Organism

Taxonomy, morphology and physiology

The bacterial blight pathogen was originally named *Pseudomonas malvacearum* by Smith (1901), who later referred to it as *Bacterium malvacearum* (E.F. Smith) (Smith, 1920). Then Dowson (1939) classified the bacterium as *Xanthomonas malvacearum* (E.F. Smith) Dow. However, the name now generally accepted is *Xanthomonas campestris* pv. *malvacearum* (E. F. Smith) Dye (*Xcm*) (Dye *et al.*, 1980).

The bacterium is an anaerobic Gram-negative rod, 0.3–0.6 × 1.3–2.7 μm in size with a single polar flagellum. It produces a yellow, copiously mucoid, wet, shining growth on 2% peptone–sucrose agar. Gelatin, casein and soluble starch are strongly hydrolysed. It does not produce nitrite from nitrate and metabolizes carbohydrates oxidatively. Acid is produced from glucose, sucrose, fructose, arabinose, galactose, maltose, cellobiose and glycerol but not from dulcitol, inulin or salicin. It is strongly lipolytic. The optimum temperature for growth is 25–30 °C, with a maximum at about 38 °C and a minimum at 10 °C (Hayward and Waterson, 1964).

Nutritional requirements

The bacterium can utilize simple sugars as a carbon source and grows well on nutrient agar (NA). Optimal growth in culture was obtained by Nayudu (1972) at pH 6.0 with a sucrose concentration of 0.2%. Amino acids were found to be a good source of nitrogen (but not of carbon) and better than inorganic nitrogen such as ammonium sulphate. Alanine, glycine, aspartic acid and glutamic acid can be utilized as the sole source of nitrogen, while alanine and glutamic acid can serve as a source of both carbon and nitrogen (Verma and Singh, 1974a).

Isolation and growth in culture

Xcm can be isolated from infected host tissue using standard bacteriological procedures. Fresh angular leaf spots are preferred for making isolations but fresh lesions from any part of the host will suffice. The pathogen can also be isolated from dried leaf tissue, after several years in store. Nutrient agar is a suitable medium for initial isolation. After pouring the medium into a Petri dish, the surface should be allowed to dry before use. If this is not done, discrete colonies cannot be obtained because the bacterium is motile in surface moisture. It is often the case that several yellow phylloplane bacteria (e.g. *Erwinia* spp. and *Flavobacterium*) grow on the isolation plate together with

Xanthomonas spp., but the saprophytic species tend to grow more quickly. Colonies visible to the naked eye after 24 hours at 30 °C on NA are unlikely to be the pathogen, which becomes visible after 36–48 hours on the plate as spherical, yellow, glistening colonies. Individual colonies which appear to have the correct characteristics should then be subcultured on to fresh NA plates and incubated for 48–72 hours before further testing.

IDENTIFICATION AND STORAGE

Obtaining pure cultures of a yellow bacterium is not a guarantee that the isolate is the blight pathogen. The identity of the isolate must be established by conducting a Gram stain test and other biochemical tests or by pathogenicity tests on seedlings of a fully susceptible variety such as Acala 44.

Once the identity of the isolate has been confirmed, the bacterium can be stored for several months on slants of a suitable medium placed in the refrigerator. The pathogen may remain viable for longer if the storage medium is different in composition from the isolation medium. Peptone sucrose agar (PeSA) is suitable and a small quantity of calcium carbonate added may be beneficial in preventing acid accumulation.

Variation in the Pathogen

There had been earlier speculation concerning the possibility of pathogenic specialization in the blight pathogen (e.g. Knight, 1946), but it was the breakdown in resistance of material produced in Sudan and Uganda when tested in India which led to the conclusion that biological races of the pathogen existed which were capable of overcoming resistance conditioned by two major genes (Balasubramanyan and Raghavan, 1950).

Soon after the cultivation of blight-resistant cottons began in the USA, new and virulent races of the pathogen appeared. In 1952, the cultivar Stoneville 20 became severely infected at three locations (Simpson, 1953). Hunter and Blank (1954) then isolated a strain capable of attacking all lines with resistance derived from Stoneville 20. Later, Bird and Hunter (1955) designated the original strain as race 1 and the new strain as race 2.

Brinkerhoff (1963) extended the work of Bird and Hunter to identify 12 races, using a set of differential hosts. In order to overcome the problem of inconsistencies in results obtained from different parts of the USA when these hosts were inoculated with the 12 races, a set of eight differential hosts was developed, each with a similar genetic background. The use of these differentials minimized the effect of genotype–environment interactions to give more consistent results at different sites, allowing previous results to be confirmed and a further three races to be identified (Hunter *et al.*, 1968; Brinkerhoff, 1970). Verma and Singh (1970) used the same hosts to identify two more races of the pathogen in India and another race was found in

Table 2.1. Races of the bacterial blight pathogen according to the system of Hunter *et al.* (1968).

Race no.	A	B	C	D	E	F	G	H	I	Ja
					Host differentials and their response to races					
1	+	+	−	−	−	−	−	−	−	−
2	+	+	+	−	−	−	−	−	−	−
3	+	+	−	−	+	−	−			
4	+	+	−	−	−	+	−			
5	+	+	−	−	+	+	−			
6	+	+	−	+	+	−	−	+	−	−
7	+	+	−	+	+	+	−	+	−	−
8	+	+	+	+	+	−	−	+	−	−
9	+	−	+	−	−	+	−			
10	+	+	+	+	+	+	−	+	−	−
11	+	+	−	−	−	−	−	+	−	−
12	+	+	+	−	−	−	−	+	−	−
13	+	−	−	−	−	−	−			
14	+	+	+	−	+	+	−	+	−	−
15	+	+	+	−	+	−	−			
16	+	+	+	+	−	+	−			
17	+	+	+	−	−	+	−			
18	+	+	+	+	+	+	−	+	+	+
19	+	−	−	−	+	+	−	+	−	−

[a] Host differentials: A, Acala 44; B, Stoneville 2B; C, Stoneville 20; D, Mebane B-1; E, 1-10B; F, 20-3; G, 101-102B; H, Gregg; I, Empire B4; J, DPX P4.

Pakistan (Hussain and Brinkerhoff, 1978), bringing the total number of races to 18, until race 19 was reported from central America (Ruano and Mohan, 1982) (Table 2.1).

Arnold and Brown (1968) put forward the view that the pathogen exhibits continuous variation and that there is therefore little value in attempting to define races. However, investigations by Gabriel *et al.* (1986) led to the conclusion that race specificity did indeed exist in *Xcm* just as in some fungal pathogens, and that the interaction between races of the pathogen and cotton genotypes fitted the conventional gene-for-gene model. It therefore appears that distinct races exist which differ from each other in pathogenic capability, but there is little doubt that symptom expression in any one interaction between pathogen and host is heavily influenced by environmental factors. In Zimbabwe, for instance, the predominant race exhibited a pathogenic capability corresponding to race 6 or race 10, depending on factors such as crop moisture status and prevailing weather conditions at and following inoculation.

The use of resistant varieties, especially if cultivated over large areas, has led to the appearance of new races at a number of locations. With the release in

Table 2.2. Races of the bacterial blight pathogen according to the system of Verma and Singh (1974b).

| Race no. | Host differentials and their response to races | | | | | | | Alternative race no.[b] |
	A	B	C	D	E	F	G[a]	
1	+	−	−	−	−	−	−	13
2	+	+	−	−	−	−	−	1, 11
3	+	−	+	−	−	−	−	−
4	+	−	−	+	−	−	−	−
5	+	−	−	−	+	−	−	−
6	+	−	−	−	−	+	−	−
7	+	+	+	−	−	−	−	2, 12
8	+	+	−	+	−	−	−	−
9	+	+	−	−	+	−	−	3
10	+	+	−	−	−	+	−	4
11	+	−	+	+	−	−	−	−
12	+	−	+	−	+	−	−	−
13	+	−	+	−	−	+	−	9
14	+	−	−	+	+	−	−	−
15	+	−	−	+	−	+	−	−
16	+	−	−	−	+	+	−	19
17	+	+	+	+	−	−	−	−
18	+	+	+	−	+	−	−	15
19	+	+	+	−	−	+	−	17
20	+	+	−	+	+	−	−	6
21	+	+	−	+	−	+	−	−
22	+	+	−	−	+	+	−	5
23	+	−	+	+	+	−	−	−
24	+	−	+	+	−	+	−	−
25	+	−	+	−	+	+	−	−
26	+	−	−	+	+	+	−	−
27	+	+	+	+	+	−	−	8
28	+	+	+	+	−	+	−	16
29	+	+	+	−	+	+	−	14
30	+	+	−	+	+	+	−	7
31	+	−	+	+	+	+	−	−
32	+	+	+	+	+	+	−	10, 18

[a] Host differentials A–G as in Table 2.1.
[b] Alternative numbering system of Hunter *et al.* (1968).

New Mexico of Acala 1517Br, resistant to race 1, there was a rapid build-up of race 2, which soon accounted for 40–60% of isolates tested. In 1961, Acala 1517Br-2, which was resistant to both races, was released and its cultivation led to a decrease in the occurrence of race 2 (Chew *et al.*, 1969). Five years later, varieties with resistance to race 1 and race 7 were being attacked (Chew and Booth, 1974). In Arkansas, the cultivation of Rex varieties with resistance to race 1 also led to a severe disease problem (McCutchen and Fulton, 1962). Breakdown of resistance occurred in Sakel cultivars of *G. barbadense* in Sudan (Gunn, 1961) and Upland cultivars in Tanzania (Crosse, 1963), which was attributed in both cases to race changes in the pathogen. In Central Africa, a new race (race 20) has appeared which is capable of causing symptoms on varieties previously resistant to the 19 other races (Follin, 1983b).

Verma and Singh (1974c) have suggested an alternative system of numbering the races. Their system numbers the races from 1 to 32, following a mathematical progression, according to the number of differential hosts (or resistance genes) attacked. This leaves room for the addition of further races as they are identified (Table 2.2). Race numbers under the original American system of Hunter *et al.* (1968) were allocated according to the chronological order in which the races were identified and do not relate directly to a gene-for-gene model. Both systems use the same set of differential hosts except that, in the Indian system, Gregg 8 is reserved for distinguishing biotypes within races. Race 10 and race 18 in the American system both cause symptoms in all the differentials except 101–102B. They were distinguished only after the addition of two further differential varieties, Empire B4 and DPXP4, which are susceptible to race 18 (Bird, 1981). Verma and Singh (1974c) prefer to reserve the additional differentials for distinguishing biotypes and in their system race 10 and race 18 are therefore identified as biotypes of race 32.

Shortly after the beginning of large-scale cultivation of cotton in Australia, only race 1 was identified among collected isolates (Brinkerhoff, 1966). Later, race 10 was also identified using the basic eight differentials. But, when the isolates were retested with the extra hosts added to the set, those reported as race 10 were redesignated as race 18 (Fahy and Cain, 1987). Of eight isolates tested from Queensland and Western Australia, five were found to be race 1 and three were race 18 (Allipi and Hayward, 1987).

The set of differential hosts assembled by Hunter *et al.* (1968) has been used in a number of countries to determine the races present in the local population of the pathogen. In most cases, more than one race has been identified (see Table 2.3). More than one race has been isolated from a single lesion on leaves of *G. hirsutum* (Chowdhury *et al.*, 1979). Also, race changes have occurred during passage through the plant. By 56 days after inoculation of Stoneville 20 with race 1, 77% of bacterial cells recovered from the plant were found to be race 2 (Schnathorst, 1970). A similar observation was made by Hande and Rane (1983) in India, where five out of 18 reisolates exhibited virulence towards one more host differential than did the parent isolate with which the plant was inoculated.

Phage Typing

Bacteriophages associated with bacterial blight lesions have been known for many years (e.g. Matsumoto and Husioka, 1939). Phages which attack the genus *Xanthomonas* show considerable specificity at the species and sub-species level (Singh *et al.*, 1970) and variation in the susceptibility of *Xantho-monas* isolates to lysis by phages was first noted by Okabe (1939). Such selectivity was used by Hayward to distinguish between strains of the bacterial blight pathogen (Hayward, 1963). Two types of the bacterium were distinguished on the basis of their susceptibility to phage lysis and these types

Table 2.3. Races of the bacterial blight pathogen and some countries where they have been identified.

Race no.[a]	Countries where identified	Reference
1	Australia, USA, India	Allipi and Hayward, 1987; Bird and Hunter, 1955; Chowdhury *et al.*, 1979
2	USA, India	Bird and Hunter, 1955; Chowdhury *et al.*, 1979
3	USA, India	Miller, 1968; Chowdhury *et al.*, 1979
4	USA, India	Miller, 1968; Verma and Singh, 1975b
5	USA, India	Miller, 1968; Srinivisan and Taneja, 1974
6	Nigeria, Zimbabwe, India	Poswal, 1988; Hillocks and Chinodya, 1988; Chowdhury *et al.*, 1979
7	Venezuela, Nigeria, USA	Alcala and Camino, 1987; Poswal, 1988; Miller, 1968; Singh *et al.*, 1981
8	Venezuela, Peru, Pakistan, India	Alcala and Camino, 1987; Cedano and Delgado, 1986; Hussain, 1984; Sing *et al.*, 1981
9	No information	
10	Pakistan, Nigeria, Zimbabwe, India	Hussain, 1984; Poswal, 1988; Hillocks and Chinodya, 1988; Chowdhury *et al.*, 1979
11	USA, India	Miller, 1968; Singh *et al.*, 1981
12	Pakistan, USA, India	Hussain, 1984; Miller, 1968; Chowdhury *et al.*, 1979
13	India	Singh *et al.*, 1981
14	Philippines, USA, India	Dizon & Reyes, 1987; Miller, 1968; Singh *et al.*, 1981
15	USA, India	Miller, 1968; Singh *et al.*, 1981
16	West and Central Africa, India	Follin, 1983a; Verma and Singh, 1970
17	India	Verma and Singh, 1970
18	Australia, West and Central Africa, Pakistan, Nicaragua, India	Allipi and Hayward, 1987; Follin, 1983a; Hussain, 1984; Zachowski and Rudolph, 1988; Singh *et al.*, 1981
19	Brazil, India	Ruano and Mohan, 1982; Chowdhury *et al.*, 1979
20	Sudan, Chad, Upper Volta	Follin, 1983b; Ruano *et al.*, 1988

[a] Race numbers in accordance with the system of Hunter *et al.* (1968).

were later shown to differ pathogenically, although pathogenic differences existed also within cultures designated as Type 2 (Crosse and Hayward, 1964).

Disease Cycle

The blight pathogen survives poorly in soil in the absence of plant debris (Massey, 1930) and the main source of primary inoculum is contaminated seed and crop residues (Brinkerhoff and Hunter, 1963; Brinkerhoff and Fink, 1964). During seed germination, the basal cap adheres to the cotyledons, causing lesions to appear on them as the seedling emerges (Verma, 1986). The cotyledon can also become infected as it emerges through the soil, due to contact with infested trash.

Once the seedling is above ground, it may become infected as a result of secondary inoculum spread from other infected plants, or from infested debris by wind and runoff from rainfall or irrigation (Rolfs, 1935; King and Brinkerhoff, 1949).

The secondary cycle of infection on seedlings, following cotyledon infection from seed or crop debris, occurs with the spread of bacteria to the lower leaves, by rain splash or contact with the cotyledons as they move in the wind (Hansford *et al.*, 1933). The pathogen enters the leaf through stomata, abundant infection occurring only when the stomata are open and the intercellular spaces filled with water (Weindling, 1948). There is also some evidence that the bacterium is capable of direct penetration of epidermal cells at high inoculum rates due to enzyme activity (Dharmarajulu, 1966). Once within the leaf, the bacteria spread from the substomatal cavity, breaking down the walls of the spongy mesophyll cells. Pectic enzymes are produced by the pathogen but only in the presence of a pectinaceous substrate (Aboel-Dahab, 1964; Vidhyasekaran *et al.*, 1971; Papdiwal and Deshpande, 1983). The production of cellulase and protease enzymes has been demonstrated in culture and they play an important role in pathogenesis (Verma and Singh, 1971c, 1975a). The palisade cells are also broken down and the mesophyll tissue is completely destroyed by the time the visible lesions on the surface become dry and dark in colour (Massey, 1929; Thiers and Blank, 1951).

Histological studies have shown that tissue destruction is confined to the parenchyma, seldom spreading to the vessels. The bacterium spreads through the plant in the intercellular spaces of the cortical parenchyma. When the cotyledon is infected from contaminated seed, the bacteria spread to the main stem along the cortical parenchyma, affecting each developing organ as the plant grows (Thiers and Blank, 1951). Tissues surrounding the stele are destroyed, allowing free passage to the bacterium, and the endodermis breaks down some distance in advance of the visible lesion (Massey, 1929).

Under some conditions, not clearly defined, true vascular infection occurs, allowing systemic spread of the bacteria within the plant, even under conditions too dry for disease spread in the more superficial tissues (Rolfs,

1935; Bhagwat and Bhide, 1962). Jahkanwar and Bhagwat (1971) have demonstrated by microscopy the presence of the bacterium in the conducting tissues and in the cotyledonary tissue and resin canals of the seed. They conclude from this latter observation that the seed can become infected by bacteria carried along vascular bundles.

The pathogen can survive at least as long as the period between crops, in dry leaf trash and infected seed (Schnathorst, 1964), so that the disease cycle is completed with the return of bacteria to the soil in falling leaves, and when infected seed is used for planting the following season's crop.

Epidemiology

Massey (1929, 1934) demonstrated the importance of soil temperature in primary infection of the seedling. Infection of germinating seeds did not occur above 30 °C, with little infection at 28 °C. Disease development was slow below 15 °C and only mild infection occurred at 16–20 °C. Severe infection was limited to soil temperatures in the range 26–28 °C. In contrast to these findings, Stoughton (1930) reported that soil temperature had little effect on secondary infection.

Free water was regarded by Massey (1931) as essential for cotyledon infection from seed-borne inoculum. The incidence of seedling infection is therefore higher in poorly draining soils or if heavy rain occurs during germination and emergence. While the presence of free water and moderate temperature are optimal for primary infection of the seedling, secondary spread within the crop is favoured by high relative humidity (RH) and high temperatures. Work conducted by Stoughton (1931, 1933) showed that, at high RH, the disease was most severe at 36 °C, infection decreasing at progressively lower temperatures. The optimum temperature for disease development in the USA was also reported to be about 36 °C, especially with moderately low night temperatures (19 °C), under which conditions blight resistance breaks down in G. hirsutum (Brinkerhoff and Presley, 1967). If the temperature is too high, humidity tends to be lower and the leaf surface dries out. Under these conditions the leaf tissues are resistant to the disease (Brinkerhoff, 1963).

Massey found that lesions formed on leaves when the RH was as low as 25%, provided that inoculum levels were high and the leaf surface wet. This is probably why outbreaks of the disease in California have been associated with the use of sprinkler irrigation (Schnathorst et al., 1960). Stoughton (1932, 1933) found that at 25 °C infection was greatest at 85% RH but only slight at 70% RH. In northern India, blight incidence was severe at locations where RH was 75% or more for at least ten weeks of the growing season. Rain at four to six weeks after planting was regarded as a warning of possible epidemic conditions (Verma and Singh, 1971b).

Although the cotton plant can be infected by the disease at any stage, young leaves of young plants are more susceptible than older leaves or young leaves on older plants and the disease is more likely to reach epidemic proportions if the infection is well established in the crop by six weeks after planting.

Optimal conditions for the development of an epidemic are therefore as follows: (i) the establishment of primary infection at the seedling stage; (ii) early rainfall to distribute the disease through the crop by six weeks after planting; (iii) periods of heavy wind-driven rain after canopy has formed, interspersed with periods of sunshine to raise the RH within the crop to 85% or over; and (iv) high temperature during the secondary phase of the disease, with temperatures in the range of 32–38 °C during the day and 17– 20 °C at night.

Role of Insects in Dissemination of the Pathogen

Xcm, in common with many other phytopathogenic bacteria, produces a polysaccharide slime which oozes out when infected host tissue is damaged. The slime readily adheres to the body parts of insects, which disseminate the pathogen mechanically in moving between plants. The common sucking pests are more likely to spread the infection by making numerous punctures in the leaves through which the bacteria can enter the host. In the case of the tarnished bug (*Lygus vosseleri*), the bacterium may be carried internally and transmitted to the leaf through the feeding stylet (Logan and Coaker, 1960). The presence of the pathogen has also been demonstrated within the larvae of the spiny bollworm (*Earias* spp.), which had acquired it after feeding on a pure culture or from infected leaf pieces (Borkar *et al.*, 1980). The pathogen is carried externally and internally by the stainer bug (*Dysdercus kregigii* Fabr.). Healthy plants developed blight symptoms after they were infested with bugs which had been placed for 30 min on a culture of the bacterium (Verma *et al.*, 1981). In the USA, the disease is also transmitted by the cotton fleahopper (*Pseudatomascelis seriatus*) (Martin *et al.*, 1988).

In Africa, the stainer bug appears more often towards the end of the season and bacterial blight symptoms on the boll may be associated with their feeding punctures. They introduce the bacterium into the young boll, which they pierce in order to reach the seed; this can result in internal boll rot. They also feed on the seeds after boll split and may be responsible for introducing the bacterium into the seed.

Control

CHEMICAL SEED TREATMENT

It was established early in the history of bacterial blight that contaminated seed was a source of primary inoculum (Massey, 1930, 1931). Various seed treatment methods were developed to reduce this source of seedling infection and the practice was established in Africa and the USA by the early 1950s. The methods most commonly used were acid delinting with concentrated sulphuric acid, steeping in 1% formaldehyde and dusting with mercuric compounds (Wickens, 1953).

Acid delinting is now widely practised and is a prerequisite where seed is mechanically planted. It is very effective in eliminating surface contamination by micro-organisms and provides adequate control of the disease in California, where conditions are too dry at harvest for internal infection of the seed (Schnathorst *et al.*, 1960). The procedure is not effective against internal inoculum (Schnathorst, 1968) and, in areas where internal infection occurs, the seed must be chemically treated after or, in some countries, instead of acid delinting.

Seed-dressing with mercuric compounds was widely adopted in the Sudan (Tarr, 1953; El-Nur, 1970a), India (Verma *et al.*, 1974) and Nigeria (Wickens, 1958). Cuprous oxide was less effective than the mercurials when the two were compared in Uganda (Logan, 1960), but was preferred in East Africa (Brown, 1976) and gradually replaced mercurial compounds in many other areas because of their toxicity. Despite their toxicity, the use of organomercurial compounds as cotton seed-dressings still finds widespread acceptance. They are still favoured in India when alternatives are not available (Bhandari *et al.*, 1969; Singh *et al.*, 1970) and are the main seed treatment compounds in Francophone Africa (Innes, 1983). The high price of copper in the 1960s provided the impetus for research into alternatives to copper formulations (Table 2.4).

The bactericidal compound Bronopol was recommended in Nigeria (Dransfield, 1969) and in Tanzania, where all seed intended for planting was treated with the compound before leaving the ginnery (Hillocks, 1984). In

Table 2.4. Some chemicals used as seed-dressings and/or foliar sprays to control bacterial blight.

Common name	Application method	Reference
Antibiotics, e.g. streptomycin sulphate	Usually foliar	Verma *et al.*, 1974
Bronopol	Seed-dressing	Dransfield, 1969
TCMTB	Seed-dressing	Pulido and Bolton, 1974
Carboxin	Seed-dressing and foliar	Jalali and Grover, 1974
Copper compounds	Seed-dressing	Logan, 1960
Mercury (inorganic)[a]	Seed-dressing	El-Nur, 1970a
Organomercurial[a]	Seed-dressing	Shivanathan and Gunasingham, 1966
Oxycarboxin	Seed-dressing and foliar	Jalali and Grover, 1974

[a] Compounds containing mercury have been withdrawn from most countries.

India and the USA, TCMTB was found to give good control (Verma and Singh, 1971a; Pulido and Bolton, 1974), and it inhibits growth of the bacterium in culture at lower concentrations than either mercurial compounds or copper oxychloride (Verma *et al.*, 1975). The oxanthiin compounds, carboxin and oxycarboxin, also appear to be very effective against the blight pathogen. They are recommended for use in India at a rate of $2 \, g \, kg^{-1}$ of seed (Jalali and Grover, 1974) and in Colombia at $8-10 \, g \, kg^{-1}$ of seed (Vina and Granada, 1986).

Antibiotics are used in bacterial blight control in the former USSR (e.g. Kravchenko and Ponmarev, 1964) and in India, where compounds such as streptomycin and tetracyclines are sometimes employed as seed-dressings (Verma *et al.*, 1974).

Seed treatment has its limitations as a control measure. The organomercurial compounds still used in parts of India do not completely eliminate internal inoculum unless combined with antibiotics (Verma, 1986). There is a danger that the use of seed-dressing chemicals tempts the farmer into ignoring other control measures such as crop rotation. Unless the chemical is also applied to the soil surrounding the seed, little protection is provided against trash-borne infection (Arnold and Arnold, 1961). In experiments conducted in Uganda by Logan (1961), seed treatment with Agrosan reduced primary infection of the seedling and also reduced the incidence of stem infection, the main cause of yield loss. But, under conditions favouring secondary spread of the disease, the beneficial effects of seed treatment were not apparent at the boll stage. Seed treatment can only be an effective means of control when the crop is not exposed to secondary infection after emergence, from infected volunteers or crop residues. For instance, the pathogen can be carried considerable distances by irrigation runoff to cause severe infection in seedlings previously protected from primary infection by seed-dressing (King and Parker, 1939).

FOLIAR SPRAYS

When seed treatment has failed to control the disease beyond the seedling stage, the crop can be protected from an epidemic by chemicals applied to the foliage. This is an expensive practice and rarely worthwhile in crops of Upland cotton, but it may be justified in long-staple cottons. Foliar sprays for blight control have not been used in Africa but have been advocated in Brazil (Frenhani *et al.*, 1969). They are most widely used in India (Verma *et al.*, 1974), particularly where they can be combined with sprays for control of insect pests and other foliar diseases such as Alternaria leaf spot (Padaganur and Basavaraj, 1983).

Some control can be achieved with copper-based fungicides, although compounds such as copper oxychloride act only as protectants. Much better results have been obtained with systemic compounds, mainly antibiotics and oxanthiin compounds (Jalali and Grover, 1974). Nayak *et al.* (1976) tested a number of protectant and systemic compounds against bacterial blight in the field. Carboxin and oxycarboxin were the most effective of those tested and were recommended at 1.5–2 kg ha⁻¹. Agrimycin and Busan-72 also gave good control. Although copper oxychloride does not give particularly good control on its own, it is much more effective in combination with antibiotics or oxanthiin compounds. Singh and Verma (1973) obtained effective control with foliar sprays of 0.2% copper oxychloride and 1% oxycarboxin. Ekbote (1985) recommended agrimycin at 5 ppm with 0.25% copper oxychloride.

In some cases, antibiotics have been used on their own to control bacterial blight. Mathur *et al.* (1973) were able to obtain a 64% increase in yield over the untreated control by using seed treatment with agrimycin (3 g 10 kg⁻¹ of seed), followed by foliar sprays containing 25 ppm of agrimycin.

Verma (1986) suggests that the decision to spray should be based on examination of the crop six weeks after planting. Spraying should then begin within the next two weeks at 10–20-day intervals, up to a maximum of three sprays. In Haryana State (India), disease severity was much reduced and yields increased by a mixture of agrimycin (0.01%) and copper oxychloride (0.2%), applied to the foliage at three growth stages; 40–50, 70–80 and 85–95 days after planting (Chauhan *et al.*, 1983).

Verma and Singh (1971b) noted that the antibiotics had the lowest minimum inhibitory concentrations *in vitro*, but were not very effective at controlling the disease in the field. The most effective seed-dressings were TCMTB, carboxin and oxycarboxin. Antibiotics have the added disadvantage that resistant strains of the pathogen have developed in culture (Nafade and Verma, 1985) and might do so in the field.

Verma (1986) concluded in his review of chemical control that the most effective practice was to use one of the oxanthiin compounds as a seed-dressing to reduce primary infection, and apply foliar sprays of the same compound at six to eight weeks after planting, if necessary. The use of the oxanthiin compounds for seed treatment has the additional advantage of controlling damping-off caused by *Rhizoctonia solani*.

CULTURAL PRACTICE

Measures intended to reduce seasonal carry-over of the pathogen in crop residues can be very effective in controlling bacterial blight. Soon after the importance of crop residues in survival of the pathogen in the Sudan Gezira was recognized, flooding the land after harvest was advocated as a control measure (Andrews, 1937). In the USA, burning surface residue at the end of the season was recommended (King and Parker, 1939).

The pathogen may survive for long periods in dry leaf trash (Schnathorst, 1964) but viability is lost more quickly if the residues are buried. Ploughing-in of residues was recommended as a control measure in East Africa (Arnold and Arnold, 1961) and also contributes to control of the disease in other countries, such as Zimbabwe. Brinkerhoff and Fink (1964) noted that the survival period of the bacterium was shorter in moist than in dry soil, and then recommended chopping crop debris and burying it in moist soil at the end of the season. It has been suggested that the significance of wet conditions in bringing about a rapid decline in pathogen viability may be related to the action of antagonistic bacteria under anaerobic conditions (Habash, 1968). In areas such as California, which experience dry conditions at the end of the cotton season, the rate of decay of crop residues can be accelerated by irrigating the land after the trash has been ploughed in (Schnathorst, 1966).

In India, the pathogen was found to remain viable in crop residues for six months on the surface and three months in buried trash. The survival period was reduced to one month by flooding the land or planting a crop of wheat after the cotton harvest (Verma *et al.*, 1977). Except under the driest conditions, therefore, the pathogen rarely survives in the field for longer than 12 months, so that crop rotation can provide effective control. In Zimbabwe, a break of one season between cotton crops, during which the land is planted to maize or sorghum, is sufficient to virtually eliminate crop residues as a source of primary inoculum.

Heavy rain at planting is recognized as an important factor in the establishment of early infection of the crop and, where this occurs, delayed planting may reduce primary infection. However, it was noted in both the USA (Hare and King, 1940) and India (Patel and Kulkarni, 1950) that seed cotton left in the field can germinate to produce infected volunteer plants as a result of contact with infected residues or because the seed itself was contaminated. Therefore, where delayed planting or crop rotation are practised to reduce early infection, it is advisable also to remove volunteer plants.

INTEGRATED CONTROL

In areas where conditions are favourable to the development of epidemics, none of the control measures advocated are likely to provide complete control on their own. Most countries in which the disease is an immediate or potential threat to production operate some kind of integrated control programme, although some components of the system may have been instigated for reasons other than blight control. Destruction of crop residues at the end of the season is widely practised to control a number of pests and diseases (Pothecary and Ofield, 1968). In Tanzania, the blight control programme combines the use of resistant varieties with chemical seed treatment and there is a closed season, aimed primarily at exclusion of red bollworm (*Diparopsis castenea*), which also reduces the carry-over of bacterial blight (Hillocks, 1981). In Zimbabwe, where planting seed is acid delinted, it is not considered necessary to further treat the seed due to the success of resistant varieties and careful management of a sanitation programme, involving destruction of crop residues and enforcement of a closed season.

Bacterial blight first appeared in California in 1951, becoming widespread on the susceptible Acala varieties by 1957, when a sanitation programme was initiated in an attempt to control the disease (Schnathorst, 1966).

An integrated approach was adopted which included the use of acid delinted seed, obtained from blight-free areas, crop rotation, ploughing-in of crop residues and a preference for furrow irrigation, rather than sprinklers. Strict adherence to this programme resulted in eradication of the disease by 1962. Blight did not reappear in the state until 1967, when it was reintroduced on contaminated seed brought in from another state for variety trials (Schnathorst, 1968).

BIOLOGICAL CONTROL

Verma and Singh (1975c) reported that pre-inoculation of leaves of a susceptible variety with avirulent strains and heat-killed cells of the blight bacterium, or with phylloplane bacteria, 20 hours before inoculation with a virulent strain, delayed symptom expression by three days. Although no practical method of biological control has been developed up to the present, much work has been done with phylloplane bacteria since that first report (e.g. Chowdhury and Verma, 1980a; Sinha and Verma, 1983; Verma *et al.*, 1983). Verma (1986) considers that, properly exploited, they show considerable potential as biocontrol agents.

The main phylloplane bacteria identified on cotton by Verma *et al.* (1982) were *Aeromonas* (29%), *Flavobacterium* (29%), *Corynebacterium* (7%), *Micrococcus* (21%) and *Pseudomonas* (14%). Those showing activity against *Xcm* represent about 14% of the total bacterial population, compared with 34% for the pathogen. Dilute sprays with certain phylloplane bacteria resulted in a reduction in the number of lesions per leaf, although not in the number of infected leaves. The greatest protective effect was given by two species of *Flavobacterium*, one species of *Pseudomonas* and one of *Aeromonas* (Verma *et al.*, 1982).

RESISTANT VARIETIES

Genetics and sources of resistance

Considerable variability for resistance to blight exists in the genus *Gossypium*. Among the tetraploid cottons, the full range of disease expression from fully susceptible to resistant is found in the Upland group. The highest degree of resistance is found in *G. hirsutum* var. *punctatum* and many Upland varieties exhibit some degree of field resistance or tolerance. Little resistance occurs naturally in *G. barbadense*, most being either susceptible or highly susceptible. Where some resistance has been encountered, it is probably due to introgressive hybridization. Resistant varieties of both tetraploid species have been produced in a number of countries, mainly by the transfer of resistance genes originally derived from other species of *Gossypium*.

None of the wild or cultivated species of diploid cottons are as susceptible as some varieties of *G. barbadense*. The more susceptible species are *G. thurberi* and *G. stocksii*. Indian races of *G. arboreum* and *G. herbaceum* which have been cultivated on the subcontinent for centuries are highly resistant or immune to the disease. However, some members of both species are susceptible, especially the perennial types. *G. anomalum* and *G. raimondii* species groups are either resistant or highly resistant (Knight and Hutchinson, 1950; Brinkerhoff, 1970; Verma, 1986).

Heritable resistance to bacterial blight was first demonstrated by Knight and Clouston (1939) in the Sudan, where severe losses to the disease were a regular occurrence on the highly susceptible *G. barbadense* cultivars grown in the Gezira. Knight (1944, 1948a, 1950, 1953a,b, 1954c, 1963) screened over 1000 accessions of diploid and tetraploid species of *Gossypium*. He identified ten major genes for blight resistance (B-genes) to which he ascribed the symbols B_1–B_{10}. Eight of these genes were dominant or partially dominant in their expression. Of the remaining two genes, B_8 from *G. anomalum* was recessive and B_6 from *G. arboreum* was originally described as a modifier gene with no effect by itself but was later shown to be a recessive gene, conferring intermediate resistance in the homozygous state (Saunders and Innes, 1963). In some cases, the major genes (oligogenes) were accompanied by minor genes (polygenes) which enhanced their expression (Knight, 1948b).

Similar surveys were carried out in the USA, where resistant plants were encountered at low frequencies in local populations of cultivated cottons (Brinkerhoff *et al.*, 1952; Mooseberg, 1953). In the Soviet Union too, blight-resistant plants were identified in the local population of Upland cotton (e.g. Verderevsky and Voitovitch, 1957).

The gene B_7 was originally identified in the variety Stoneville 20, a blight-resistant selection made in the USA by Simpson and Weindling (1946). Bird and Blank (1951) concluded that resistance in Stoneville 20 was recessive, but in the Sudan it was dominant in its expression (Knight, 1953b). This gene formed the basis of blight resistance in some of the early breeding pro-grammes in the USA and later work established that the expression of B_7 resistance is influenced by the presence of minor genes. The gene could therefore behave as dominant or recessive, depending on the genetic back-ground of the recipient variety (Bird and Hadley, 1958; El-Zik and Bird, 1970).

Six more B-genes were subsequently added to Knight's original list. The gene B_{11} was found in *G. herbaceum* by Knight, but it is of low potency (Innes, 1966). Two genes for resistance were found in the Upland variety Allen and given the symbols B_9 and B_{10}. They were not homologous with Knight's B_9 and B_{10} and were therefore given the symbols B_{9L} and B_{10L} (Innes, 1965a).

Green and Brinkerhoff (1956) studied the inheritance of three more major genes, which they identified in American Upland varieties. These were given the symbols B_{1n} from the variety 1–10B, B_N from Northern Star and B_S from Stormproof 1. Resistance in another three varieties from the USA was shown by Bird and Hadley (1958) to be controlled by polygene complexes. These were given the symbols B_{Sm}, found in Stoneville 2B and Empire, and B_{Dm}, found in Deltapine (Table 2.5).

Breeding Programmes

SUDAN

The first successful breeding programme for resistance to bacterial blight was conducted in the Sudan, where the Sakel cultivars (*G. barbadense*), which had desirable fibre qualities, were highly susceptible. Having identified major genes for resistance, Knight (1954a, b) and later Innes (1966) transferred all the genes to Sakel by back-crossing (Knight, 1945). The genes were also transferred to Upland varieties in the Sudan (Knight, 1954a; Kheiralla, 1970).

The genes B_2, B_3 and B_6, proved to be the most useful in the Sakel background. B_2 gave a degree of resistance on its own which was greatly enhanced when combined with B_3. B_6 greatly enhanced leaf resistance conferred by B_3 (Knight, 1953a). $B_2 + B_6$ gave field immunity and the combination of the three genes gave complete immunity under Sudanese conditions

Table 2.5 Major genes and polygene complexes conferring resistance to bacterial blight (Innes, 1983).

Gene symbol	Description and original source
B_1	Weak dominant gene from Uganda B31 (*G. hirsutum*)
B_2	Strong dominant gene from Uganda B31 and others
B_3	Partially dominant gene from Schroeder 1306 (*G. hirsutum* var. *punctatum*)
B_4	Partially dominant gene from Multani strain NT 12/30 (*G. arboreum*)
B_5	Partially dominant gene from Grenadine White Pollen, (a perennial *G. bardadense*)
B_6	Recessive gene from Multani Strain NT/12/30
B_7	Gene from Stoneville 20 and others in USA (*G. hirsutum*)
B_8	Recessive gene from *G. anomalum*
B_{9K}	Strong dominant gene from Wagad 8 (*G. herbaceum*)
B_{10K}	Weak, partially dominant gene from Kufra Oasis (*G. hirsutum* var. *punctatum*)
B_{9L}	Strong dominant gene from Allen 51-296 (*G. hirsutum*)
B_{10L}	Weak gene from same source as B_{9L}
B_{11}	Weak gene from same source as B_{9K}
B_{1N}	Dominant gene from unknown *G. hirsutum* in USA
B_N	Dominant gene from Northern Star, a variety of *G. hirsutum* from USA
B_S	Dominant gene from Stormproof 1, a variety of *G. hirsutum* from USA
B_{Dm}	Polygene complex found in a Deltapine variety (*G. hirsutum*) from USA
B_{Sm}	Polygene complex found in Stoneville 2B and Empire varieties (*G. hirsutum*) from USA

(Knight, 1956). The first resistant commercial Barbadense variety in Sudan was Bar XL1 (B_2), followed by Bar 14/25 (B_2B_3) and later by Barakat (B_2B_6). Among the successful Upland varieties were Bar SP84 (B_2), Bar 7/8 (B_3) and Bar 11/7 (B_2B_3). The gene combination B_2B_6 was transferred to the susceptible variety Acala 4-44 to produce Barac(67)B which became widely grown in the Gezira. Instead of back-crossing, a different approach was used in the production of Albar(57)12. Because the population exhibited considerable variability in reaction to the disease, line breeding with heavy selection pressure for resistance was used to produce this variety, which is still grown in the Nuba Mountains (Innes, 1983).

A new isolate has been identified in Sudan which shows virulence towards cultivars carrying B_2B_6 resistance. But a line has been developed (S295) which is resistant to this isolate of the bacterium (Wallace and El-Zik, 1990).

EAST AND CENTRAL AFRICA

The origin of many of the Upland varieties developed by the Cotton Research Corporation in the countries of East and Central Africa which are now members of the Commonwealth can be traced back to the landrace Allen (Fig. 2.4). Allen was originally taken from the USA to Nigeria, where the blight-resistant selection Nigerian Allen was the source of the Samaru seed issues in Nigeria, the BPA and SATU varieties in Uganda and the Albars of Central and Southern Africa (Innes, 1965b; Innes and Jones, 1972).

Resistance in Allen is conditioned by the genes B_2 and B_3 derived by introgression from *G. hirsutum* var. *punctatum* in Nigeria (Knight and Hutchinson, 1950). Blight resistance in the early years of cotton expansion in Africa was therefore due to B_2 and possibly B_2B_3 in some Allen derivatives.

Many other American Uplands were introduced into Africa and became widely distributed in East and Central Africa. They became mixed and interpollinated, so losing their individual identities. Over the years, these cottons became adapted to conditions in the locality where they were grown and new names were adopted to describe cottons grown in a particular area (Arnold, 1970). The cultivated cotton in Tanzania known as Mwanza local suffered severe attacks from the jassid bug (*Empoasca* spp.), causing severe damage to the crop. A successful selection programme for improved jassid resistance was conducted at Ukiriguru Research Station. As Mwanza local became more resistant to jassid, it was apparent that the material was susceptible to bacterial blight, and crosses were made between jassid-resistant Mwanza selections and blight-resistant Albar 51 from Uganda. The first commercial varieties to emerge from the hybridization with Albar were UK 64 and, later, UK 74. Further crosses were made to introduce more B-genes into the Ukiriguru population. Crosses with Reba W296 gave rise to the commercial variety UK 71. Selections from crosses with Bar 12/16 are involved in the background of the successful commercial variety UK 77.

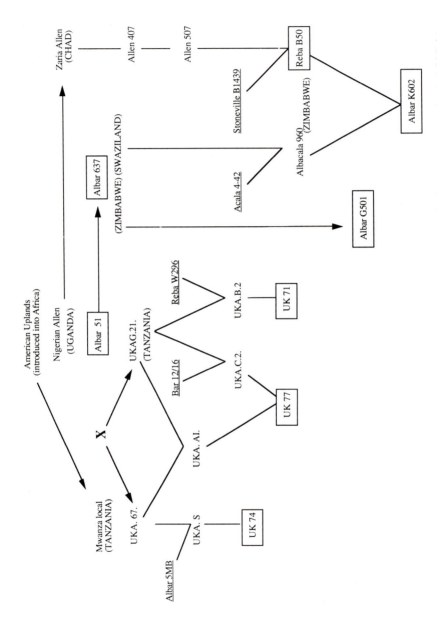

Fig. 2.4. Origin of some cotton varieties grown in East and Central Africa which are resistant to bacterial blight.

Until the early to mid-1980s, commercial cultivars in Zimbabwe were derived by reselection from within original populations of blight-resistant Albar selections, obtained from Uganda (see Fig. 2.4). One of these varieties was Albar G501, which exhibited good blight resistance unless exposed to inoculum under warm, wet conditions early in the growing season. Beginning with Albar K602, the more recent commercial varieties Albar K502 and Albar K603 are selections from crosses between Albar material and the blight-resistant variety Reba B50. They have shown a level of blight resistance superior to that of Albar G501.

Francophone Africa

In the French-speaking African countries, successful blight-resistant varieties have been developed by the Institut de Recherche du Coton et des Textiles Exotiques (IRCT) from two sources, Zaria Allen and the American variety Triumph with introgression from N'Kourala (Innes and Jones, 1972; Roux, 1977). The genes which might contribute to resistance in varieties derived from these sources are B_2 and B_{9L} with B_3 and minor genes (Lagière, 1960; Innes, 1963, 1965a,b). Reba B50 (Allen × Stoneville) has been particularly successful both as a commercial variety in Central Africa, South America and Thailand (Roux, 1977) and as a donor of blight-resistance genes for commercial varieties in Zimbabwe (Hillocks and Chinodya, 1988), Paraguay and Argentina (IRCT, 1977a).

One of the most widely distributed varieties from the IRCT programme was BJA-592, derived from the cross Triumph × N'Kourala. It is grown in Chad, Mali, Cameroon, Ghana and Senegal (Roux, 1977). Although the blight-resistant varieties grown in some of these countries were susceptible to a new race (race 20) of the pathogen, which was found in the early 1980s, a resistant line (S295) was produced in Chad within a few years of the appearance of the race 20 (Girardot and Follin, 1987).

There is considerable interest being shown in triple hybrids, originally produced in the Ivory Coast from *G. hirsutum* × *G. arboreum* × *G. raimondii* (HAR) (Innes, 1975). The HAR variety L229-10 is grown in the Ivory Coast, Togo and Benin and L142-9 is grown in Cameroon, both of them having only moderate blight resistance, derived from Allen (Innes, 1983).

India

G. arboreum and *G. herbaceum* have a long history of cultivation in India and a high level of resistance is to be found within both species. The widespread cultivation of tetraploid cottons is much more recent and many of the varieties grown are blight-susceptible (Innes, 1971; Verma and Singh, 1974c). There is therefore increasing interest in breeding for resistance. Race 18 of the pathogen is widely distributed in the country and is capable of overcoming resistance derived from B_2, B_4, B_7, B_{1n} and B_N (Verma *et al.*, 1980). Doubt has been cast on the worth in India of resistance controlled by the gene combination B_2B_3 in the absence of minor gene complexes (Balasubramanyan and Ragha-van, 1950).

Although so far not meeting the standards of yield and quality found in the more susceptible varieties, resistant lines have been produced in India. The earliest resistant Upland types were produced from crosses between *G. hirsutum* and *G. herbaceum* (Jagannatharao *et al.*, 1952, 1953). Beginning with Upland material containing the genes B_2 and B_3, breeding for resistance has been conducted at the Indian Agricultural Research Institute in New Delhi. Inoculation was carried out, using mainly race 18 but also mixtures of both races to produce the resistant line BJR-734. This line has not been accepted as a commercial variety but has been utilized as a donor of resistance to blight and the jassid bug in further breeding programmes (Singh *et al.*, 1973). A series of lines resistant to races 1, 3, 5, 7, 10 and 13 of the pathogen have been produced at the Central Institute for Cotton Research by crossing some local and introduced Uplands with 101-102B, Reba B50 and BJA-592 (Singh *et al.*, 1988).

USA

The first resistant varieties produced in the USA, such as Acala 1517Br, Austin and Blightmaster, were selected from crosses with Stoneville 20, which contains the gene B_7. Selected against race 1 of the pathogen, resistance broke down in these varieties with the arrival of race 2. They were replaced by Acala 1517Br2, resistant to both races due to the incorporation of the B_2 gene, and this variety was in turn replaced in New Mexico and west Texas by Acala 3080 and Acala 1517-70 (Brinkerhoff, 1970). In the Rex varieties grown in Missouri, resistance is conditioned by the gene B_2 from Stoneville 2B together with tolerance contributed by Empire (Mooseberg, 1965; Brinkerhoff, 1970).

It was noted in the USA that high levels of resistance can be obtained by transferring B-genes to lines with low levels of resistance conditioned by polygenes (Bird, 1960, 1962). B-genes were transferred to Empire, Deltapine and others from Knight's strains of *G. barbadense*, by back-crossing and rigorous screening of the segregating population with virulent mixtures of races. Immune plants were obtained after several back-cross generations.

Brinkerhoff *et al.* (1984) have defined immunity to bacterial blight as a response to all 18 known races of the pathogen which shows no macroscopic symptoms under field levels of inoculum. Macroscopic symptoms of cellular necrosis, typical of the hypersensitive response, do not occur unless inoculum levels exceed 1×10^6 cells ml^{-1} (Hopper *et al.*, 1975), which is at least twice the highest level normally found in the field.

The first immune line developed by Bird (1960, 1962; see also Brinkerhoff *et al.*, 1984) was 101-102B, obtained from a cross between Empire WR and Bar 4/16, a *G. barbadense* line containing the gene combination B_2B_3. Two back-crosses to Empire were followed by a cross to the Upland line MVW.

Resistance in 101-102B is considered to be due to B_2B_3 together with the polygene complex B_{Sm} from Empire. At the time it was developed it was resistant to all known races of the pathogen.

Once homozygous immune strains have been developed, the immunity can be transferred by crossing. Brinkerhoff and Hunter (1964, 1965, 1966) transferred immunity from 101-102B to three Uplands which already contained B_2. In each case, a single cross was made, followed by selection within the segregating population in response to inoculation with a mixture of virulent races, and then inbreeding. The gene complex B_{Sm} acts as a modifier which greatly enhances the resistance conferred by B_2B_3 (El-Zik and Bird, 1970). The combination $B_2B_3B_{Sm}$ behaves as a stable unit or 'super-gene' which can be readily transferred as a completely dominant character (Innes, 1969; Innes *et al.*, 1974).

Another immune line, Im-216, was produced by several generations of selection and inbreeding from a segregating population of Bird's B_2B_3 Empire. The population was screened using inoculum consisting of a mixture of races capable of overcoming resistance conferred by all known single B-genes (Brinkerhoff, 1970).

The gene B_4 in the Empire background acts as a strong gene of large effect (Bird, 1964) and confers a greater degree of resistance than the gene combinations B_2B_6, B_2B_3 or $B_2B_3B_7$ (El-Zik and Bird, 1970). Bird has produced the highly resistant Tamcot strains, which contain combinations such as $B_2B_3B_7B_4$ (Bird, 1974, 1976a,b, 1979a,b,c).

Lines developed for resistance to bacterial blight were found to contain a much higher proportion of plants resistant to Fusarium wilt than occur in blight-susceptible populations (Brinkerhoff and Hunter, 1961). Then, Bird (1970) demonstrated that resistance and escape from five major diseases were interrelated and had genes in common. Although the strongest association was with resistance to the Fusarium wilt/root-knot nematode complex, lesser associations were found between blight resistance and resistance to Verticillium wilt, Phymatotrichum root rot and seedcoat resistance to mould (Bird, 1982). These observations led to the development of the multi-adversity resistance (MAR) programme at Texas A&M University, with the objective of producing high-yielding Upland varieties with resistance to several pests and diseases by exploiting the potential of polygene complexes (Bird, 1972, 1982, 1983).

The MAR programme began in 1963 with a diverse gene pool, but with the emphasis on stocks developed for resistance to bacterial blight. The population contained the major genes B_2, B_3, B_4, B_6 and B_7 with the minor gene complexes, B_{Sm}, B_{Dm} and B_{Tm}. Material was included which exhibited a wide range of traits, such as glandless, nectariless and okra leaf, associated with resistance to insect pests. Some MAR selections have been released as varieties and have proved to be successful in parts of Texas, e.g. Tamcot SP21 and SP23 (Bird, 1982), and these have been followed by Tamcot SP21S, SP37 and CAMD E (Bird, 1979a,b,c), which show improved resistance to pests and diseases, including immunity or a high degree of resistance to bacterial blight, as well as improved yield potential.

Stability of Resistance

The capacity for changes in virulence of the pathogen, in response to selection pressure provided by host genes for resistance, became apparent once resistant varieties were cultivated on a large scale. In the USA, single gene resistance quickly broke down and varieties were produced with combinations of B-genes. Breeders now emphasize the importance of resistance based on oligogenes combined with polygene complexes (El-Zik and Bird, 1970).

Resistance dependent on more than one major gene has broken down in the Sudan where the Barakat varieties, containing the gene combination B_2B_6 have been attacked (Sippell et al., 1983). In Central Africa, strains have appeared that are capable of attacking previously immune cottons. The new strain (race 20) has been found in the Sudan, Chad and Upper Volta, where varieties with major gene resistance are widely grown, but has not been detected in neighbouring countries growing predominantly susceptible varieties. This race produces symptoms on the host differential, 101-102B, which contains the gene combination $B_2B_3B_{Sm}$, and which is resistant to races 1 to 19 of the pathogen (Follin, 1983b). The variety Reba B50 PR has shown a high level of resistance to race 20 (Ruano et al., 1988).

These new strains have already been exploited by the MAR programme to produce lines with the necessary combination of B-genes to confer resistance to them and to the original 19 races (Bird et al., 1984).

Nature of Resistance

The main antimicrobial secondary metabolites in cotton are the terpenoids. The cotton terpenoids are present in the form of aldehydes (TA), unique to the Gossypiae. They may be preformed as in the pigment glands, or synthesized de novo in response to wounding, chemical injury or infection. When the vascular system of a blight-susceptible variety (Seabrook Sea Island) and a blight-resistant variety (Rex Smoothleaf) was inoculated with Xcm, a direct relationship was found between the rate of induced TA synthesis and host resistance (Bell and Stipanovic, 1978).

During the first 48 hours after inoculation, TA accumulation was similar in both varieties. Subsequently, TA synthesis was virtually suppressed in the susceptible variety, suggesting that the pathogen actively interfered with TA synthesis. Heat-killed cells of the pathogen induced more TA synthesis in the resistant than in the susceptible cotton.

Essenberg et al. (1982) identified two phytoalexins from leaves and cotyledons of inoculated blight-resistant plants. Extracts from the plants inhibited pathogen growth and contained more 2,7-dihydroxycadalene and laciniline than were obtained from inoculated susceptible plants, or from non-inoculated resistant plants. However, the quantity of phytoalexin detected was sufficient to inhibit *in vivo* growth of the bacterium only if concentrated close to the site of infection.

Although a high degree of correlation in the expression of resistance in different parts of the plant has been demonstrated for a wide range of genotypes (e.g. Knight, 1946; Innes, 1963, 1966), certain genotypes exhibit resistance at one growth stage or organ but susceptibility at another. It may be that the resistance mechanism may vary to some extent according to the particular B-genes involved.

Bird (1954) reported that resistance in strains with the B_7 gene was due to their having a nitrogen/carbohydrate balance which was unfavourable to disease development. Resistant lines had a higher total carbohydrate content than the susceptible lines. These conclusions were later confirmed when debudded plants became less resistant and stem-girdled plants became more resistant than untreated controls. This was explained in terms of lower carbohydrate levels when plants remained in the vegetative state and accumulation of carbohydrate in stem-girdled plants. This supports the field observation that plants become less susceptible after the onset of fruiting (Bird and Joham, 1959). Hughes and Fowler (1953) also correlated resistance with higher levels of glucose in the leaves, and Verma and Singh (1974a) found higher glucose levels in all tissues except the seed of resistant varieties, compared with the same tissues in susceptible varieties.

Genetic differences in amino acid content have also been correlated with differences in susceptibility to blight. Vohra and Chand (1971) found that highly susceptible cottons contained more amino acid than moderately susceptible or resistant cottons. Resistant varieties differed from others in the absence of threonine, D–L alanine and arginine and in containing cystine and lysine. In contrast, Verma and Singh (1974a) reported higher total amino acid content in resistant varieties and that serine was present in small amounts in the resistant variety but absent from the susceptible ones.

The immune, hypersensitive response is characterized by the activity of cell wall-degrading enzymes, such as polymethyl galacturonase (Hopper *et al.*, 1975), which leads to rapid membrane breakage and cell necrosis (Cason *et al.*, 1978). In both the susceptible and immune response in the leaf, infection can be induced by a single cell which grows into a colony. Growth is logarithmic in both cases, but ceases at a lower population in the immune plant. Symptoms appear more rapidly after inoculation of immune tissues than in susceptible tissues. Borkar and Verma (1985) reported that, following leaf inoculation by infiltration of inoculum into the tissues, symptoms appeared after 72 hours in compatible host–pathogen interactions, whereas in the non-compatible interaction symptoms appeared after only 16 hours. In the studies reported by Cason *et al.* (1977) hypersensitive damage to the immune host occurred within six hours, with typical leaf lesions appearing much later on the susceptible host. Perry (1966) reported that 20 days after inoculation the bacterial population had reached 6×10^6 in the susceptible leaf but only 1.5×10^6 in a resistant leaf.

The immune response in the leaf seems to be a local one around each developing colony. A small bacteriostatic zone is created around the colony. Infiltration of water dilutes the bacteriostatic principle, allowing the bacterial population to increase again until the bacteriostatic zone is re-established (Essenberg et al., 1979a,b). This suggests a water-soluble bacteriostatic compound which accumulates close to the infection site. Al Mousawi et al. (1982) have suggested that, in the resistant and immune reaction, bacteria are physically localized by the accumulation of 'fibrillar material'.

Inoculum Preparation and Inoculation Techniques

Inoculum required for screening purposes can be prepared from fresh infected material, dried leaves or isolates maintained in store. The pathogen can remain viable on a suitable medium in the refrigerator for a year or more, but it is advisable to renew the cultures at least once during a 12-month period. NA and PeSA are suitable media for inoculum preparation. In Zimbabwe, sucrose is added to the NA (NAS) to give a more vigorous growth. When preparing inoculum from a storage culture, it is better to streak out the culture on to a dry NA plate to check for viability and purity before transferring to NAS. After 48 hours at 30 °C the culture is ready for use. Growth on the surface of the plate is suspended in 10 ml of sterile distilled water, giving a concentration of about 10^8 cells ml^{-1}. This stock suspension can be diluted $\times$ 10 for stem inoculation and $\times$ 50 for boll inoculation (Hillocks and Chinodya, 1988).

In the absence of pure culture facilities, infected leaves can be crushed and soaked in water to provide inoculum. This method is particularly suitable for leaf spray inoculation and was used successfully in breeding for resistance in the Sudan (e.g. Knight and Clouston, 1939). Approximately 2.5 kg of air-dried infected leaf trash was soaked for 2–3 hours in 180 litres of rain-water before spraying (Knight, 1946).

Susceptible cotton tissues will produce symptoms at a wide range of inoculum concentrations, provided conditions are favourable to disease development. However, the relative number of cells introduced into host tissue can affect disease expression, especially under suboptimal conditions for disease development (Weindling, 1948; Innes and Last, 1961). Woody tissue of the stem requires relatively concentrated inoculum whereas less concentrated inoculum may be used on the young boll. Disease expression in the leaf is probably most subject to environmental influences. Lesion size increases with increasing inoculum concentration, and a reduction from 2×10^6 to 2×10^2 greatly reduced symptom severity in leaves inoculated by spraying (Innes and Last, 1961). If the inoculum is too concentrated, a hypersensitive reaction of rapid necrosis and tissue collapse occurs in resistant and immune varieties

(Brinkerhoff, 1963; Verma *et al.*, 1979). The optimum inoculum concentration to differentiate resistance from susceptibility varies to some extent with the inoculation technique used, tissue to be inoculated and general level of resistance in the population to be inoculated.

There has been some controversy as to whether resistance to infection of the seedling, leaf, stem and boll are under the same genetic control. Arnold (1963) pointed out the importance of interactions between genotype and environment, and concluded that a basic mechanism controlled leaf, stem and boll resistance, but that additional factors may operate in the stem. Innes (1966) found evidence of genetic differences in certain genotypes when leaf and stem resistance were compared. Also, in a programme to select for blight resistance in the Philippines, no correlation could be found between results obtained from inoculation of different tissues (Dizon and Reyes, 1987). Knight (1946) considered that stem and leaf resistance were correlated and Lagière (1960) found evidence of positive correlation for resistance between leaf and boll, leaf and stem and stem and boll. Nevertheless, certain cottons possess resistance in some tissues while other tissues are susceptible. Stoneville 20, for instance, has little resistance in the cotyledons and leaves but good stem and boll resistance. Bar 7/8 has shown a high degree of resistance in the cotyledon, good leaf resistance and fair stem resistance, but no resistance to boll infection. It has therefore been suggested that both stem and boll resistance should be considered in breeding programmes (e.g. Balasubramanyan and Raghavan, 1950; Wickens, 1953).

The response to inoculation in different organs of the plant can also be affected by which B-genes are present in the host. B_2 and B_2B_3 give little protection from the disease under very wet conditions (Last, 1959b). A recent commercial variety in Zimbabwe, Albar G501 (B_2B_3), responds to stem inoculation as a resistant or susceptible type, depending on weather conditions immediately preceding and in the first two weeks after inoculation. It also shows resistance to natural leaf infection but behaves as a susceptible variety under heavy disease pressure early in crop development. B_2 and B_3 both reduce the severity of leaf infection under suboptimal conditions for disease development but have little effect on stem resistance in the Sakel background. B_6 also gives little protection against leaf infection but in combination with B_2 conferred leaf and stem resistance (Last, 1958).

Arnold and Brown (1968) suggest that it is desirable to screen for resistance at different stages of plant growth and in a range of environments. Programmes based on alternate testing sites are more likely to produce plants with well-balanced resistance mechanisms than those in which the genotype evolved in isolation (Innes *et al.*, 1974).

Leaf spray inoculation involves spraying the under-side of young leaves of six-week-old plants with a suspension of the bacteria. The spray is directed under pressure at the under-side of the leaf from ground level. Spraying should be carried out in the morning when the stomata are open and twice a day on two consecutive days to increase the chance of inoculating during good

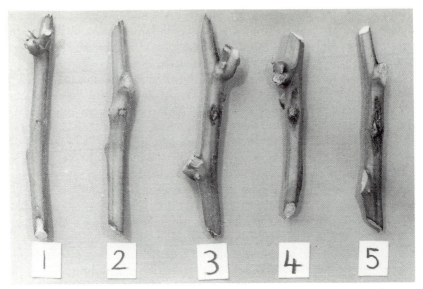

Fig. 2.5. Grading scale used in screening for resistance to bacterial blight following stem inoculation with *Xanthomonas campestris* pv. *malvacearum* (Hillocks and Chinodya, 1988).

conditions for infection (Hunter *et al.*, 1968; Verma and Singh, 1975b). When pure cultures are used to prepare inoculum, 10^6 bacterial cells ml^{-1} is a suitable concentration.

    Inoculation of the seedling hypocotyl may be done using a fine needle charged with inoculum consisting of a thick suspension in water of the bacteria or taken directly from the surface of the culture medium (Wickens, 1953; Brown, 1980). In older plants, stem inoculation is done six to eight weeks after planting. A sewing-machine needle, the eye of which is charged with bacterial suspension, is used to pierce the uppermost internode, rotating it as it is withdrawn (Hillocks and Chinodya, 1988).

    Several techniques have been used to inoculate the boll, using either atomization, brushing (Wickens, 1953; Logan, 1958), needle puncture (Logan, 1958) or abrasion (Lagière, 1960). Brush inoculation is suitable only under optimal conditions for disease development. Atomization must be done on very young bolls at the time of corrolla drop and results in extensive shedding. With abrasion it is difficult to create a wound of uniform size and depth. The method preferred in East Africa (Hillocks, 1981) is to pierce the boll surface with a short needle (2 mm) charged with inoculum. Bolls approximately two weeks old are selected (with reference to bolls tagged at flowering) and two locules per boll of two bolls per plant are inoculated. Lesions are scored when they reach maximum size, 21–25 days later.

    Seed inoculation is not of much value in screening because there is a high risk of disease escape, especially if conditions are too dry during emergence.

Fig. 2.6. Response of three cultivars to boll inoculation and the bacterial blight pathogen; S, susceptible; R, resistant; I, intermediate (Hillocks and Chinodya, 1988).

The method is more useful in the evaluation of seed-dressing chemicals. Seeds are soaked in a bacterial suspension containing 10^7 cells ml^{-1} for three hours at 25 °C and dried before planting. One litre of suspension is sufficient for 140 g of seed (Verma, 1986).

Disease Scoring Systems

Following inoculation of the stem, the lesion spreads along the internode and lesion development reaches a maximum at about three weeks after inoculation. Lesion development can be rated by using a scoring system or by simply measuring lesion length. The use of a scoring system is preferred in screening for resistance, where large numbers of plants must be evaluated (e.g. Wickens, 1953; Lagière, 1960). Accurate measurement of lesion length may give more reliable results in studies of the genetics of resistance or of pathogenic variation, especially if seedling hypocotyl inoculation is used (e.g. Verma and Singh, 1970; Brown, 1980). In Zimbabwe, a scale of 1 to 5 is used (Fig. 2.5), based on the degree of spread away from the callus tissue around the inoculation puncture (Hillocks and Chinodya, 1988). Grade 6 is occasionally required for very susceptible material when the lesion spreads beyond the inoculated internode.

The response to boll inoculation may also be graded or measured quantitatively. The lesion tends to spread concentrically from the puncture (see Fig.

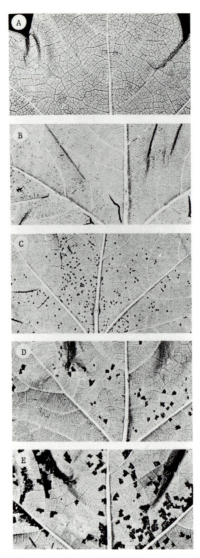

Fig. 2.7. Grading scale for scoring bacterial blight reactions following leaf spray inoculation with *Xanthomonas campestris* pv. *malvacearum*; A–B, grade 1; C, grade 2; D, grade 3; E, grade 4. Grades 1 and 2 are resistant reactions, grades 3 and 4 are susceptible reactions (Hunter *et al.*, 1968).

2.6) and lesion diameter can be measured when it reaches its maximum size at about three weeks after inoculation.

The severity of naturally occurring leaf infection or the response to leaf-spray inoculation is assessed using some form of grading scale. In the case of natural infection, the most simple method is to assess the percentage of infected plants in the plot or row. If the material under evaluation is unreplicated or more detailed information is required, a grading scale must be used which incorporates both the number of infected leaves per plant and the severity of infection. Following leaf-spray inoculation, where the reaction can be evaluated on a particular leaf, most scoring systems have been based on lesion size rather than lesion number (see Fig. 2.7). In the original system developed by Knight and Clouston (1939), leaves were graded from 0 to 12 (grade 11 was omitted). However, in the USA the grading scale was reduced to five grades (e.g. Hunter *et al.*, 1968) and, in India, a 1 to 4 scale was used in the identification of races of the pathogen (Verma and Singh, 1970, 1975b).

References

ABOEL-DAHAB, M.K. (1964) Pectic and cellulytic enzymes of *Xanthomonas malvacearum*, the incitant of bacterial blight of cotton. *Phytopathology* 54, 597–601.

ALCALA DE MERCANO, D. AND CAMINO, J.M. (1987) Characterization and identification of races of the causal agent of cotton angular leaf spot in Venezuela. *Review of Plant Pathology* 66, 406.

ALLEN, S.J. (1986) Bacterial blight strikes again. *Australian Cotton Grower* 7, 32–3.

ALLIPI, A. AND HAYWARD, A.C. (1987) Races of *Xanthomonas campestris* pv. *malvacearum* occurring in Queensland and Western Australia. *Australian Plant Pathology* 16, 16–19.

AL-MOUSAWAI, A.H., RICHARDSON, P.E., ESSENBERG, M. AND JOHNSON, W.M. (1982) Ultrastructural studies of a compatible interaction between *Xanthomonas campestris* pv. *malvacearum* and cotton. *Phytopathology* 72, 1222–30.

ANDREWS, F.W. (1937) Investigations on Blackarm disease under field conditions. 2. The effect of flooding infective cotton debris. *Empire Journal of Experimental Agriculture* 5, 204–18.

ARNOLD, M.H. (1963) The control of bacterial blight in rain-grown cotton. 1. Breeding for resistance in African Upland varieties. *Journal of Agricultural Science* 60, 415–27.

ARNOLD, M.H. (1965) The control of bacterial blight in rain-grown cotton. 2. Some effects of infection on growth and yield. *Journal of Agricultural Science* 65, 29–40.

ARNOLD, M.H. (1970) Cotton improvement in East Africa. In: Leakey, C.L.A. (ed.), *Cotton Improvement in East Africa*. Commonwealth Agricultural Bureaux, Farnham Royal, pp. 178–208.

ARNOLD, M.H. AND ARNOLD, K.M. (1961) Bacterial blight of cotton: trash-borne infection. *Empire Cotton Growing Review* 38, 258–70.

ARNOLD, M.H. AND BROWN, S.J. (1968) Variation in the host–parasite relationship of a crop disease. *Journal of Agricultural Science* 71, 19–36.

ATKINSON, G.F. (1891) The black rust of cotton. *Alabama Agricultural Experimental Station Bulletin* 27, 1–16.

BALASUBRAMANYAN, R. AND RAGHAVAN, A. (1950) Bacterial blight of cotton in Madras. *Indian Cotton Growing Review* 4, 118–20.

BELL, A.A. AND STIPANOVIC, R.D. (1978) Biochemistry of disease and pest resistance in cotton. *Mycopathologia* 65, 91–106.

BHAGWAT, V.Y. AND BHIDE, V.P. (1962) Vascular infection of some cottons by *Xanthomonas malvacearum*. *Indian Cotton Growing Review* 16, 80–3.

BHANDARI, D.R., SINGH, R. AND BHARADWAJ, R.P. (1969) Control of blackarm (*Xanthomonas malvacearum*) of cotton. *Madras Agriculture Journal* 56, 485–7.

BIRD, L.S. (1954) Genetic-controlled carbohydrate and soluble nitrogen combinations in plant tissues causing resistance to the bacterial blight disease of cotton. *Plant Disease Reporter* 38, 653–60.

BIRD, L.S. (1959) Loss measurements caused by bacterial blight disease of cotton. *Phytopathology* 49, 315 (Abstr.).

BIRD, L.S. (1960) Developing cotton immune to bacterial blight. *Proceedings Cotton Improvement Conference* 12, National Cotton Council, Memphis, Tennessee, pp. 16–17.

BIRD, L.S. (1962) Use of races of *Xanthomonas malvacearum* for establishing levels of selection pressure in developing bacterial blight immune cottons. *Proceedings Cotton Improvement Conference* 14, National Cotton Council, Memphis, Tennessee, pp. 6–8.

BIRD, L.S. (1970) Cotton varieties which resist and escape multiple diseases. *Phytopathology* 60, 581–2 (Abstr.).

BIRD, L.S. (1972) Interrelationships of resistance and escape from multi-disease and other adversities. *Proceedings of the Beltwide Cotton Production Research Conference*, National Cotton Council, Memphis, Tennessee, pp. 92–7.

BIRD, L.S. (1974) Improving cotton for immunity from bacterial blight. *Proceedings of the Beltwide Cotton Production Research Conference*, National Cotton Council, Memphis, Tennessee, p. 23.

BIRD, L.S. (1976a) Registration of Tamcot SP21, Tamcot SP23 and Tamcot SP37 cottons. *Crop Science* 16, 884.

BIRD, L.S. (1976b) Status of progress in developing cottons immune from bacterial blight. *Texas Agricultural Experimental Station Progress Report* PR-3420, Texas A & M University College Station, p. 3.

BIRD, L.S. (1979a) Registration of Tamcot SP21S cotton. *Crop Science* 19, 410–11.

BIRD, L.S. (1979b) Registration of Tamcot CAMD-E cotton. *Crop Science* 19, 411–12.

BIRD, L.S. (1979c) Registration of Tamcot SP37H cotton. *Crop Science* 19, 412.

BIRD, L.S. (1981) Report of the bacterial blight committee. *Proceedings of the Beltwide Cotton Production Research Conference*, National Cotton Council, Memphis, Tennessee p. 6.

BIRD, L.S. (1982) The MAR (multi-adversity resistance) system for genetic improvement of cotton. *Plant Disease* 66, 172–6.

BIRD, L.S. (1983) Multi-adversity resistance for developing durable host protection from disease and insects. *Phytopathology* 73, 769 (Abstr.).

BIRD, L.S. AND BLANK, L.M. (1951) Breeding strains of cotton resistant to bacterial blight. *Texas Agricultural Experimental Station Bulletin* 736, 1–25.

BIRD, L.S. AND HADLEY, H.H. (1958) A statistical study of the inheritance of Stoneville 20 resistance to the bacterial blight disease of cotton in the presence of *Xanthomonas malvacearum* races 1 and 2. *Genetics* 43, 750–67.

BIRD, L.S. AND HUNTER, R.E. (1955) Report of the bacterial blight committee. *Proceedings of the Beltwide Cotton Production Research Conference*, National Cotton Council, Memphis, Tennessee, p. 20.

BIRD, L.S. AND JOHAM, H.E. (1959) The influence of nitrogen source and carbohydrate change by debudding and girdling on bacterial blight resistance caused by the B_7 gene in cotton. *Plant Disease Reporter* 43, 86–9.

BIRD, L.S., EL-ZIK, K.M., THAXTON, P.M., HOWELL, M. AND PERCY, R.G. (1984) Maintaining immunity and high resistance in cotton to races of *Xanthomonas campestris* pv. *malvacearum*. *Phytopathology* 74, 818.

BLASINGAME, D. (1990) Disease loss estimate committee report. *Proceedings of the Beltwide Cotton Production Research Conference*. National Cotton Council, Memphis, Tennessee, p. 4.

BORKAR, S.G. AND VERMA, J.P. (1985) Population dynamics of *Xanthomonas campestris* pv. *malvacearum* in compatible and incompatible reactions of cotton cultivars. *Acta Phytopathologica Academiae Scientiarum Hungaricae* 20, 31–4.

BORKAR, S.G., VERMA, J.P. AND SINGH, R.P. (1980) Transmission of *Xanthomonas malvacearum* the incitant of bacterial blight of cotton through spotted bollworm. *Indian Journal of Entomology* 42, 390–4.

BRINKERHOFF, L.A. (1963) Variability of *Xanthomonas malvacearum* the cotton bacterial blight pathogen. *Oklahoma Agricultural Experimental Station Technical Bulletin* T98, 1–96.

BRINKERHOFF, L.A. (1966) Tests for races of *Xanthomonas malvacearum* from Australia. *Plant Disease Reporter* 50, 323–4.

BRINKERHOFF, L.A. (1970) Variation in *Xanthomonas malvacearum* and its relation to control. *Annual Review of Phytopathology* 8, 85–110.

BRINKERHOFF, L.A. AND FINK, G.B. (1964) Survival and infectivity of *Xanthomonas malvacearum* in cotton plant debris and soil. *Phytopathology* 54, 1198–201.

BRINKERHOFF, L.A. AND HUNTER, R.E. (1961) Frequency of cotton plants resistant to Fusarium wilt in some lines of cotton resistant or susceptible to bacterial blight. *Plant Disease Reporter* 45, 126–7.

BRINKERHOFF, L.A. AND HUNTER, R.E. (1963) Internally infected seed as a source of inoculum for the primary cycle of bacterial blight of cotton. *Phytopathology* 53, 1397–401.

BRINKERHOFF, L.A. AND HUNTER, R.E. (1964) Immunity to 10 races of *Xanthomonas malvacearum* and inheritance of the gene B_7 in Acala 44. *Proceedings of the Beltwide Cotton Production Research Conference*, National Cotton Council, Memphis, Tennessee, p. 10.

BRINKERHOFF, L.A. AND HUNTER, R.E. (1965) The development of blight immunity and isogenic blight differentials. *Proceedings of the Beltwide Cotton Production Research Conference*, National Cotton Council, Memphis, Tennessee, p. 30.

BRINKERHOFF, L.A. AND HUNTER, R.E. (1966) Current research on bacterial blight: transfer and stability of immune reaction. *Proceedings of the Beltwide Cotton Production Research Conference*, Memphis, Tennessee, p. 15.

BRINKERHOFF, L.A. AND PRESLEY, J.T. (1967) Effect of 4 day and night temperature regimes on bacterial blight reaction of immune, resistant and susceptible strains of Upland cotton. *Phytopathology* 57, 47–51.

BRINKERHOFF, L.A., GREEN, J.M., HUNTER, R. AND FINK, G. (1952) Frequency of bacterial blight resistant plants in twenty cotton varieties. *Phytopathology* 42, 98–100.

BRINKERHOFF, L.A., VERHALEN, L.M., JOHNSON, W.M., ESSENBERG, M. AND RICHARDSON, P.E. (1984) Development of immunity to bacterial blight of cotton and its implications for other diseases. *Plant Disease* 68, 168–73.

BROWN, S.J. (1976) Plant pathology. In: Arnold, M.H. (ed.), *Agricultural Research for Development*. Cambridge University Press, Cambridge, pp. 151–73.

BROWN, S.J. (1980) The relationship between disease symptoms and parasitic growth in bacterial blight of cotton. *Journal of Agricultural Science* 94, 305–12.

BRYAN, M.K. (1932) An atypical lesion on cotton leaves caused by *Bacterium malvacearum*. *Phytopathology* 22, 263–4.

CASON, E.J., RICHARDSON, P.E., BRINKERHOFF, L.A. AND GHOLSON, R.K. (1977) Histopathology of immune and susceptible cotton cultivars inoculated with *Xanthomonas malvacearum*. *Phytopathology* 67, 195–8.

CASON, E.T., RICHARDSON, P.E., ESSENBERG, M., BRINKERHOFF, L.A., JOHNSON, W.M. AND VENERE, R.J. (1978) Ultrastructural cell wall alteration in immune cotton leaves inoculated with *Xanthomonas malvacearum*. *Phytopathology* 68, 1015–21.

CEDANO, C.E. AND DELGADO, J.M.A. (1986) Race 8 of *Xanthomonas campestris* pv. *malvacearum* present on cotton (*G. hirsutum* and *G. barbadense*) in the San Martin Department, Peru. *Review of Plant Pathology* 67, 256.

CHAUHAN, M.S., KAIRON, M.S. AND KARWASRA, S.S. (1983) Determination of minimum number of sprays of agrimycin plus blitox for the control of bacterial blight of cotton under Haryana conditions. *Indian Journal of Mycology and Plant Pathology* 13, 187–91.

CHEW, C.F. AND BOOTH, J.A. (1974) Report of the bacterial blight committee. *Proceedings of the Beltwide Cotton Production Research Conference*, National Cotton Council, Memphis, Tennessee, p. 6.

CHEW, C.F., PRESLEY, J.T. AND STATEN, G. (1969) Survey, screening and breeding for bacterial blight resistance in cotton. *Plant Disease Reporter* 53, 390–1.

CHOWDHURY, H.D. AND VERMA, J.P. (1980a) Multiplication of *Xanthomonas malvacearum* and a phylloplanae bacterium in leaves of *Gossypium hirsutum*. *Indian Phytopathology* 33, 245–8.

CHOWDHURY, H.D. AND VERMA, J.P. (1980b) Phylloplane bacteria of *Gossypium hirsutum* associated with *Xanthomonas malvacearum*, the incitant of bacterial blight of cotton. *Indian Phytopathology* 33, 347–9.

CHOWDHURY, H.D., SINGH, R.P. AND VERMA, J.P. (1979) Presence of more than one race of *Xanthomonas malvacearum* in lesions on leaves of *Gossypium hirsutum*. *Indian Phytopathology* 32, 110–12.

CICCARONE, A. (1959) Angular leaf spot of cotton in Italy. *FAO Plant Protection Bulletin* 7, 147.

CROSSE, J.E. (1963) Pathogenicity differences in Tanganyika populations of *Xanthomonas malvacearum*. *Empire Cotton Growing Review* 40, 125–30.

CROSSE, J.E. AND HAYWARD, A. (1964) Relationship between pathogenicity and phage types in *Xanthomonas malvacearum*. *Empire Cotton Growing Review* 41, 49–50.

DHARMARAJULU, K. (1966) A study of pathological history of infection of *Xanthomonas malvacearum* (E.F. Smith) Dowson in the leaf petiole and stem of *Gossypium hirsutum* L. *Indian Cotton Journal* 20, 143–8.

DIZON, T.O. AND REYES, T.T. (1987) Screening of cotton (*Gossypium hirsutum*) for resistance to bacterial blight and blackarm caused by *Xanthomonas campestris* pv. *malvacearum*. *Review of Plant Pathology* 67, 373.

DONSON, W.J. (1939) On the systemic position and generic names of the gram negative bacterial plant pathogens. *Zentralblatt für Bakteriologie, Parasitenkunde, Infektionskrankheiten und Hygiene* 100, 177–93.

DRANSFIELD, M. (1969) Report from Northern States, Nigeria. *Empire Cotton Growing Corporation Reports from Experimental Stations*, pp. 24–6.

DYE, D.W., BRADBURY, J.F., GOTO, M., HAYWARD, A.C., LELLIOT, R.A. AND SCHROTH, M.N. (1980) International standards for naming pathovars of phytopathogenic bacteria and a list of pathovar names and pathotype strains. *Review of Plant Pathology* 59, 153–68.

EDGERTON, C.W. (1912) The rots of the cotton boll. *Louisiana Agricultural Experimental Station Bulletin* 137, 133pp.

EKBOTE, M.V. (1985) Chemical control of bacterial blight of cotton. *Journal of Maharashtra State University* 10, 26–7.

EL-NUR, E. (1970a) Bacterial blight of cotton in Sudan. *Pest Articles and News Summaries* 16, 132–41.

EL-NUR, E. (1970b) Bacterial blight of cotton. In: Siddig, M.A. and Hughes, L. (eds), *Cotton Growth in the Gezira Environment*. Sudan Agricultural Research Corporation, pp. 179–88.

EL-ZIK, K.M. AND BIRD, L.S. (1970) Effectiveness and specific genes and gene combinations in conferring resistance to races of *Xanthomonas malvacearum* in Upland cotton. *Phytopathology* 60, 441–7.

ESSENBERG, M., CASON, E.T., HAMILTON, B., BRINKERHOFF, L.A., GHOLSON, R.K. AND RICHARDSON, P.E. (1979a) Single cell colonies of *Xanthomonas malvacearum* in susceptible and immune cotton leaves and the local resistant response to colonies in immune leaves. *Physiological Plant Pathology* 15, 53–68.

ESSENBERG, M., HAMILTON, B., CASON, E.T., BRINKERHOFF, L.A., GHOLSON, R.K. AND RICHARDSON, P.E. (1979b) Localized bacteriosis indicated by water dispersal of colonies of *Xanthomonas malvacearum* within immune cotton leaves. *Physiological Plant Pathology* 15, 69–78.

ESSENBERG, M., DOHERTY, M.A., HAMILTON, B.K., HENNING, V.T., COVER, E.C., MCFAIL, S.J. AND JOHNSON, W.M. (1982) Identification and effects on *Xanthomonas campestris* pv *malvacearum* of two phytopalexins from leaves and cotyledons of resistant cotton. *Phytopathology* 72, 1349–56.

FAHMY, T. (1930) The angular leaf spot of cotton in Egypt. *Empire Cotton Growing Review* 7, 30–6.

FAHY, P.C. AND CAIN, P. (1987) Races of bacterial blight, *Xanthomonas campestris* pv *malvacearum* present in Australian cotton crops in 1964–1980. *Australian Plant Pathology* 16, 17–18.

FAULWETTER, R.C. (1917) Wind-blown rain, a factor in disease dissemination. *Journal of Agricultural Research* 10, 639–48.

FINDLAY, W.P.K. (1928) Some conditions influencing the development of bacterial disease of cotton (*Bacterium malvacearum*). *Empire Cotton Growing Review* 5, 29–39.

FOLLIN, J.C. (1983a) New races of *Xanthomonas campestris* pv. *malvacearum* virulent to the B_2B_3 gene combination in *Gossypium hirsutum*. *Proceedings of the Beltwide Cotton Production Research Conference*, National Cotton Council, Memphis, Tennessee, p. 5.

FOLLIN, J.C. (1983b) Races of *Xanthomonas campestris* pv *malvacearum* (Sm.) Dye in Western and Central Africa. *Coton et Fibres Tropicales* 38, 274–9.

FRENHANI, A.A., DA SILVEIRA, A.P., ABRAHO, J., CRUZ, B.P.B. AND DA SILVEIRA-SALIMA, G.P. (1969) Chemical control of angular leaf spot (*Xanthomonas malva-*

cearum) of cotton (*Gossypium* spp.) in Sao Paulo State. *Review of Plant Pathology* 49, 183.

GABRIEL, D.W., BURGES, A. AND LAZO, G.R. (1986) Gene for gene interaction of five cloned avirulence genes from *Xanthomonas campestris* pv. *malvacearum*. *Proceedings of the National Academy of Sciences of the USA* 83, 6415–19.

GIRARDOT, B. AND FOLLIN, J.C. (1987) First results on gene determination for cotton resistance to the new race of *Xanthomonas campestris* pv *malvacearum*. *Coton et Fibres Tropicales* 42, 139–40.

GREEN, J.N.M. AND BRINKERHOFF, L.A. (1956) Inheritance of 3 genes for bacterial blight resistance in Upland cotton. *Agronomy Journal* 48, 481–5.

GUNN, R.E. (1961) Bacterial blight of cotton: loss of disease resistance in certain combinations in Sakel cotton. *Empire Cotton Growing Review* 38, 284–6.

HABASH, H.A. (1968) Role of antagonistic bacteria in the decline of *Xanthomonas malvacearum* in wet cotton debris. *Cotton Growing Review* 45, 36–41.

HANDE, Y.K. AND RANE, M.S. (1983) Stability of the races of *Xanthomonas campestris* pv *malvacearum* (E.F. Smith) Dowson. *Journal of Maharashtra State University* 8, 229–30.

HANSFORD, G.G. (1932) Blackarm disease of cotton in Uganda. *Empire Cotton Growing Review* 9, 21–31.

HANSFORD, G.G., STOUGHTON, R.H. AND YATES, F. (1933) An experiment on the incidence and spread of angular leaf spot disease of cotton in Uganda. *Annals of Applied Biology* 20, 404–20.

HARE, J.F. AND KING, C.J. (1940) The winter carry-over of angular leaf spot infection in Arizona cotton fields. *Phytopathology* 30, 679–84.

HAYWARD, A.C. (1963) A note on bacteriophage typing of African strains of *Xanthomonas malvacearum*. *Empire Cotton Growing Review* 40, 216–18.

HAYWARD, A.C. AND WATERSON, J.M. (1964) *Xanthomonas malvacearum*. *CMI Descriptions of Pathogenic Fungi and Bacteria* No. 12. Commonwealth Mycological Institute, Kew, UK.

HILLOCKS, R.J. (1981) Cotton disease research in Tanzania. *Tropical Pest Management* 27, 1–12.

HILLOCKS, R.J. (1984) Production of cotton plants resistant to *Fusarium* wilt with special reference to Tanzania. *Tropical Pest Management* 30, 234–46.

HILLOCKS, R.J. AND CHINODYA, R. (1988) Current status of breeding for resistance to bacterial blight in Zimbabwe. *Tropical Pest Management* 34, 303–8.

HOPPER, D.G., VENERE, R.J., BRINKERHOFF, L.A. AND GHOLSON, R.K. (1975). Necrosis induction in cotton. *Phytopathology* 65, 206–13.

HUGHES, L.C. AND FOWLER, H.D. (1953) Resistance to blackarm disease associated with high glucose content of the leaf. *Nature* 172, 316–17.

HUNTER, R.E. AND BLANK, L.M. (1954) Pathogenicity differences of *Xanthomonas malvacearum* isolates. *Phytopathology* 44, 332 (Abstr.).

HUNTER, R.E. AND BRINKERHOFF, L.A. (1964) Longevity of *Xanthomonas malvacearum* on and in cotton seeds. *Phytopathology* 54, 617.

HUNTER, R.E., BRINKERHOFF, L.A. AND BIRD, L.S. (1968) Development of a set of Upland cotton lines for differentiating races of *Xanthomonas malvacearum*. *Phytopathology* 58, 830–2.

HUSSAIN, T. (1984) Prevalence and distribution of *Xanthomonas campestris* races in Pakistan and their reaction to different cotton lines. *Tropical Pest Management* 30, 159–62.

HUSSAIN, T. AND BRINKERHOFF, L.A. (1978) Race 18 of the cotton bacterial blight pathogen, *Xanthomonas malvacearum* identified in Pakistan in 1977. *Plant Disease Reporter* 62, 1085–7.

INNES, N.L. (1963) Resistance to bacterial blight of cotton in Albar. *Nature* 200, 387–8.

INNES, N.L. (1965a) Resistance to bacterial blight of cotton: the genes B_9 and B_{10} *Experimental Agriculture* 1, 189–91.

INNES, N.L. (1965b) Inheritance of resistance to bacterial blight of cotton. 1. Allen (*G. hirsutum*) derivatives. *Journal of Agricultural Science* 64, 257–71.

INNES, N.L. (1966) Inheritance of resistance to bacterial blight of cotton. 3. *Herbaceum* resistance transferred to tetraploid cotton. *Journal of Agricultural Science* 66, 433–9.

INNES, N.L. (1969) Inheritance of resistance to bacterial blight of cotton. 4. Tanzania selections. *Journal of Agricultural Science* 72, 41–57.

INNES, N.L. (1971) Impressions of cotton production and research in India. *Cotton Growing Review* 48, 163–74.

INNES, N.L. (1975) Upland cotton of triple hybrid origin. *Cotton Growing Review* 52, 46–58.

INNES, N.L. (1983) Bacterial blight of cotton. *Biological Reviews* 58, 157–76.

INNES, N.L. AND JONES, G.B. (1972) Allen: a source of successful African cotton varieties. *Cotton Growing Review* 49, 201–15.

INNES, N.L. AND LAST, F.T. (1961) Cotton disease symptoms caused by different concentrations of *Xanthomonas malvacearum*. *Empire Cotton Growing Review* 38, 27–9.

INNES, N.L., BROWN, S.J. AND WALKER, J.T. (1974) Genetical and environmental variation for resistance to bacterial blight of Upland cotton. *Heredity* 32, 53–72.

IRCT (1977) Recent cotton varieties (*G. hirsutum*). *Coton et Fibres Tropicales* 32, 233–5.

JAGANNATHARAO, C., RAGHAVAN, A. AND APPARAO, P. (1952) A review of recent progress in the work for the evolution of jassid and blackarm resistant strains of Cambodia for the Tungabhodra project area. *Indian Cotton Growing Review* 6, 147–55.

JAGANNATHARAO, C., MARAR, K.S. AND SANTHANAN, V. (1953) A brief review of cotton breeding problems in Madras State with special reference to improvement in yield and quality. *Indian Cotton Growing Review* 7, 48–52.

JAHKANWAR, P.L. AND BHAGWAT, Y.W. (1971) Vascular infection by *Xanthomonas malvacearum* and internal seed infection. *Cotton Growing Review* 48, 304–6.

JALALI, B.L. AND GROVER, R.K. (1974) Control of *Xanthomonas malvacearum* on cotton by agrimycin and oxanthiin compounds. *Indian Journal of Agricultural Science* 44, 664–6.

KHEIRHALLA, A.I. (1970) Varieties in the Gezira environment. 2. Medium staple varieties. In: Siddig, M.A. and Hughes, C.C. (eds), *Cotton Growth in the Gezira Environment*. Sudan Agricultural Research Corporation, pp. 173–8.

KING, C.J. AND BRINKERHOFF, L.A. (1949) The dissemination of *Xanthomonas malvacearum* by irrigation water. *Phytopathology* 39, 88–90.

KING, C.J. AND PARKER, R.B. (1939) Factors influencing the distribution and persistence of angular leaf spot in irrigated cotton fields. *Phytopathology* 29, 754 (Abstr.).

KNIGHT, R.L. (1944) The genetics of blackarm resistance. 4. *Gossypium punctatum* (Sch. and Thom.) crosses. *Journal of Genetics* 46, 1–27.

KNIGHT, R.L. (1945) The theory and application of the backcross technique to cotton breeding. *Journal of Genetics* 47, 76–86.

KNIGHT, R.L. (1946) Breeding cotton resistant to blackarm disease (*Bacterium malvacearum*). *Empire Journal of Experimental Agriculture* 14, 153–74.

KNIGHT, R.L. (1948a) The genetics of blackarm resistance. 6. Transference of resistance from *Gossypium arboreum* L. to *G. barbadense*. *Journal of Genetics* 48, 359–69.

KNIGHT, R.L. (1948b) The role of major genes in the evolution of economic characters. *Journal of Genetics* 48, 370–87.

KNIGHT, R.L. (1950) The genetics of blackarm resistance. 8. *Gossypium barbadense*. *Journal of Genetics* 50, 67–76.

KNIGHT, R.L. (1953a) The genetics of blackarm resistance. 9. The gene B_{6m} from *Gossypium arboreum*. *Journal of Genetics* 51, 270–5.

KNIGHT, R.L. (1953b) The genetics of blackarm resistance. 10. The gene B_7 from Stoneville 20. *Journal of Genetics* 51, 515–19.

KNIGHT, R.L. (1954a) Cotton breeding in the Sudan. Part 1. Egyptian cotton. *Empire Journal of Experimental Agriculture* 22, 68–80.

KNIGHT, R.L. (1954b) Cotton breeding in the Sudan. Part 3. American Upland cotton. *Empire Journal of Experimental Agriculture* 22, 176–84.

KNIGHT, R.L. (1956) The genetical approach to disease resistance in plants. *Empire Cotton Growing Review* 33, 191–6.

KNIGHT, R.L. (1963) The genetics of blackarm resistance. 12. Transference of resistance from *G. herbaceum* to *G. barbadense*. *Journal of Genetics* 58, 324–46.

KNIGHT, R.L. AND CLOUSTON, T.W. (1939) The genetics of blackarm resistance. 1. Factors B_2 and B_3. *Journal of Genetics* 38, 133–59.

KNIGHT, R.L. AND HUTCHINSON, J.B. (1950) The evolution of blackarm resistance in cotton. *Journal of Genetics* 50, 36–58.

KOTASTHANE, S.R. AND AGRAWAL, S.C. (1970) Efficacy of four seed protectants for the control of bacterial blight of cotton. *Pest Articles and News Summaries* 16, 334–5.

KRAVCHENKO, V.S. AND PONMAREV, E.K. (1964) Antibiotics against gummosis. *Review of Applied Mycology* 44, 273.

LAGIÈRE, R. (1960) Bacterial blight of cotton (*Xanthomonas malvacearum* (E.F. Smith) Dowson worldwide and in the Central African Republic. IRCT, Paris (French).

LAST, F.T. (1958) Stem infection of cotton by *Xanthomonas malvacearum* (E.F. Smith) Dowson. *Annals of Applied Biology* 46, 321–35.

LAST, F.T. (1960) Effect of *Xanthomonas malvacearum* (E.F. Smith) Dowson on cotton yields. *Empire Cotton Growing Review* 37, 115–17.

LEYENDECKER, P.J. (1950) Plant disease survey for New Mexico. *Plant Disease Reporter* 34, 39–44.

LOGAN, C. (1958) Bacterial boll rot of cotton. 1. Comparison of two inoculation techniques for the assessment of host resistance. *Annals of Applied Biology* 46, 230–42.

LOGAN, C. (1960) An estimate of the effect of seed treatment in reducing cotton crop losses caused by *Xanthomonas malvacearum* (E.F. Smith) Dow. in Uganda. *Empire Cotton Growing Review* 37, 241–55.

LOGAN, C. AND COAKER, T.H. (1960) The transmission of bacterial blight of cotton (*Xanthomonas malvacearum* (E.F. Smith) Dow.) by the cotton bug, *Lygus vosseleri* Popp. *Empire Cotton Growing Review* 39, 26–9.

McCUTCHEON, B.E. AND FULTON, N.D. (1962) Bacterial blight of cotton in Arkansas. *Plant Disease Reporter* 46, 168–9.

MARTIN, W.R., STERLING, W.L., KENERLY, C.M. AND MORGAN, P.W. (1988) Transmission of bacterial blight of cotton, *Xanthomonas campestris* pv *malvacearum*, by feeding of the cotton fleahopper and implications for stress ethylene-induced square loss. *Journal of Entomological Science* 23, 161–8.

MASSEY, R.E. (1929) Blackarm disease of cotton: the development of *Pseudomonas malvacearum* E.F. Smith within the cotton plant. *Empire Cotton Growing Review* 6, 124–46.

MASSEY, R.E. (1930) Studies on blackarm disease of cotton. 1. *Empire Cotton Growing Review* 7, 181–95.

MASSEY, R.E. (1931) Studies on blackarm disease of cotton. 2. *Empire Cotton Growing Review* 8, 187–213.

MASSEY, R.E. (1934) Botany and plant pathology. *Annual Report Gezira Agricultural Research Service* 1932–1933, p. 126.

MATHUR, R.L., DAFTARI, L.N. AND JHAMARIA, S.L. (1973) Controlling blackarm of cotton by fungicides and antibiotics. *Indian Journal of Mycology and Plant Pathology* 3, 107–8.

MATSUMOTO, G. AND HUSIOKA, Y. (1939) Bacteriophage in relation to *Bacterium malvacearum*. *Annals of the Phytopathological Society of Japan* 8, 193–8.

MILLER, J.W. (1968) Race build-up of *Xanthomonas malvacearum* on cotton in Southeast Missouri. *Plant Disease Reporter* 52, 739–41.

MOOSEBERG, C.A. (1953) Breeding cottons resistant to bacterial blight disease. *Arkansas Agricultural Experimental Station Bulletin* 534, 1–21.

MOOSEBERG, C.A. (1965) Rex Smoothleaf cotton. *Arkansas Farm Research* 14, 5–9.

NAFADE, S.D. AND VERMA, J.P. (1985) Drug resistant mutants of *Xanthomonas campestris* pv *malvacearum*. *Indian Phytopathology* 38, 74–80.

NAYAK, M.L., SINGH, R.P. AND VERMA, J.P. (1976) Effective chemical sprays to control bacterial blight of cotton. *Zeitschrift für Pflanzenkrankheiten und Pflanzenschutz* 83, 407–15.

NAYUDU, M.V. (1972) Interaction of L glutamic acid and L serine in the nutrition of *Xanthomonas malvacearum*. *Indian Journal of Microbiology* 12, 93–7.

OKABE, N. (1939) Bacteriophage in relation to *Bacterium malvacearum*. 2. Relation between variants and phage. *Review of Applied Mycology* 18, 249.

PADAGANUR, G.M. AND BASAVARAJ, M.K. (1983) Spray schedule for the control of important foliar diseases and bollworms of cotton in the transition belt of Karnataka. *Indian Journal of Agricultural Science* 53, 725–9.

PAPDIWAL, P.B. AND DESHPANDE, K.B. (1983) Pectin methylesterase of *Xanthomonas campestris* pv *malvacearum*. *Indian Phytopathology* 36, 134–5.

PATEL, M.K. AND KULKARNI, Y.S. (1950) Bacterial leafspot of cotton. *Indian Phytopathology* 3, 51–63.

PERRY, D.A. (1966) Multiplication of *Xanthomonas malvacearum* in resistant and susceptible cotton leaves. *Empire Cotton Growing Review* 43, 37–40.

POSWAL, M.A.T. (1988) Races of *Xanthomonas campestris* pv *malvacearum* (Smith) Dye, the causal organism of bacterial blight of cotton in Nigeria. *Journal of Phytopathology* 123, 6–11.

POTHECARY, B.P. AND OFIELD, R.J. (1968) Destruction of old cotton for pest and disease control. *World Crops* 20, 39–43.

PULIDO, M.L. AND BOLTON, R.A. (1974) A non-mercurial seed treatment for cotton seed. *Pest Articles and News Summaries* 20, 251–4.

ROLFS, F.M. (1935) Dissemination of the bacterial organism. *Phytopathology* 25, 971 (Abstr.).

ROUX, J.B. (1977) Sélection de variétés résistantes à certains maladies du cotonnier. *Séminaire Cotonnier d'Alep*, October 1977, IRCT, Paris.

RUANO, D. AND MOHAN, S.K. (1982) A new race of *Xanthomonas campestris* pv *malvacearum* (Smith) Dye in Panama State. *Fitopatologia Brasiliera* 7, 439–41.

RUANO, O., ALMEIDA, W.P., FOLLIN, J.C. AND GIRARDOT, B. (1988) Identification of sources of resistance to races 18 and 20 of *Xanthomonas campestris* pv. *malvacearum* in cotton genotypes. *Review of Plant Pathology* 69, 298.

SAUNDERS, J.H. AND INNES, N.L. (1963) The genetics of bacterial blight resistance in cotton: further evidence on the gene B_{6m}. *Genetical Research* 4, 382–8.

SCHNATHORST, W.C. (1964) Longevity of *Xanthomonas malvacearum* in dried cotton plants and its significance in dissemination of the pathogen on seed. *Phytopathology* 54, 1009–11.

SCHNATHORST, W.C. (1966) Eradication of *Xanthomonas malvacearum* from California through sanitation. *Plant Disease Reporter* 50, 168–71.

SCHNATHORST, W.C. (1968) Introduction of *Xanthomonas malvacearum* into California in acid-delinted and fumigated cotton seed. *Plant Disease Reporter* 52, 981–3.

SCHNATHORST, W.C. (1970) Altered host specificity in race 1 of *Xanthomonas malvacearum* by passage through a resistant variety of *Gossypium hirsutum*. *Phytopathology* 60, 258–60.

SCHNATHORST, W.C., HALISKY, P.M. AND MARTIN, R.D. (1960) History, distribution, races and disease cycle of *Xanthomonas malvacearum* in California. *Plant Disease Reporter* 44, 603–8.

SHIVANATHAN, P. AND GUNASINGHAM, V. (1966) Effect of seed protectants on the control of bacterial blight of cotton. *Tropical Agriculture* 122, 61–3.

SIMPSON, D.M. (1953) Report of the bacterial blight committee for 1952. *Proceedings of the Beltwide Cotton Production Conference*, Memphis, Tennessee, pp. 4–8.

SIMPSON, D.M. AND WEINDLING, R. (1946) Bacterial blight resistance in a strain of Stoneville cotton. *Journal of the American Society of Agronomy* 38, 630–5.

SINGH, M., VERMA, J.P., AGRAWAL, R.A., SINGH, V.P., SINGH, R.P. AND KATIYAR, K.N. (1973) Breeding for resistance to jassids, pink bollworm and bacterial blight in cotton. *Proceedings of the 2nd SABRAO General Congress*, New Delhi, 190–200.

SINGH, R.P., VERMA, J.P. AND RAO, Y.P. (1970) Eradication of seed infection of blackarm of cotton. *Current Science* 39, 330–1.

SINGH, R.P., VERMA, J.P. AND PRASAD, R. (1981) Distribution of races of *Xanthomonas malvacearum* on tetraploid cottons. *Indian Phytopathology* 34, 399–401.

SINGH, V.V., MESHRAM, S.R. AND NARAYANAN, S.S. (1988) Evolution of upland cotton lines with multi-racial resistance to bacterial blight caused by *Xanthomonas campestris* pv *malvacearum*. *Indian Journal of Agricultural Science* 58, 255–8.

SINHA, P.P. AND VERMA, J.P. (1983) Role of phylloplane bacteria in bacterial blight of cotton. *International Journal of Tropical Plant Disease* 1, 125–8.

SIPPELL, D.W., KHALIFA, H., EL-HIO OMER, M. AND BINDRA, O.S. (1983) A method to accumulate horizontal resistance in *Gossypium hirsutum* to *Xanthomonas malvacearum*. *Proceedings of the International Congress of Plant Pathology*, Abstract 4, p. 830.

SMITH, E.F. (1901) The cultural characteristics of *Pseudomonas hyacinthi*, *P. campestris*, *P. phaseoli* and *P. stewarti* from one flagellate yellow bacterium parasitic on plants. *Bulletin of the Division of Vegetable Physiology and Pathology*, US Department of Agriculture, p. 153.

SMITH, E.F. (1920) *Bacterial Disease of Plants*. Saunders Co., Philadelphia.

SRINIVISAN, K.V. AND TANEJA, N.K. (1974) The occurrence of race 5 of *Xanthomonas malvacearum* in India. *Current Science* 43, 450.

STOUGHTON, R.H. (1930) The influence of environmental conditions on the development of the angular leaf spot disease of cotton. 2. The influence of soil temperature on primary and secondary infection of seedlings. *Annals of Applied Biology* 17, 493–503.

STOUGHTON, R.H. (1931) The influence of environmental conditions on the development of the angular leaf spot disease of cotton. 3. The influence of air temperature on infection. *Annals of Applied Biology* 18, 523–34.

STOUGHTON, R.H. (1932) The influence of environmental conditions on the angular leaf spot disease of cotton. 4. The influence of atmospheric humidity on infection. *Annals of Applied Biology* 19, 370–7.

STOUGHTON, R.H. (1933) The influence of environmental conditions on the development of the angular leaf spot disease of cotton. 5. The influence of alternating and varying conditions on infection. *Annals of Applied Biology* 20, 590–611.

TARR, S.A.J. (1953) Seed treatment against blackarm disease of cotton in the Anglo-Egyptian Sudan. 2. Steeping and slurry methods of seed treatment and the effects of water steeping on germination of cotton seed. *Empire Cotton Growing Review* 30, 117–32.

TARR, S.A.J. (1956) Seed treatment against blackarm disease of cotton in the Sudan. IV. Recent research on wet treatments. *Empire Cotton Growing Review* 33, 98–104.

TARR, S.A.J. (1958) Seed treatment of cotton against diseases and insect pests. *Outlook on Agriculture* 2, 168–77.

TARR, S.A.J. (1961) Seed treatment against blackarm diseases of cotton in the Sudan. V. Results with seed carrying light and moderate infection. *Empire Cotton Growing Review* 38, 30–5.

TENNYSON, G. (1936) Invasion of cotton seed by *Bacterium malvacearum*. *Phytopathology* 26, 1083–5.

THIERS, H.D. AND BLANK, L.M. (1951) A histological study of bacterial blight of cotton. *Phytopathology* 41, 499–510.

VERDEREVSKY, D. AND VOITOVITCH, K. (1957) A method of producing cotton varieties resistant to gummosis. *Review of Applied Mycology* 36, 644–5.

VERMA, J.P. (1986) *Bacterial Blight of Cotton*. CRC Press, Inc., Boca Raton, Florida.

VERMA, J.P. AND SINGH, R.P. (1970) Two new races of *Xanthomonas malvacearum* the cause of blackarm of cotton. *Cotton Growing Review* 47, 203–5.

VERMA, J.P. AND SINGH, R.P. (1971a) Busan: an effective chemical to control bacterial blight of cotton. *Cotton Growing Review* 48, 60–2.

VERMA, J.P. AND SINGH, R.P. (1971b) Epidemiology and control of bacterial blight of cotton. *Proceedings of the Indian National Science Academy* Part B 74, 326–31.

VERMA, J.P. AND SINGH, R.P. (1971c) Pectic and cellulolytic enzymes of *Xanthomonas malvacearum* and the incidence of bacterial blight of cotton. *Current Science* 40, 21–2.

VERMA, J.P. AND SINGH, R.P. (1974a) Studies on the nutrition of *Xanthomonas malvacearum* and its relationship to the free amino acids, organic acids and sugars of resistant and susceptible cotton cultivars. *Acta Phytopathologica Academiae Scientiarum Hungaricae* 9, 55–63.

VERMA, J.P. AND SINGH, R.P. (1974b) Recent studies on the bacterial diseases of fibre and oil seed crops in India. In: Raychauhuri, S.P. and Verma, J.P. (eds), *Current Trends in Plant Pathology*. Lucknow University, India, pp. 134–45.

VERMA, J.P. AND SINGH, R.P. (1975a) Cellulases, pectinases and proteases of Indian isolates of *Xanthomonas malvacearum*. *Indian Phytopathology* 28, 379–83.

VERMA, J.P. AND SINGH, R.P. (1975b) Studies on the distribution of races of *Xanthomonas malvacearum* in India. *Indian Phytopathology* 28, 459–63.

VERMA, J.P. AND SINGH, R.P. (1975c) Races of *Xanthomonas malvacearum*, loss of their virulence and the protective effort of avirulent strains, heat killed cells and phylloplane bacteria. *Journal of Plant Diseases and Protection* 83, 748–57.

VERMA, J.P., SINGH, R.P. AND NAYAK, M.L. (1974) Bacterial blight threat to the cultivation of cotton in India and steps necessary to eradicate it. *Cotton Development* 4, 23–7.

VERMA, J.P., SINGH, R.P. AND NAYAK, M.L. (1975) Laboratory evaluation of chemicals against *Xanthomonas malvacearum* the incitant of bacterial blight of cotton. *Indian Phytopathology* 28, 170–3.

VERMA, J.P., NAYAK, M.L. AND SINGH, R.P. (1977) Survival of *Xanthomonas malvacearum* under North Indian conditions. *Indian Phytopathology* 30, 361–5.

VERMA, J.P., CHOWDHURY, H.D. AND SINGH, R.P. (1979) Interaction between different races of *Xanthomonas malvacearum* in leaves of *Gossypium hirsutum*. *Zeitschrift für Pflanzenkrankheiten und Pflanzenschutz* 86, 460–4.

VERMA, J.P., SINGH, R.P., BORKAR, S.G., SINHA, P.P. AND PRASAD, R. (1980) Management of bacterial blight of cotton with particular reference to variation in the pathogen, *Xanthomonas malvacearum* (Smith) Dowson. *Annals of Agricultural Research* 1, 98–107.

VERMA, J.P., CHOWDHURY, H.D. AND SINGH, R.P. (1982) Biological control of bacterial blight of cotton. In: Chattapadhay, S.B. and Samajpati, N. (eds), *Advances in Mycology and Plant Pathology*. Oxford and IBH Publishing, Calcutta, pp. 279–85.

VERMA, J.P., SINGH, R.P., CHOWDHURY, H.D. AND SINHA, P.P. (1983) Usefulness of phylloplane bacteria in the control of bacterial blight of cotton. *Indian Phytopathology* 36, 574–7.

VIDHYASEKARAN, P., PARAMBARAMAM, C. AND DURAIRAJ, P. (1971) Pectolytic enzymes of *Xanthomonas malvacearum*. *Indian Journal of Microbiology* 11, 93–6.

VINA, D. AND GRANADA, G.A. (1986) Evaluation of the production of certified cotton seed in relation to the incidence of angular leaf spot. *Review of Plant Pathology* 67, 142.

VOHRA, S. AND CHAND. J.N. (1971) Relation of free amino acids to the development of angular leaf spot of cotton. *Phytopathologische Zeitschrift* 70, 177–80.

WALLACE, T.P. AND EL-ZIK, K.M. (1990) Quantitative analysis of resistance in cotton to three new isolates of the bacterial blight pathogen. *Theoretical and Applied Genetics* 79, 443–8.

WEINDLING, R. (1948) Bacterial blight of cotton under conditions of artificial inoculation. *US Department of Agriculture Technical Bulletin* No. 956.

WICKENS, G.M. (1953) Bacterial blight of cotton: a survey of present knowledge, with particular reference to possibilities of control of the disease in African rain-grown countries. *Empire Cotton Growing Review* 30, 81–103.

WICKENS, G.M. (1956) Vascular infection of cotton by *Xanthomonas malvacearum* (E.F. Smith) Dow. *Annals of Applied Biology* 44, 129–37.

WICKENS, G.M. (1958) Present practice in the treatment of cotton seed against bacterial blight, *Xanthomonas malvacearum* (E.F. Smith) Dow. *Empire Cotton Growing Review* 35, 9–12.

WICKENS, G.M. (1961) Bacterial blight of cotton: its influence on yield. *Empire Cotton Growing Review* 38, 241–57.

ZACHOWSKI, A. AND RUDOLPH, K. (1988) Characterization of isolates of bacterial blight of cotton from Nicaragua. *Journal of Phytopathology* 123, 344–52.

3 Verticillium Wilt

A.A. BELL*

Introduction

Verticillium wilt is one of two vascular wilt diseases affecting cotton, the other being Fusarium wilt (Chapter 4). Vascular pathogens have the ability to colonize plant roots and penetrate to the vascular tissues, where they are contained and proliferate within the xylem vessels, eventually becoming distributed throughout the plant. Although these fungi may grow out of the vascular tissues in the advanced stages of infection and, after the death of the host (at least in the case of Fusarium wilt), may sporulate on the surface of crop residues, they can be considered to have no secondary cycle of infection. Vascular wilt fungi are specialized soil-invading pathogens with narrow host ranges. They have the ability to survive for long periods in the soil in the absence of a host by producing sclerotized or thick-walled resting structures which resist desiccation and lysis. Under optimal conditions for infection, the susceptible host is normally killed by a combination of toxic fungal metabolites, accumulated fungal material and host responses to infection, leading to vascular occlusion and moisture deficit.

History and Distribution

Verticillium wilt of cotton was first reported in 1914 after Carpenter isolated *Verticillium dahliae* from a few diseased plants of Upland cotton (*Gossypium hirsutum* L.) growing in Arlington, Virginia. In 1918 Carpenter showed that *V. dahliae* from okra in South Carolina caused wilt of cotton plants inoculated

* Cotton Pathology Research Unit, United States Department of Agriculture, Agricultural Research Service, Southern Plains Area, Southern Crops Research Laboratory, Route 5, Box 805, College Station, Texas 77840, USA.

in the greenhouse. Bewley (1922) later caused wilt in the greenhouse by inoculating Asiatic cotton (*G. herbaceum* L.) with *V. dahliae* from tomato in Europe.

Sherbakoff (1928) first recognized that *V. dahliae* could cause an economically important disease of cotton, when he found the disease in Lake County, Tennessee, in 1927. The next year he found the disease extensively in counties along the Mississippi River in both Tennessee and Arkansas. Farmers in this area had observed the disease in cotton planted in 'Gumbo' (sedimentary loam) soils for several years. In some fields, 100% of the plants were infected (Sherbakoff, 1929). In 1932, Miles and Persons reported that Verticillium wilt of cotton also occurred extensively in Mississippi, in heavy loam soils along the Mississippi River, and it was occasionally found in other river-bottom soils in the state. They first observed the disease at the Delta Branch Experimental Station, Stoneville, in experimental plots that had been used as a Fusarium wilt nursery since 1927. Cultivars grown on the plots, including those resistant to Fusarium wilt, had not shown any appreciable resistance to wilt over a period of three years. Isolations were made from plants grown on the plot, and *V. dahliae* was found to be responsible for most of the wilt. It is probable that *V. dahliae* was distributed throughout the cotton-growing areas of the upper Mississippi River Delta already in the early 1920s.

Diseases first diagnosed as Fusarium wilt near St David, Arizona, in 1921 and near Waxahatchie, Texas, in 1929 were later identified as Verticillium wilt (Taubenhaus, 1936; Anon., 1949). Shapovalov and Rudolph (1930) first reported the disease in California, after it was observed near Wasco in Kern County in 1927. By 1930, wilt was present in Kern, Tulare, Kings, and Madera Counties in California with the most severe infections in Kern County near Wasco and Shafter (Herbert and Hubbard, 1932). In 1936, H.D. Barker and C.D. Sherbakoff (unpublished report) surveyed cotton wilts throughout the USA and confirmed the causal agents by making hundreds of isolations. They found that Verticillium wilt was common in certain areas of Texas, Arizona, New Mexico and Missouri as well as Tennessee, Arkansas, Mississippi and California. They speculated that much of the 'Fusarium wilt' in Oklahoma was probably also due to *V. dahliae*. Taubenhaus also found the disease in 1936 in Ellis County and the El Paso area of Texas and near Anthony, New Mexico.

V. dahliae is indigenous to soils of the South West and Far West of the USA. When desert soils were first cleared for irrigation agriculture in California, Arizona and New Mexico, wilt appeared in the initial cotton crop of some fields, as scattered irregular patches, varying from single plants to 6 m in diameter. Within a few years, the disease had spread throughout such fields, and the patches were no longer evident (Presley, 1950). The fungus seems to have been associated with certain plant communities in the desert before cultivation was begun. Similar outbreaks of wilt occurred in Australia when cotton was first grown on land newly cleared for irrigated agriculture (Evans, 1967).

Shortly after the disease was described in the USA, it was reported from Central Asia, Bulgaria and Greece (Butler, 1933; Miles, 1934). In the next few years it was confirmed in Peru, Brazil, Uganda, China and various republics of the Soviet Union. Cotton wilt was a serious problem in Peru as early as 1911 but was incorrectly attributed to Fusarium wilt until the 1930s (Boza Barducci and Garcia Rada, 1942). Cotton wilt was recognized by Zaprometov as a serious problem in the former Soviet Union, especially Uzbekistan, as early as 1927, but *V. dahliae* was not identified as the major cause until the 1930s (Mukhamezhanov, 1966).

By the 1940s the disease was recognized as a serious threat to cotton production in Peru, the Soviet Union, Uganda and parts of the western and south-western USA. Ezekiel and Dunlap (1940) estimated that the wilt destroyed 15% of the cotton crop in the El Paso Valley, Texas, in 1939, and by this time many fields in California were also severely damaged by *V. dahliae*. By the end of the 1940s, losses from the disease in California were between 10 and 15% annually. Surveys in the Mesilla Valley, New Mexico, revealed the disease in 75% and 88% of the fields in 1944 and 1946, respectively. Losses ranging from 5 to more than 20% occurred in 27–30% of the fields (Leyendecker *et al.*, 1947).

Early breeding programmes led to the development and release of tolerant cultivars, such as Acala 1517WR by Leding in New Mexico in 1946, 108-F in Russia in the late 1940s, and Acala 4-42 by Harrison in California in 1954 (Wilhelm, 1981). Acala 4-42 and 108-F at first reduced wilt damage dramatically; losses from wilt in California dropped from 13% in 1953 to 2% in 1955 and 1956 when Acala 4-42 was planted throughout the wilt-affected areas of the state. However, within ten years, each of these cultivars lost its effectiveness. This was caused by three factors: (i) cotton monoculture increased inoculum density of the fungus in soils; (ii) intensified use of irrigation water and nitrogen fertilizers greatly increased disease severity; and (iii) the most virulent strain of the pathogen spread and increased in percentage of the total population of the pathogen.

Irrigated cotton monoculture also caused wilt to become a major problem in other locations. For example, the disease became very serious in the Comarca Lagunera region in the states of Durango and Coahuila in northern Mexico by 1953, following the construction of the Presa Lazaro Cardenas reservoir in 1946 and the drilling of numerous deep wells, which allowed much more intensive irrigation and encouraged cotton monoculture (Duffield *et al.*, 1953). Similarly, Verticillium wilt quickly became a major problem following the development of irrigated monoculture of cotton on the High Plains and Rolling Plains of Texas, New Mexico and Oklahoma and in higher elevations of Arizona in the 1950s. Wilt also became the major disease problem of irrigated cotton in Turkey, Syria, Iran, Zimbabwe, South Africa, Swaziland, northern Iraq, the Namoi River Valley in New South Wales, Australia, the Negev region of Israel, the Madras and Mysore States of India and the Guadalquivir Valley in southern Spain. With the expansion and intensification of cotton culture in China, the incidence of cotton wilt increased from 10% of the acreage in 1973 to 33% in 1982 (Liu, 1985). The exact proportion of wilt due to *V. dahliae* in China is uncertain, because Verticillium and Fusarium wilts commonly occur together.

Symptoms

Symptoms of Verticillium wilt depend on the cultivar, the virulence of the fungal isolate, the developmental stage of the plant and the environmental conditions, especially temperature. Infection and symptoms generally develop only when mean temperatures are below 30 °C. During a growing season, wilt may first appear in relatively young plants before the maximum temperatures have been reached, and then disappear in midsummer, reappearing when temperatures decline and plants are heavily fruited. Plants infected when young (see Plate 2A), show general yellowing, epinasty and defoliation of leaves and may be killed quickly or remain stunted as they recover during warmer weather. Healthy plants compensate for the early diseased plants. Consequently, early infections usually cause little loss of yield or quality, if plants later recover under higher temperatures.

Plants infected during fruiting develop characteristic mosaic patterns on the infected leaves, beginning at the base of the plant and progressing towards the top. These leaf symptoms first appear as yellowing of tissues along the margins and between the major veins (see Plate 2B, C). As symptoms progress, these areas become more intensely yellow, and occasionally red, before the tissues become necrotic and turn white, tan or brown. By the time leaf symptoms appear, a cross-section of the stem will reveal uniformly distributed tan-to-brown flecks in the wood (see Plate 2D), which are due to the formation of melanized products in the infected xylem vessels and surrounding cells. As the disease progresses, many of the leaves and the young bolls are defoliated, and plants take on a generally unthrifty appearance but are rarely killed. With the P-1 strain, defoliation often becomes complete, and terminal dieback may occur at the tips of the stems, which become black in colour. Lateral buds at the base of the plant may break dormancy, even before defoliation is complete, and may form a number of new side-branches which give the plant a bushy appearance. The pathogen should be isolated from petioles or veins of diseased leaves to confirm the cause of any cotton wilt.

Losses

Verticillium wilt (or Verticillium and Fusarium wilts in the case of China) is now the most important disease causing losses to the cotton crop in the three major cotton-producing countries (China, the former Soviet Union and the USA) and in eight of the other top 20 cotton-producing countries (Turkey,

Australia, Greece, Syria, Zimbabwe, Peru, South Africa and Spain). It is also the major pathogen in certain states of India, Brazil, Argentina and Mexico.

Yield losses due to Verticillium wilt in the USA are shown in Table 3.1. The peak loss in the United States occurred in 1961, when 580 000 bales were lost to the disease. The highest percentage loss (4.4%) occurred in 1967. Since then, losses have decreased somewhat due to a decline in the use of nitrogen fertilizer and irrigation water and the growing of new cultivars with improved resistance.

Losses from the disease are most severe in the former Soviet Union, where many potentially high-yielding farms annually lose 25–30% of their crop. Losses were estimated as 500 thousand metric tonnes of raw cotton (*c.* 760 000 bales) in 1966, with 80% of the losses occurring in Uzbekistan (Mukhamezhanov, 1966). Losses have declined since then, with the release of the 'Tashkent' and other resistant cultivars and with improved cultural management. However, losses could become greater if the P-1 strain spreads throughout Uzbekistan and other republics. The extreme importance of wilt in the former Soviet Union is shown by the tremendous research effort that has been dedicated to it. In 1966 there were 220–250 scientific and scientific-technical personnel working on Verticillium wilt of cotton at the Institute of Experimental Biology of Plants (Mukhamezhanov, 1966), and in 1990 at the Fifth International *Verticillium* Symposium in Leningrad, 48 Russian scientists gave 54 papers concerned with *V. dahliae* and the wilt of cotton.

Estimated loss of yield due to cotton wilt in China in 1982 was 100 million kg (*c.* 460 000 bales) (Liu, 1985; Shen, 1985). Loss estimates are not available from most other countries and regions but, if losses are similar in percentage

Table 3.1. Losses from potential cotton production caused by Verticillium wilt in the USA from 1952 to 1990[a].

Years	% Loss	Bales loss (1000s)	Years	% Loss	Bales lost (1000s)
1952–54	1.62	559.2	1973–75	2.60	861.3
1955–57	1.50	623.3	1976–78	2.59	896.6
1958–60	1.93	907.9	1979–81	2.36	1037.3
1961–63	2.57	1238.8	1982–84	1.50	521.6
1964–66	3.03	1028.6	1985–87	1.90	740.0
1967–69	3.48	966.6	1988–90	1.46	857.0
1970–72	2.41	839.1			
				Total	11 077.3

[a] Average percentage loss by state from 1952 to 1990 was: New Mexico, 5.73; California, 3.87; Arizona, 2.92; Texas, 2.43; Oklahoma, 2.28; Missouri, 1.79; Mississippi, 1.62; Arkansas, 1.55; Tennessee, 1.45; North Carolina, 0.34; Louisiana, 0.22; Alabama, 0.06; Georgia, 0.01; and South Carolina, 0.01%. Estimates are calculated from the annual publications of the US Cotton Disease Council.

to those in the USA and China, the combined loss would be about 700 000 bales. Thus, the current losses of yield in the world are about 1.5 million bales, worth more than US$1 billion.

In addition to its effect on yield, severe wilt also causes a significant reduction in fibre quality. This is due to an increase in the percentage of immature fibres which decreases fibre length, uniformity, strength and grade. Yarns spun from fibre produced on wilted plants are lower in grade and inferior in appearance, and the number of neps and the amount of manufacturing waste are increased considerably. Seed weight and vigour are also decreased by the disease.

Causal Organism

TAXONOMY

Verticillium Nees ex Link 1824, a genus of the Hyphomycetes, has predominantly hyaline vegetative hyphae and colonies with moderate growth rates. Conidiophores produced on solid media are usually well differentiated and erect, branched over most of their length, bearing whorls of slender flask-shaped or aculeate divergent phialides with inconspicuous collarettes. Conidia are hyaline or brightly coloured, mostly one-celled, and usually borne in mucilaginous heads. Hyphal cells, phialides and conidia are uninucleate. while hyphal tips may be multinucleate.

Prior to 1970, the causal organism of Verticillium wilt of cotton was often given as *V. albo-atrum*. This was due to a controversy over the taxonomy of *Verticillium*. Some scientists did not accept the validity of establishing *V. dahliae* as a new species and, consequently, referred to the pathogen from cotton as *V. albo-atrum*. Studies since 1970 have validated the two species.

Five species of *Verticillium*, *V. albo-atrum* Reinke & Berthold 1879, *V. dahliae* Klebahn 1913, *V. nigrescens* Pethybridge 1919, *V. nubilum* Pethybridge 1919, and *V. tricorpus* Isaac 1953, cause wilt diseases of plants. These five species are distinguished from other *Verticillium* species by the production of black resting structures (usually in a reverse pattern from near the margin of the colony) and by the ability to cause vascular infections of plants. These species are separated from each other by the morphology of the dark resting structure. *V. dahliae* colonies produce only discrete black microsclerotia as resting structures; mycelium, conidiophores and conidia are hyaline. *V. tricorpus* also produces microsclerotia, but it is easily distinguished from *V. dahliae* by its light orange mycelium and large numbers of dark chlamydospores in colonies. The taxonomy and morphology of *Verticillium* species have been reviewed (Smith, 1965; Isaac, 1967; Domsch *et al.*, 1980).

Under field conditions, only *V. dahliae* damages cotton. *V. albo-atrum* also causes a severe wilt of cotton in the greenhouse at low temperatures (20–24 °C). However, this species is unable to cause disease above 25 °C, whereas *V. dahliae* can cause cotton wilt at temperatures as high as 30 °C. In the cotton belt of the USA, mean temperatures below 25 °C do not normally occur in cotton fields until the crop is almost mature. Cotton fields are also not conducive to the survival of the dormant structures of *V. albo-atrum*. *V. tricorpus* colonizes cotton roots extensively and sometimes occurs in soils in much higher concentrations than *V. dahliae*. However, *V. tricorpus* does not colonize the xylem of cotton thoroughly enough to reduce yields. Likewise, *V. nigrescens* can be isolated from cotton roots, leaves and bolls, and it is very common in some cotton fields, but it also does not extensively colonize xylem or reduce yield. *V. nubilum* has not been isolated from cotton or cotton fields.

MORPHOLOGY

Colonies of *V. dahliae* grow moderately fast (2.0–3.5 mm day^{-1} at 20–25 °C) on PDA and initially appear white or light cream, but later become black, with the formation of microsclerotia. Patterns of microsclerotia formation vary considerably among isolates. In some cases, microsclerotia may not form until cultures are several weeks old. Sectors with varied patterns of microsclerotia formation often occur in colonies, and they may be due to either nuclear or mitochondrial mutations. The hyaline conidiophores are erect or prostrate and have several whorls of three to four phialides that are subulate and mostly 1.0–2.5 × 16–35 μm. Conidia are ellipsoidal to short-cylindrical, hyaline and mainly one-celled (Fig. 3.1). Haploid conidia are 1.4–3.2 × 2.5–6.0 μm and diploid conidia are 1.6–3.5 × 5–12.5 μm. Stable diploid isolates are designated

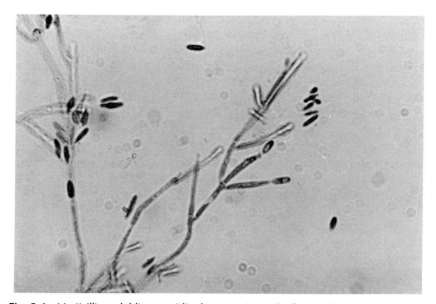

Fig. 3.1. *Verticillium dahliae*: conidiophores and conidia from culture growing on PDA (× 380).

as *V. dahliae* var. *longisporum* and are relatively rare. Unstable homozygous and heterozygous diploids can be formed with proper cultural techniques (Tolmsoff, 1983). Microsclerotia are dark brown to black and torulose, and consist of almost globose cells that originate by repeated budding and multi-lateral septation of a single or a few adjacent hyphae. They may be either elongate or globose, with dimensions of 15–100 × 50–200 µm. On water agar, microsclerotia of the P-1 strain are normally elongate, whereas those of the P-2 strain are globose. Other variations in microsclerotium morphology involve size of individual cells and emergence of hyphae or conidiophores from among the dark melanized globose cells.

PHYSIOLOGY

The nutrient requirements of the wilt pathogen consist simply of a carbon source and the mineral elements commonly found in soils and plant tissues. The fungus grows well on the simple sugars and amino acids that are normally found in root exudates and xylem sap. However, the fungus also produces a large number of hydrolytic enzymes that allow digestion of carbohydrates, proteins and nucleic acids in plant tissues. Oxygen concentrations and water availability rarely affect growth. Maximal mycelial growth and sporulation occur at 1.9–2.7% oxygen and at a water potential of −10 to −30 bars.

The optimal temperature for growth is 22–27 °C, and some isolates will grow slowly at 32 °C. The P-1 strain grows better than the P-2 strain at 28 °C and above. Temperature is usually the greatest limiting factor to the growth of the fungus in the cotton plant during most of the growing season. Tempera-ture tolerance of the fungus is increased somewhat by decreased water poten-tial. Consequently, the fungus may grow at 35 °C with water potentials of −30 to −40 bars, but it will not grow above 33 °C when water is readily available. The fungus cannot normally cause disease of cotton above 30 °C. Either light or negative pressure potential stimulates sporulation and inhibits formation of microsclerotia.

The dark pigments in the microsclerotia have been identified as melanins formed from 1,8-dihydroxynaphthalene via the pentaketide pathway. The melanins are deposited as granules in the outer cell wall and surrounding matrix. They apparently protect the microsclerotia against dehydration, para-sitism by micro-organisms and toxic chemicals in soil (Bell and Wheeler, 1986).

Hydrolytic enzymes (pectinase and cellulase), plant growth regulators (IAA and ethylene) and complex polymers containing protein, lipid and polysaccharide (PLP) are produced by *V. dahliae*, and these have been impli-cated in the production of disease symptoms. However, the importance of any of these compounds is still controversial. Detailed reviews of the physi-ology of *V. dahliae* are available (Puhalla and Bell, 1981; Heale, 1988; Bell, 1992).

Host range

The known host range of *V. dahliae* now exceeds 400 plant species. Severe disease is caused on food and fibre crops, such as cotton, sunflower, peanut, potato, tomato, eggplant, pepper, broad bean, runner bean, pea, squash, cucumber, melon, watermelon, onion, oilseed rape, hop, olive, pistachio, apricot, plum, peach, cherry, avocado, mango and strawberry; on ornamentals and shade trees, such as maple, rose, chrysanthemums and snapdragons; and on common agricultural weeds, such as velvetleaf, ragweed, pigweed and cocklebur. The host range of a strain, and especially a single isolate, will be considerably smaller than that of the species. For example, the P-1 strain has very weak virulence to tomato, potato and melons, and has rarely been isolated from these plants in the field. In contrast, the P-4 strain is very virulent to potato but has the least virulence of any strain to cotton.

Isolation and growth in culture

The pathogen is readily isolated from petioles or small veins of leaves that show symptoms but are still attached to the plant (Llosa, 1938). After the leaf is thoroughly washed with tap-water, sections of major leaf veins between symptomatic interveinal areas are cut out with a No. 3 cork-borer, dipped in 1% hypochlorite (20% bleach) for about 15 seconds, rinsed with sterile water and placed on PDA amended with 200 ppm streptomycin sulphate, 200 ppm penicillin G and 0.2–0.5% ethanol. Petioles are treated with hypochlorite for 30–60 seconds, rinsed with sterile water, cut into 0.5–1.0 cm lengths, split and placed on media. In some cases, surface disinfection with mercuric chloride in aqueous ethanol may be more effective than hypochlorite. Cultures may be placed on prune extract agar, Czapek's agar or potato–carrot dextrose agar to stimulate production of microsclerotia. Alternatively, 1.0 mM catechol can be added to the PDA to accomplish this purpose (Bell, 1992).

The fungus can also be isolated from the woody tissue of the stem and root. However, there is a much greater risk of obtaining saprophytes, especially *Fusarium* species, or less virulent variants of *V. dahliae* from these tissues. The leaf isolates will occasionally yield *V. nigrescens*, which apparently colonizes insect wounds. Thus, caution must be taken not to confuse these two species.

Fungal strains can be maintained genetically pure on PDA by making transfers as hyphal tips while cultures are still growing. The strains may be preserved by mixing conidia from hyphal fronts in advance of mature microsclerotia with milk and freeze-drying them. The sealed freeze-dried preparations may be stored in the refrigerator or freezer. Cultures on agar slants can also be preserved under mineral oil, or air-dried microsclerotia may be stored over desiccant in the refrigerator or freezer. Revival of these latter cultures will give rise to variants; therefore, the wild type must be reselected.

Variation and Genetics of the Pathogen

Isolates of *V. dahliae* frequently show variation in morphology and virulence to different plant species. Serological and auxotrophic variants have also been described. The most frequent morphological variants show decreased production of black pigment in the culture and have been referred to as white, hyaline or albino variants. Three distinct types of these variants occur: (i) true albinos, designated as *alm* mutants, which make normal numbers of albino microsclerotia; (ii) largely hyaline colonies, designated as *rms* or *hyal* mutants, which readily form dark microsclerotia only if treated with catechol; and (iii) hyaline colonies, designated as *nms* or *hyal*+ mutants, which have fluffy mycelium and do not make microsclerotia, even in response to catechol. The *nms* variation is due to mutations in mitochondrial DNA, whereas the *alm* and *rms* variations are due to mutations or changes in nuclear DNA. The *rms* and *nms* variants grow more rapidly than the wild-type fungus and can overgrow the wild mycelium. Consequently, they often replace the wild parent during culture transfers unless precautions are taken. Transfers must be made from hyphal tips, while colonies are still actively growing, if genetic purity is to be maintained. Other morphological mutants show differences in the colour or morphology of microsclerotia and in the colour of the mycelium. Most of these were developed with mutagens and were used to elucidate the pathways of biosynthesis of melanins and carotenoids.

In 1966, Schnathorst and Mathre proved that different levels of virulence to cotton occur among strains of *V. dahliae*. They designated the most virulent strain as a defoliating strain because it completely defoliated young Acala 4-42 plants in the greenhouse. Subsequently, Puhalla and Hummel (1983) and Joaquim and Rowe (1990) showed that defoliating and non-defoliating strains belong to genetically isolated, subspecific populations. Isolates belonging to the two populations (or heterokaryon compatibility groups) as described by Joaquim and Rowe will be designated as P-1 and P-2 strains in this paper. All defoliating isolates belong to the P-1 strain.

The P-1 strain is indigenous to the Mississippi River Delta, the South West and the Far West in the United States. All cotton isolates tested from Missouri, Arkansas, Oklahoma, New Mexico and Arizona were P-1 isolates, and most isolates from Texas were P-1 isolates (Mathre *et al.*, 1966; Puhalla and Hummel, 1983). Thus, the highly virulent P-1 strain has probably been responsible for most of the Verticillium wilt in these areas since the disease was first discovered. The P-1 strain occurs sporadically in California, predominating in a few areas such as Tulare and Kings Counties, while the less virulent P-2 strain originally predominated in other areas of California.

The P-1 strain may have been introduced into other countries with the import of US seed. The P-1 strain has been found in the Rimac valley of Peru along with the P-2 strain (Schnathorst, 1969). Isolates that cause severe wilt in Mazandaran in northern Iran (Puhalla and Hummel, 1983), in the lower Guadalquivir Valley of Andalucia in southern Spain (Blanco-Lopez *et al.*,

1989) and in the Yangtze River Valley in the Jiangsu Province in China (Oingii and Chiy, 1990) also belong to the P-1 strain. Isolates designated as races 2 and 3 in the former Soviet Union defoliate Tashkent varieties, which have resistance similar to Acala 4-42 (Portenko and Kas'yanenko, 1988). Thus, races 2 and 3 are probably equivalent to the P-1 strain. Races 2 and 3 predominate in the Hissar Valley, Tajikistan. The disease can be expected to become a greater problem in the eastern hemisphere with spread of the P-1 strain to additional geographical areas.

Isolates showing variation in virulence have been designated as defoliating (P-1) and non-defoliating (P-2) isolates in the USA and races 0, 1, 2, 3 and 4 in the CIS. The defoliating isolates are distinguished by their virulence to the Deltapine 15 and Acala 4-42-77 cultivars of *G. hirsutum*, the Tanguis 2885 cultivar of *G. barbadense*, and the Pearson Improved cultivar of tomato (Schnathorst and Mathre, 1966). The races are distinguished by the reactions of the 108-F and Tashkent-1 cultivars of *G. hirsutum* and of a strain of *G. arboreum* L. (Portenko and Kas'yanenko, 1988). Races 2 and 3 are probably equivalent to the defoliating (P-1) isolates, while races 1 and 4 are equivalent to the non-defoliating (P-2) isolates. Hyaline variants of the *nms* type generally show a marked decrease in virulence, whereas *alm* and *rms* variants generally retain the virulence of the wild parent.

Variants may arise as a consequence of mutation, parasexual (mitotic) recombination or heteroploidy. Spontaneous mutations occur frequently for morphological types and nicotinamide requirement (Puhalla, 1980). The frequencies of mutations can be greatly increased with ultraviolet light or toxic chemicals (Puhalla, 1973). Potassium chlorate (1.5–2.5%) is added to media to select nitrate reductase mutants (Joaquim and Rowe, 1990), and glycerol is added to select for other auxotrophic requirements (Puhalla, 1976). Acriflavine is used to induce mutations in mitochondrial DNA (Typas and Heale, 1976). Mutants have been used in genetic studies to show that the parasexual cycle operates and to establish five linkage groups, which apparently correspond to five chromosomes (Heale, 1988). Tolmsoff (1983) concluded that most variation in cultures originates from microsclerotial cells, as a consequence of the development of heteroploids in these cells. Several detailed reviews of the variation and genetics of *V. dahliae* are available (Puhalla and Bell, 1981; Tolmsoff, 1983; Tolmsoff and Howell, 1984; Hastie and Heale, 1986; Heale 1988).

Disease Cycle

The disease cycle may be summarized as follows (Schnathorst, 1981; Huisman and Gerik, 1989; Bell, 1992).

1. The pathogen survives as dormant microsclerotia that occur mostly free, but also in plant debris in the soil, persisting for many years at depths down to 40 cm.

2. Microsclerotia germinate in response to root exudates and establish colonies on the cotton root surfaces, usually between 3 and 10 mm back from the root tip.

3. Hyphae from the root colonies penetrate deep into the cotton root cortex, spread along and around the endodermis of the stele, and eventually enter the xylem vessels by direct penetration or through wounds. Wounds are not required for infection but may increase the incidence and severity of wilt.

4. The pathogen proliferates within the cotton xylem vessels by forming conidia that are carried by the xylem 'stream' to end walls where they germinate, penetrate into the next vessel segment and form new conidia. Hyphae from some conidia also penetrate through pit membranes and form conidia in adjoining vessels. These steps are repeated many times during the process of complete systemic invasion of the xylem. Symptoms do not generally appear until most of the vessels in a leaf become totally occluded by a combination of the fungus, its catabolic products and the resistance responses of the cotton plant.

5. Hyphae of *V. dahliae* penetrate from xylem vessels into surrounding parenchyma cells and, eventually, throughout the necrotic tissue of dead leaves, stems and roots, where new microsclerotia are formed after several weeks or months, depending on availability of moisture. With decay of the tissue and cultivation, the microsclerotia are again dispersed in the soil.

Epidemiology and Ecology

The severity of Verticillium wilt in cotton depends on inoculum potential and virulence of the pathogen, resistance potential of the cultivar, temperature (especially in the soil), soil conditions (pH, moisture content and nutrient content) and concentrations of microbial antagonists or synergists. Losses in yield and quality are caused primarily by damage to, and loss of, foliage, especially during fruiting and early stages of boll development. In the northern hemisphere, yield is generally correlated negatively with the percentage of plants showing foliar symptoms from late July to early September. For example, Waddle and Fulton (1955) in Arkansas found that yield of seed cotton was decreased by 25 and 17 kg ha^{-1} for each 1% increase in the percentage of plants showing foliar symptoms on 1 August 1953 and 1955, respectively. The earlier that symptoms occur in the critical period, the greater the damage will be to the affected plant. In the USA, correlations between yield and foliar symptoms generally become progressively smaller after August until harvest, but they are still higher than those between yield and systemic invasion, as determined by observing vascular browning in the wood at the base of the stem.

Relationships between inoculum density and incidence and severity of wilt have been studied extensively in California (Ashworth, 1983; DeVay and

Pullman, 1984). The incidence of susceptible plants showing systemic infection at harvest increases progressively as the inoculum density increases from 0 to 10 propagules g^{-1} of soil with 100% incidence at the highest density. The rate at which the percentage of plants infected increases after disease onset in July (calculated from disease progress curves) also increases progressively with increases of inoculum density from 0 to 50 propagules g^{-1} of soil. Similarly, mean disease severity, expressed as reductions in plant height and in numbers of internodes, retained squares and bolls, is increased with increasing inoculum density up to 50 propagules g^{-1} of soil. The critical numbers of propagules required for given damage levels will vary with virulence of the pathogen and resistance of the host. Much greater inoculum densities are required for the P-2 than for the P-1 strain to cause the same degree of damage. Greater inoculum densities are required for resistant than for susceptible cultivars to cause the same damage. Relationships between inoculum density and disease severity also are influenced by other factors: methods used to measure inoculum density, environmentally induced dormancy in microsclerotia and microbial antagonists (Bell, 1992).

Most of the microsclerotia that serve as inoculum in soil are formed in undecomposed plant residues, during the cool winter months following harvest. Grishechkina (1990) in Russia found considerable variation in numbers of microsclerotia formed in different tissues: 82 000–7 000 000 g^{-1} in dry leaves, 2000–827 000 g^{-1} in stems and 300–172 000 g^{-1} in roots. The quantities were much greater in plant residues of susceptible *G. hirsutum* Tashkent 1 and Andizhan 5 than in resistant *G. barbadense* Termez 7. In New Mexico, Blank and Leyendecker (1951) found that incorporating one infected stalk 0.8 m^{-2} into soil free of the pathogen gave nearly 100% infection in the following cotton crop. Incorporating gin trash from wilt-infected cotton at the rate of 25 tonnes ha^{-1} for three years on 'cut and filled' areas of a levelled field, in the High Plains of Texas, increased wilt incidence from 32 to 81% and 53 to 86%, respectively (Waker and Onken, 1969). The increased severity of wilt caused by the gin trash persisted for at least seven years.

The resistance of the cotton plant to infection depends on its genetic constitution, age and physiological condition. Various levels of tolerance or resistance to the disease have been found and incorporated into modern cultivars. None of these sources of resistance prevents infection of the vascular system, although they may slow the rate and incidence of such infection. The primary benefit of resistance seems to be slowing the disease progress rate for any given inoculum density. This gives increased yield and quality but may allow 100% incidence of symptoms by harvest. Young plants are generally more susceptible than older plants, at least until a full fruit load is developed and the plant becomes determinant. As a consequence, disease is often more severe if plants are infected while young. Plants that are growing rapidly and are succulent are much more susceptible to wilt than those that are growing slowly and are hardened by unfavourable growing conditions. Cultural practices can strongly affect the relative resistance of any given cultivar.

Temperature is the most important environmental determinant of wilt and is usually a factor limiting disease expression, in most years and localities (Bell and Presley, 1969; Barrow, 1970b; DeVay and Pullman, 1984). Verticillium wilt does not occur until mean soil temperatures are below 30 °C and remains mild until temperatures decrease below 28 °C. As temperatures decrease, the severity of the disease increases progressively and measurably with each degree of temperature decrease down to 22 °C. Consequently, a decrease in temperature of only 2 or 3 °C from the normal mean during late July and early August can greatly increase losses from the disease. The effects of temperature are due to effects on both the growth and sporulation rate of the fungus (optimal at 22–27 °C) and the rate of active defence response in the host (optimal at 30–35 °C) (Bell and Presley, 1969). Because of these distinctly different temperature optima, the rate of pathogen colonization/rate of host response increases rapidly as temperatures are decreased from 30 to 22 °C. In the USA, temperatures in late July and early August over much of the cotton belt are often restrictive to wilt development, so that both early infections before this time and late infections after this time are often seen during the course of the growing season. Plants infected early may recover during warm weather or, if plants are severely diseased or killed, adjoining healthy plants may compensate for them. The critical period for wilt development, therefore, is during the period of temperature decline following mid-season. The years in which wilt has been most devastating have occurred when the normal high temperatures failed to develop during midsummer.

Soil conditions may directly affect the pathogen, alter host resistance or change relationships with other micro-organisms (Bell, 1989). Verticillium wilt normally occurs in neutral or slightly alkaline soils (pH 6–9). Growth, microsclerotia production and survival of the fungus are inhibited in soil at pH 5.5 and below. This may be due to the manganese and aluminium ions, which are progressively increased in concentration in soil solution as the pH of soil is lowered. Cotton plants grown on acid soils also accumulate much higher concentrations of these ions in leaves than do plants grown on neutral soils. Potassium deficiency or liberal nitrogen fertilization also significantly increases the incidence and severity of wilt. High levels of soil moisture due to rainfall or irrigation generally decrease soil temperature and increase disease incidence and severity, whereas prolonged flooding of soil generally decreases inoculum density and subsequent disease, by favouring antagonists or causing accumulation of toxic metabolites in soil.

V. dahliae is a soil invader rather than a soil inhabitant, because it is unable to grow appreciably as a saprophyte on tissue other than that which it originally parasitized. The fungus cannot colonize healthy plant leaves or stems placed in soil, even though large numbers of microsclerotia are produced in infected leaves or stems incorporated into soil (Wilhelm, 1965). The fungus is vulnerable to many antagonists in soil and is a poor competitor. It will grow extensively in sterile soils, but, once the soils are inoculated with mixtures of other micro-organisms, the fungus no longer grows as a saprophyte. Practices that increase the activity of saprophytes, without other-wise favouring disease, often decrease wilt severity. Various mycolytic bacteria, actinomycetes and mycoparasitic fungi suppress *V. dahliae* in cul-ture, and are also associated with wilt-suppressive soils. A few micro-organ-isms, such as endomycorrhizal fungi and the reniform and root-knot nematodes, increase wilt severity in greenhouse tests (Bell, 1991a). Mycorrhi-zal fungi apparently increase host susceptibility by increasing phosphorus availability, whereas nematodes provide wounds that facilitate the pene-tration of roots by the wilt pathogen. These interactions have not been demonstrated to be important under normal field conditions.

Control

No single method is highly effective in controlling the disease. Consequently, an integrated management system is necessary to minimize losses from the disease (El-Zik, 1985). Control begins with the selection of a cultivar that has some degree of resistance to wilt, good agronomic characteristics and adap-tation to the geographical location. Then a combination of cultural practices are adapted for this cultivar to minimize the losses from wilt. Chemical controls can be effective, but are generally not used because of their prohib-itive cost. Biological controls are used to a limited extent only in Russia.

RESISTANT CULTIVARS

In the field, cultivar resistance to Verticillium wilt is expressed as a delay in the onset of visible symptoms, a decrease in the rate of the disease progress curve and a decrease in symptom severity, especially during the second half of the growing season. Consequently, resistant cultivars will have lower percentages of plants with foliar symptoms, after bolls first set and temperature begins to decline, usually from 1 August to 15 September in the northern hemisphere. Foliar symptoms will also be less severe in resistant than in susceptible cultivars during this period. Cross-sections of stems of resistant, compared with susceptible, cultivars show a lower percentage of the xylem vessels with browning, and petioles of leaves contain fewer fungal propagules when they are homogenized and plated on nutrient medium.

Perhaps the greatest gains in the control of wilt have come from the elimination of susceptible and highly susceptible cultivars and lines from production and breeding programmes. In the USA in 1954, nine of the ten most planted cultivars were susceptible or highly susceptible to the P-1 strain. In contrast, the ten leading cultivars in 1990 all had greater resistance to *Verticillium* than at least seven of the ten cultivars grown in 1954. Marked increases in wilt severity have resulted from the few introductions of new, otherwise agronomically desirable, cultivars that were considerably more susceptible to wilt than previous cultivars. When grown in monoculture, such varieties cause rapid build-up of inoculum concentrations in soil. Yields, which initially may be high, consequently decline progressively and rapidly.

When selecting a resistant cultivar, it is important to know the inoculum density (or previous disease severity) in the field, the strain or strains present, the prevailing environmental conditions and the intensity of cultural practices for obtaining high yields. Various methods are available for determining inoculum densities and virulence levels of *V. dahliae* isolates (Bell, 1991b). Generally, the level of resistance needed increases with increases in inoculum density, the virulence of prevailing isolates, the number of days with soil temperatures below 30 °C at 25 cm depth during boll development and the amount of irrigation and nitrogen fertilization used. Low or moderate levels of resistance, combined with good cultural practices, often give good disease control when conditions for wilt development are marginal.

Several cultivars released in the USA in recent years have high levels of resistance to Verticillium wilt. These include Acala Prema, Acala Royale and Acala Maxxa in California; Acala 1517-91 in New Mexico; Paymaster 147, Paymaster 303, and Paymaster 404 in Texas; and Delcot 344 in Missouri. In my tests, Acala Prema and Acala 1517-91 have been equally as resistant as the resistant standard Seabrook Sea Island 12B2 (*G. barbadense*), at the initial flower stage of growth. Several other cultivars in the Acala 1517 series released in New Mexico also have moderate to high levels of resistance to wilt and resistance to bacterial blight. Moderate to high levels of resistance to both Fusarium and Verticillium wilt occur in Acala SJ-5, Deltapine Acala 90 and Delcot 344. Cultivars that have moderate to high levels of resistance to Fusarium wilt and low to moderate levels of resistance to Verticillium wilt include Deltapine 20, Deltapine 50, DES 119, Germain's GC-510, Stoneville 112 and Stoneville 506.

In the former Soviet Union, about 200 cultivars have wilt resistance derived from a single wild strain of *G. hirsutum* ssp. *mexicanum* var. *nervosum*. The first and most popular of these were the Tashkent cultivars, such as Tashkent 1, Tashkent 2, etc. More recent resistant cultivars include Andizhan 9, Andizhan 60, Kirgizsky 3, 175F, AN-510, C-6524 and C-9070 (Kravtsova, 1990). These cultivars are highly resistant to race 1 (P-2 strain) but have only slight resistance to race 2 (P-1 strain). Cultivars with moderate to high levels of resistance to Verticillium wilt have also been developed in several other countries: Laoyang 5, 8004, 8010 and Zhong Mien 12 in China (Shen, 1985); Sakel in Iran (Moshir-Abadi, 1981); and Albar G501 in Zimbabwe (Hillocks, 1991). The reaction of these cultivars to the P-1 strain is not known. Shaanxi 1155 and Liaomiao T were developed for resistance to both Fusarium and Verticillium wilts in China (Li and Shen, 1987).

CULTURAL PRACTICES

Cultural practices used to control Verticillium wilt have several objectives: (i) to prevent the introduction, spread and build-up of inoculum of the pathogen in soil; (ii) to eradicate, or reduce the inoculum potential of the pathogen in soil; and (iii) to optimize the expression and use of cultivar resistance. These objectives are achieved by manipulating crop sequences, nutrients, soil moisture, planting practices and tillage practices, and by solarization. Benefits derived from different control practices will depend on the cultivar, the inoculum concentration, the prevalence of different strains of *V. dahliae*, environmental conditions and the physics, chemistry and biology of the soil.

Rotation

Almost any disruption of continuous culture of susceptible cotton cultivars will reduce the incidence and severity of wilt, compared with continued monoculture. Fallow, rotational crops or highly resistant cultivars prevent the continued spread and large annual build-up of inoculum potential that occurs with monoculture of a susceptible or slightly resistant cotton cultivar. DeVay and Pullman (1984) found that monoculture of Acala SJ-2 in California caused average increases in the propagule densities of *V. dahliae* of 13–15 propagules g^{-1} $year^{-1}$. Increases continued each year until plateau concentrations of 60–70 or up to 300 were reached, in soils with mixes of the P-1 and P-2 strains or with the P-1 strain only, respectively. Only ten propagules g^{-1} were needed for 100% incidence of wilt, and 40 generally gave the maximum severity of foliar symptoms. Thus, the inoculum build-up from only a few years of monoculture of a susceptible variety can have a devastating effect on wilt severity.

Although most crops prevent the annual build-up of inoculum potential that would occur with cotton, many do not appreciably reduce the existing inoculum potential, especially in one-year rotations. Plant groups that reduce inoculum potential as well as wilt incidence and severity in the following cotton crop include grasses, legumes and crucifers. Rotation with paddy rice for a single season decreases inoculum potential to undetectable levels and gives lasting benefits for many years (Butterfield *et al.*, 1978; Pullman and DeVay, 1982). Flooding alone is much less effective in reducing inoculum potential. The most effective of the other grasses are the forage grasses, such as perennial ryegrass and fescue, and grain crops, such as barley, wheat and sorghum, which reduce inoculum potential 50–75% during a single year. Rye, maize, Sudan grass and hegari have also been beneficial in some locations and years. The most used legume is alfalfa, which is often grown for two to three years in rotation with cotton. Other legumes that reduce inoculum density and wilt following a one-year rotation include peas, beans, sweet clover, lespedeza, Hubam clover, white clover and soyabeans. Autumn plantings of mustard and rape, as green manure crops, have reduced inoculum potential by 80–95% in Russia (Grishechkina, 1990).

The most effective rotations use crops from at least two different plant groups. Autumn planting of mustard following summer planting of rye or maize gives good wilt control in Russia (Marupov, 1990). Similarly, soyabeans followed by barley or sorghum and then cotton in the third year have substantially reduced wilt in cotton in Arkansas (Hinkle and Fulton, 1963). Planting wheat, peas and vetch over a period of six years in Israel reduced inoculum potential by 96% but did not eradicate the pathogen (Ben-Yephet *et al.*, 1989).

Crops that can be susceptible to cotton strains of *V. dahliae* should be avoided in rotations that include olive, pistachio, sesame, safflower, peanut, cowpea, celery, tomato, potato, okra, beet, eggplant, chili pepper, mint, castorbean and sweet potato. California poppy, snapdragon and maple are also susceptible to cotton strains of the wilt pathogen.

Fertilizer use

The effects of macro- and micro-element amendments on Verticillium wilt have recently been reviewed in detail (Bell, 1989). Deficiency of potassium increases the severity of wilt, presumably by increasing host susceptibility. Correcting potassium deficiency decreases wilt and increases yield. The percentage of affected plants and severity of foliar symptoms are generally proportional to the rate of nitrogen fertilization, and this effect is amplified by deficiency of potassium. However, low to moderate rates of nitrogen also give substantial increases in yield in spite of the increased wilt severity. Therefore, it is not economically feasible, in the short term, to reduce drastically nitrogen fertilization rates in order to control wilt. The adverse effects of nitrogen can be offset by using balanced fertilizers which prevent other elemental deficiencies and by avoiding excessive nitrogen fertilization. Organic forms of nitrogen, such as urea, manure and alfalfa meal, usually increase wilt less than inorganic forms. Using lower nitrogen rates with nitrification inhibitors has also decreased wilt and increased yield. Combining organic amendments with inorganic fertilizers decreases wilt compared with the inorganic fertilizer alone.

Soil moisture

Excess moisture beyond what is minimally necessary for desired yields increases both disease incidence and severity. Excesses in moisture are especially critical during initial flowering and boll set, when soil temperatures are still increasing toward the peak reached in midsummer. Moisture levels can be regulated by controlling irrigation and providing better drainage. Decreasing irrigation frequencies or amounts, especially prior to flowering and during initial flowering, decreases the severity of wilt. Prolonged moisture stress during this period may also have detrimental effects on yield. Therefore, it is important to determine optimal timing and amounts of irrigation for wilt control that are consistent with desirable yields. Sprinkler irrigation increases wilt less than furrow irrigation (Hillocks, personal communication). In fields that have high water tables or little slope, improved drainage has decreased wilt, presumably by increasing soil temperature and providing better root aeration.

Irrigation treatments show interactions with both nitrogen fertilization rates and plant density on wilt severity. Excess moisture is even more damaging in the presence of excess nitrogen. In the High Plains of Texas, losses from Verticillium wilt decreased progressively from about 10% of the potential crop in the late 1960s to about 1% of the crop in the late 1980s. This decline has paralleled a similar progressive decline in the use of irrigation water and nitrogen fertilizer, which has resulted from the high cost of pumping irrigation water from wells in this area (Wanjura and Barker, 1987). This situation dramatically shows the strong combined effects of nitrogen fertilization and irrigation on wilt severity. A plant density of 190 000 plants ha^{-1}, with three cycles of irrigation, yielded 40% more than the conventional 50 000 plants ha^{-1}, with four cycles of irrigation. A plant density of 150 000 plants ha^{-1} with one cycle of irrigation yielded as much as the conventional control (Palomo Gil and Quirarte, 1976). Thus, optimal effects of limiting irrigation are achieved with reduced nitrogen fertilization and increased plant densities. High plant beds may also be desirable to improve drainage and increase soil temperature where poor drainage exists.

Planting and tillage

Planting and tillage practices that affect wilt incidence include seed bed preparation, seed treatment, planting dates and densities, weed control practices and treatment of cotton residues. Vigorous seedlings should be established as early as possible, so that bolls are set and developed as far as possible before temperatures become favourable for wilt late in the growing season. In many areas, high seed beds increase soil temperature and seedling growth, and they reduce wilt severity. In New Mexico, beds built to heights of 20 and 38 cm, compared with flat seed beds, raised soil temperatures at the 15 cm depth from 69.8 to 75.2 and 80.6 °C, respectively (Leyendecker, 1950).

Reducing seed transmission

Evidence for the seed transmission of *V. dahliae* has been reviewed by Sackston (1983). Microsclerotia are deposited on the surface of fuzzy seed during harvest and ginning, and in a few instances internal infestation has been found (Evans *et al.*, 1966). Acid delinting of seed and treatments with carbendazim or other suitable fungicides prevent seed transmission (Shen, 1985). Seed germination and seedling vigour can also be increased by other treatments: density sorting, systemic insecticides and acid neutralizers (Minton, 1978). These treatments decrease wilt by increasing plant population density and earliness of flowering and fruit set.

Planting density

The percentage of plants affected by wilt is related inversely to the amount of planting seed ha^{-1} and to the resulting number of plants ha^{-1}. Plant densities of 100 000–150 000 ha^{-1}, compared with more conventional numbers of 50 000–60 000, generally result in reduced wilt severity and increased yields (Leyendecker, 1950). Plant densities higher than 150 000 ha^{-1} decrease wilt even more, but usually give lower yields than the optimal densities. Increasing plant density by using either narrower rows or greater densities of plants m^{-1} in standard row widths have been similarly effective in reducing wilt (Minton *et al.*, 1972). Both raised seed beds and high plant densities have been more effective when most of the water is provided by irrigation compared with predominantly by rainfall. Similarly, these practices are most effective when fields are well drained.

Weed control

Both weeds and practices used for weed control influence wilt severity. Deep cultivation facilitates infection of plants and increases wilt incidence and severity. Therefore, shallow cultivation should be used minimally for weed control (Young *et al.*, 1959). Failure to control red root pigweed (*Amaranthus retroflexus*) in sorghum caused a significant increase in wilt in the following cotton crop, compared with cotton following weed-free sorghum (Minton, 1972). Weeds associated with wheat culture were held responsible for the high incidence of wilt in many fields when cotton was first planted in the Namoi Valley in Australia (Evans, 1971). Weeds that are highly susceptible to wilt and allow seed transport of the fungus in Australia include *Xanthium pungens, Xanthium spinosum* and *Carthamus lanatus*. Other weeds that occur in cotton fields and are susceptible to cotton isolates of *V. dahliae* include *Geranium carolinianum, Xanthium commune, Lamium amplexicaule, Abutilon theophrasti, Sida spinosa, Physalis alkekengi, Hibiscus trionum, Portulacca oleraceae, Datura stramonium, Ageratum conyzoides, Tribulus terrestris, Sphaeralcea angustifolia, Solanum rostratum, Lepidium viginicum, Ipomoea purpurea, Amaranthus palmeri* and *Chenopodium album*. These should be controlled in crop sequences involving cotton (Brown and Wiles, 1970; Johnson *et al.*, 1980).

Removal of crop residues

Cotton residues from fields infested with *V. dahliae* need to be managed in a manner that minimizes spread and build-up of inoculum potential in the soil. In Israel, cotton gin waste cannot be used for cattle bedding or other purposes that might introduce it into newly cultivated areas (Krikun and Susnoski, 1971). Composting gin trash is effective in killing most of the fungal propagules, especially when composting is performed in trenches (Staffeldt, 1959;

Sterne *et al.*, 1974). Flaming also destroys fungal propagules in cotton residues. Heating cotton tissues to 70 °C for 30 sec or 100 °C for 25 sec eliminates the pathogen even in wet tissue (Ausher *et al.*, 1973). Treatments that allow earlier and more complete decomposition of residues generally decrease production of microsclerotia. Thus, shredding of stalks and complete ploughing in should occur as soon after harvest as possible (Wilhelm and Sagen, 1985). Spraying residues with urea or other nitrogen fertilizers before discing and irrigating after discing can facilitate residue breakdown and decrease formation of inoculum in residues (Grishechkina and Sidorova, 1984).

Solarization

Solarization refers to treatments that use irradiation from the sun to heat soil. A thin, clear plastic sheet is placed over the soil and left there for several weeks. Irrigating just before the plastic is spread, or under the plastic sheet, greatly increases the efficiency of solarization. Solarization treatments have been very effective in eradicating most *V. dahliae* propagules from soils where cotton is grown under irrigation (Pullman *et al.*, 1981a; Conway and Martin, 1983). The procedure also reduces populations of *Pythium*, *Thielaviopsis* and *Fusarium*, and it gives large yield increases. Solarization normally causes soil temperatures to rise to 40–50 °C to depths of 25 cm. Cotton isolates of *V. dahliae* are killed after 26–29 days at 37 °C and after 23–27 min at 50 °C (Pullman *et al.*, 1981b). Most of the propagules were killed after 14–66 days in solarized soils in California. Solarization for one month with irrigation has also cured the potassium-disease complex and increased yields more than 60% (Weir *et al.*, 1987). In Spain, solarization has given yield increases of 230% in some fields infested with *V. dahliae* (Melero-Vara *et al.*, 1990).

CHEMICAL CONTROL

The chemical control of Verticillium wilt was reviewed by Minton (1973) and Erwin (1981). Certain chemicals eradicate the fungus from soil, or prevent its development in the cotton plant, but these are not usually used because the cost is prohibitive. There are two exceptions. When cotton is grown in sequence with a high-value crop, such as potato or strawberry, the alternate crop can justify the cost of the chemical and its application (Ben-Yephet *et al.*, 1989). Also, fungicides such as carbendazim and ethylene thiosulphonate can control seedling pathogens, as well as prevent seed transmission of the pathogen (Shen, 1985).

Fumigants that reduce populations of *V. dahliae* in soil include methyl bromide, ethylene dibromide, chloropicrin, telone (1,3-dichloropropene) and metham-sodium (sodium methyldithiocarbamate). Mixtures of chloropicrin and methyl bromide are synergistic (Munnecke *et al.*, 1982). Fumigation with 55% chloropicrin : 45% methyl bromide at 280–400 kg ha^{-1} eradicates the pathogen from soil, but it also stunts cotton plants, causing severe yield reduction, presumably due to the phytotoxicity of bromine (Wilhelm *et al.*,

1966). Metham-sodium is used in Israel before planting potatoes, which may be grown in sequence with cotton, and it reduces the disease in both crops (Ben-Yephet et al., 1989). The chemical is generally applied in irrigation water and is most effective when incorporated into the first 5–10% of the total water during an irrigation. Effectiveness of metham-sodium increases as temperature increases from 15 to 35 °C, and it is greater with slow than with fast rates of water application. Solarization further increases effectiveness of metham-sodium and speed of kill. Raising temperatures also increases the toxicity of ethylene dibromide and telone to the wilt fungus (Ben Yephet et al., 1981).

Benzimidazole fungicides are systemic in cotton and give complete control of Verticillium wilt when applied as drenches to infected cotton plants grown in pots. For example, 100 ppm of benlate in water drenches can be used to rescue infected plants in the greenhouse. In the field, 10–20 kg ha^{-1} of benlate must be incorporated into soil to reduce wilt and, even then, control is incomplete (Buchenauer and Erwin, 1971). Almost complete control and restoration of yield have been obtained with 50–100 kg ha^{-1} of benomyl or uzgen (Yunusov et al., 1973). The benzimidazoles are not taken up through foliage unless these chemicals are acidified. Benomyl and TBZ adjusted to pH 1.7 and 2.7, respectively, and applied as spray at 2500 ppm to cotton leaves and stems, twice at 1–3-day intervals, decrease wilt, but do not give complete control (Buchenauer and Erwin, 1971). Uptake through foliage is facilitated further by adding paraffin oils to spray formulations (Erwin et al., 1974) and by maintaining high humidity (Buchenauer and Erwin, 1971). Even with these improvements, the expense of using benzimidazole fungicides still offsets most or all of the yield increases obtained.

Growth retardants are used in cotton cultivation to facilitate maturity and increase yields. These chemicals also enhance cultivar resistance to Verticillium wilt. The growth retardants, TTMP at 40–80 g ha^{-1}, CCC at 10–25 g ha^{-1} and DPC at 25–50 g ha^{-1}, applied in June and July in California, mitigated wilt symptoms and markedly reduced internal numbers of V. dahliae propagules in petioles (Erwin et al., 1979). Yields were increased by 10–29%. Kinetin, IAA and various herbicides have also increased cultivar resistance to wilt (Erwin, 1981). Growth retardants did not reduce wilt or increase yields in Greece, where only the P-2 strain occurs. Thus, the beneficial effects of these chemicals may be restricted to wilt caused by the more virulent P-1 strain. Cultivar susceptibility may also affect results, because cycocel increased yield of susceptible Coker 310 but not moderately resistant Deltapine 16 in Mexico, where the P-1 strain predominates (Palomo Gil and Hernandes, 1976).

BIOLOGICAL CONTROL

Preparations of *Trichoderma viride* (= *T. lignorum*) have been used in Russia on a limited scale to control wilt for more than 20 years (Fedorinchik, 1964). The antagonist which parasitizes *V. dahliae* is grown on peat or lignin, supplemented with nutrients, and the dried preparation is referred to as Trichodermin. The fungus may be grown on barley which is broadcast and disced into the soil. Trichodermin has been applied as a seedcoat, preplant broadcast and side-dressing with fertilizer. Treatments have generally reduced wilt in soils with high organic matter content. Substantial organic matter must be added with the biocide in soils low in organic matter. Recently, the biocontrol agent has been added as part of a complex organo-fertilizer that contains *T. viride*, *Chlorella vulgaris*, hydrolytic lignin and inorganic fertilizer (COMF) (Azimkhodzbayeva and Ramasanova, 1990). From 1985 to 1989, COMF was used on farms in the Tashkent, Fergana and Andizhan regions of Uzbekistan. It reduced wilt by 22–28% and increased yield 0.3–0.5 t ha^{-1}. The preparation was especially effective in semidesert grey soils, where reduction in wilt persisted in the second and third year after application. The preparation enhances the activity of other antagonists in soil, in addition to *T. viride*.

In soils low in organic matter, *Gliocladium roseum* may be better than *T. viride* as an antagonist of *V. dahliae* (Globus and Muromtsev, 1990). Bioprep-arations of this fungus, containing 200–300 million propagules g^{-1} and ap-plied at the rate of 5–10 kg ha^{-1}, with fertilizer, before seeding and with the first side-dressing, reduced wilt 30–40% and increased yield 12–34% in Russia. Many isolates of *G. roseum* can kill the microsclerotia in soil, when the antagonist is added in a vermiculite : bran (3 : 1) preparation, to give 0.01% of the preparation thoroughly mixed in soil (Keinath *et al.*, 1990). Species of *Talaromyces*, *Chaetomium*, *Stachybotrys*, *Fimetaria*, *Podospora* and *Aspergil-lus* are also antagonistic to the wilt pathogen, in culture and in greenhouse tests. These fungi might have potential as biocontrol organisms (Wilson and Porter, 1958; Fedorinchik, 1964; Wilhelm, 1965; Marois *et al.*, 1984).

Resistance Breeding

SOURCES OF RESISTANCE

Resistance can sometimes be selected within adapted, desirable cultivars. The Tanguis cultivar of *G. barbadense* in Peru was selected from a field of susceptible *G. hirsutum* Suave (Boza Barducci and Garcia Rada, 1942). *Gossy-pium barbadense* had been cultivated for hundreds of years along the coast of Peru before the introduction of Upland cotton. Thus, Tanguis probably originated from volunteer plants of improved strains cultivated earlier in the same area. Acala 1517WR and Acala 4-42 cultivars of *G. hirsutum* originated from single plant selections from Acala 1517. Also, Acala 1517B was selected from Watson's Acala (Staten, 1970). Natural hybridization may have been involved in the origin of these cultivars, because they were selected from nurseries containing diverse germplasm, including *G. barbadense*.

The first wilt-resistant cultivars in China were selected from the adapted, desirable cultivars, Delfos 531 and Deltapine 15 (Shen, 1985).

Wilt resistance may be increased further in some cases by repeated reselection within the cultivar, or among siblings of the hybrid from which the cultivar was derived. This procedure was used to improve resistance within the Acala 4-42 cultivar, which reached its pinnacle of resistance in the Acala 4-42-77 family used in Acala 4-42 in 1966 (Cooper *et al.*, 1967). Similarly, resistance in Acala 1517C was increased progressively by reselection among siblings. Acala strains 7133, 8893 and 1028, released as Acala 1517C in 1951, 1954 and 1958, respectively, had progressive increases in resistance to Verticillium wilt (Staten, 1970). Selection within 108F has also improved the resistance of this cultivar in Russia (Shadmanov and Saranskaya, 1990). In other cases, strains selected for improved resistance have been released as new cultivar numbers. Acala SJ-2, which was released to replace Acala SJ-1, was selected from a sibling line with better resistance to wilt. Acala SJ-5, which replaced Acala SJ-4, also came from a sibling line (Wilhelm, 1981).

Extensive collections of *Gossypium* germplasm, including wild species, have been evaluated to identify sources of resistance for hybridization programmes. The highest levels of resistance have consistently occurred in Sea Island, Egyptian, American–Egyptian and South American cultivars of *G. barbadense*. Rudolph and Harrison (1939) showed that selections from the Fusarium wilt-resistant cultivars, Tuxtla, Kekchi, Missdel, Mexican Big Boll and Cook 307-6, were moderately resistant to *Verticillium*, and Acala P18 and Stoneville yielded well under heavy infestation by *V. dahliae* in California. Later, Harrison also identified moderate resistance in Moencopi Hopi, which was cultivated by North American Indians (Cotton, 1965).

Extensive evaluations in New Mexico and western Texas in the early 1950s further identified resistance to Verticillium wilt in the Fusarium wilt-resistant cultivars Coquette, Hartsville and Roxe; in selections from Acala 1517, Acala 1517WR and Acala 1517B; in Harrison's AHA ((Acala Q 6-2 × Moencopi Hopi) × Acala 1517); in the D × K 3131 breeding line from Arkansas; and in the KP 28 × B 181 breeding line from Uganda (Cotton, 1965). Wiles (1963) identified the Fusarium wilt-resistant cultivars Auburn 56 and Rex as sources of moderate resistance to Verticillium wilt. Resistance to Fusarium wilt in all cultivars of *G. hirsutum* was apparently derived by introgression from the Sea Island and Mit-Afifi cultivars of *G. barbadense* in the early 1900s (Sherbakoff, 1949). All of the listed cultivars and lines, along with Tanguis, Seabrook Sea Island 12-B-2, Pima and other cultivars of *G. barbadense*, have been used in programmes to improve resistance to Verticillium wilt. Cook 307-6 and its derivatives, e.g. Hartsville, Auburn 56 and Deltapine 5540, and the AHA and sister C6 hybrid lines have been especially useful sources of resistance. Coquette, Rex, Tanguis and the triple species hybrid (Coker 100W × (*G. arboreum* × *G. thurburi*)) are also sources of resistance in some modern cultivars in the USA.

In Russia, modern cultivars have their resistance derived from a single accession, 06422, of a wild strain of *G. hirsutum* ssp. *mexicanum* var. *nerv-osum* (= *G. hirsutum* var. *punctatum*) (Wilhelm *et al.*, 1974b; Kalandarov, 1990). These cultivars are highly resistant to race 1 (probably = P-2 strain), but they are susceptible to race 2 (probably = P-1 strain). Consequently, intense efforts are under way to develop improved resistance to race 2, using lines and cultivars from New Mexico and Texas (Kalandarov, 1990) and using wild strains of *G. hirsutum* ssp. *yucatanense*, ssp. *punctatum* and ssp. *purpu-rescens*, from the Oahaka State of Mexico (Atlanov and Rakhimzhanov, 1990). Efforts are also under way to utilize the resistance to race 2 from wild species: *G. thurburi*, *G. trilobum*, *G. aridum*, *G. lobatum* and *G. sturtii* (Egamberdiyev, 1990). Resistance also occurs in certain accessions of *G. arboreum*, *G. herbaceum*, *G. davidsonii*. *G. klotschianum*, *G. harknessi*, *G. raimondii* and *G. armourianum* (Wiles, 1963; Bell, 1973; Kasyanenko *et al.*, 1976; Wilhelm, 1981).

Okra leaf and super-okra leaf characters raise stem and canopy tempera-tures by 2–3 °C and 3–4 °C, respectively, and reduce wilt by 26 and 33% in susceptible Stoneville 7A in California (Schnathorst, 1975). Incorporation of these characters into resistant cultivars might further enhance resistance.

INOCULATION AND SCREENING TECHNIQUES

The most resistant cultivars in the USA have come from programmes that use wilt nurseries that are heavily and uniformly infested with the P-1 strain of *V. dahliae*. Greenhouse evaluations are sometimes used to supplement the field evaluations. Most successful breeding programmes have also included evalu-ations of germplasm in a wilt-free nursery. Wilt resistance in cotton germ-plasm is often associated with late fruiting, robust or growthy plant type, low lint percentage, low yield, inferior fibre length, strength and Micronaire, heavy foliage and pubescence of leaves (Cotton, 1965). Thus, minimum quality standards must be shown on wilt-free land, to prevent the accumu-lation of undesirable agronomic characters, during selection for resistance to wilt.

Resistance against the P-1 strain is most strongly expressed when mean temperatures are 27–28 °C (Bell and Presley, 1969), and resistance to the P-2 strain is best distinguished at 23–25 °C (Barrow, 1970b). Ideally, desirable temperatures should occur consistently from year to year, for best progress in a breeding programme. In the USA, desirable temperatures for screening in the field are present only in a few areas of the northern extreme, or at the highest elevations of the cotton belt, such as in northern California, the high elevations of Arizona and New Mexico, western and northern Texas and Missouri.

Uniform heavy loam soils that are already heavily infested with the wilt pathogen are generally best for a wilt nursery. The density and uniformity of inoculum potential in nursery soils are usually increased by growing susceptible cultivars under excess nitrogen fertilization and irrigation and by delaying the ploughing in of residues until midwinter, when microsclerotia are well differentiated in crop residues. Healthy plants may be stem-puncture inoculated, to give uniform infections in the nursery. Adding gin trash from severely diseased fields each year also helps to increase and maintain a high uniform inoculum potential. Nurseries are alternately planted with test materials and a susceptible cultivar, in consecutive years, to maintain a high inoculum potential.

Plants in the greenhouse have been inoculated using root-dip, root-ball and stem-puncture methods (Wiles, 1973). In the root-dip method, seedlings are grown in sterile river sand in flats until they have four leaves. Seedlings are then lifted gently from the sand, and the roots are washed and dipped in inoculum. Seedlings are then reset in the same soil. Inoculum is prepared by growing the fungus in a liquid nutrient, such as Czapek's solution, and blending the culture. Sufficient plain agar is added during blending to give a thick consistency. With the root-ball method, plants are grown in individual pots until they are six to eight weeks old. Then the intact root ball is removed from the pot, sprayed with a conidial suspension and replaced in the pot. Symptoms appear after two weeks. Several variations of stem-puncture inoculation have been developed, but the method of Bugbee and Presley (1967) is the most used. They used a syringe fitted with a 21- and 23-gauge needle, in the field and greenhouse, respectively. A conidial suspension ($2-3 \times 10^6$ ml^{-1}) is delivered from the syringe, to form a bead at the tip of the needle. The needle is then inserted into the stem of six- to eight-week-old plants at a 45° angle to the stem, until the bevel of the point is just visible. The drop of inoculum that forms in the axis between the stem and needle disappears rapidly when the xylem vessels are severed by the needle point, giving evidence of successful inoculation.

Disease severity is much greater from stem-puncture inoculation than from root inoculation. Cultivar responses to stem-puncture inoculation in the greenhouse correlate more closely with field responses than do those from root inoculations (Erwin *et al.*, 1965). Devey and Rosielle (1986) obtained phenotypic and genotypic correlations of 0.72 and 0.86, respectively, between field ratings and greenhouse ratings following stem inoculations.

Several types of observations have been used to estimate disease severity or resistance to disease. Most commonly, the percentage of plants showing visible symptoms is recorded one or more times during the second half of the growing season. More critical evaluations of resistance include an estimate of the loss of photosynthetic area and stunting (e.g. Staffeldt and Fryxell, 1955). Yield losses are correlated most closely with leaf damage during critical periods of boll development.

The date at which evaluations are made is very critical. Estimates of wilt severity made at harvest frequently correlate poorly, if at all, with yield losses. The best correlations between wilt severity and yield have been obtained from ratings taken between 1 August and 15 September in the northern hemisphere. The exact date for best readings will vary from year to year and location to location depending on when soil temperatures favourable to disease (i.e. below 30 °C for the P-1 strain and below 27 °C for the P-2 strain) first develop. The best approach is to take ratings at least every two weeks during the critical period, to ensure that maximum differences between lines are observed.

BREEDING METHODS

The pure-line and pedigree methods of plant breeding have been used most often to develop resistance to Verticillium wilt of cotton. The success of either method depends on having a large, genetically variable, original population. Several hundred or even more than a thousand germplasm accessions and cultivars of *G. hirsutum* and *G. barbadense* have been initially evaluated in wilt nurseries in the USA (Cotton, 1965; Wilhelm, 1981). Pure-line selections have most often been made by harvesting open pollinated seed (usually less than 10% cross-pollinated) from several individual plants within the most resistant lines and cultivars that have desirable agronomic traits. Progeny rows from the individual plants are planted in the wilt nursery the next year, and individual plants are again selected from the most resistant rows that show acceptable agronomic characteristics. This procedure may be repeated several years before one or several plants from the most resistant progeny row(s) are bulk-increased as the new cultivar. In many cases, resistance in pure-line selections may have come from natural hybridization, even though crosses were not intentionally made. Natural introgression from *G. barbadense* cultivars is often suspected as the source of resistance found in *G. hirsutum* cultivars, because accessions of both species were often maintained in the same breeding nursery.

In the pedigree method of breeding, selections with high levels of resistance are crossed with agronomically desirable lines, and the first selections are made in the F_2 populations. Generally 1–10% of the F_2 plants are selected for resistance in a nursery or in the greenhouse with strong disease pressure. F_3 and F_4 progeny rows are then evaluated in the field, and the most desirable plants are selected from the most resistant progeny rows. Selections may continue for several more generations before a few plants are bulk-increased as the new cultivar (Cotton, 1965; Cooper *et al.*, 1967).

Occasionally, back-crossing is used to combine other desirable characters, such as bacterial blight resistance, into a line or cultivar with resistance to Verticillium wilt. This procedure has not been extensively used for wilt resistance, because resistance is often recessive or only partially dominant and controlled by multiple genes. Consequently, resistance is difficult to identify in back-cross progeny, beyond the first back-cross.

Several methods have been developed for combining resistance to Verticillium wilt with resistance to other diseases. In the method most often used, resistance to individual diseases is first incorporated into agronomically desirable lines, and then these lines are crossed and the pedigree method is used to select desirable progeny that have resistance to both diseases. This method was used to incorporate and maintain resistance to both bacterial blight and Verticillium wilt in cultivars released from New Mexico since 1968 (Staten, 1970). It was also used with diallele sets of crosses to combine resistance to Verticillium wilt, bacterial blight and the Fusarium wilt/root-knot nematode complex in Oklahoma (Brinkeroff *et al.*, 1977). In the latter programme, progeny rows from F_2, F_4 and F_6 generations were evaluated in Verticillium wilt and Fusarium–nematode nurseries, where they were also inoculated with races 1, 2, 4 and 10 of the bacterial blight pathogen.

At least two approaches have been developed to breed simultaneously for resistance to several diseases, including Verticillium wilt. Both methods use extensive greenhouse evaluations as well as field nurseries. Sappenfield *et al.* (1980) used 'sequential inoculation selection' (SIS) to obtain plants with multiple disease resistance. Plants are sequentially inoculated, in the greenhouse, with root-knot nematodes, the bacterial blight pathogen, *F. oxysporum* f.sp. *vasinfectum* and a P-1 isolate of *V. dahliae*. Selections are self-pollinated and seed is produced in the greenhouse. Progeny rows are then evaluated in both a Verticillium wilt and a Fusarium wilt/root-knot nematode nursery, where they are spray-inoculated with races of the bacterial blight pathogen. Seed from individual plant selections is then recycled through the screen, or selected plants are used for crosses. Bird (1982) used a multi-adversity resistance (MAR) system, in which laboratory and greenhouse screening is directed against seed and seedling diseases and bacterial blight. Selfed seed from the greenhouse selections is then evaluated in Verticillium wilt, Fusarium wilt root-knot nematode, *Phymatotrichum* root rot, bacterial blight and insect nurseries. Selections from lines showing some resistance to all adversities are then recycled through the screens or are used in crosses. Baker and Sappenfield (1985) concluded that greater progress in developing resistance to wilts is made with the SIS than with the MAR programme. They proposed a new system that combines the best attributes of both systems.

GENETICS

Genetic studies of resistance have been complicated by the fact that resistance is incomplete and that it is variably expressed, depending on the inoculum potential and virulence of the *V. dahliae* strains, nitrogen and moisture levels in the soil and, especially, temperature (Bell and Presley, 1969; Barrow, 1970b). In addition, Barrow (1970a) found that at least some cultivars and lines are heterozygous for resistance, and true breeding lines must first be found by screening under rigid environmental conditions, using a single strain and inoculum concentration. Bell and Presley (1969) showed that all cultivars and species, regardless of genetic level of resistance, can vary from a susceptible to an immune reaction to the P-1 strains, over a temperature range of 4 °C. The critical ranges were 27–31, 25–29 and 23–27 °C for genetically susceptible, moderately resistant and highly resistant cultivars, respectively.

Thus, genetically different levels of resistance to the P-1 strain are best distinguished at a mean temperature of 27 °C. Similarly, Barrow (1970b) concluded that resistance to the P-2 strain is best distinguished at mean temperatures of 22.5 °C. Because of the temperature effects, a plant with a moderate level of resistance to the P-1 strain could be classified as highly resistant at 29 °C but susceptible at 25 °C. This probably explains why about equal numbers of investigators have concluded that resistance is dominant, susceptibility is dominant and resistance is additive. Each conclusion can be correct with the same genetic stocks, depending on the mean temperature and the virulence level of the pathogen. Under field conditions, lines with moderate resistance can also appear resistant or susceptible with low or very high inoculum concentrations, or with low and very high levels of nitrogen fertilization, respectively.

Most studies indicate that resistance in cultivars of *G. hirsutum* is multigenic and can be explained by pooled additive and dominant effects (Verhalen *et al.* 1971; Devey and Roose, 1987). In a study of combining abilities of strains and varieties in the New Mexico programme, Barnes and Staten (1961) found that, among 43 F_1 populations, five were more tolerant than either parent, 33 were intermediate between parents and five were more susceptible than either parent. Thus, transgressive segregation towards either resistance or susceptibility may occur, but most resistance appears to be quantitative. In a diallele study of ten lines and cultivars, Verhalen *et al.* (1971) also concluded that resistance was usually quantitatively inherited, but that occasional dominance towards susceptibility occurred. Roberts and Staten (1972) concluded that some of the conceived dominance towards susceptibility in field studies may be due to escapes in F_2 plants that were selected for F_3 progeny row evaluation.

In a few cases, resistance appears to be due to a single dominant gene. Barrow (1973) found that the moderate resistance of the Acala line 9519 from New Mexico to a P-2 isolate of the pathogen from California behaved as a single dominant gene when 9519 was crossed with the susceptible line Acala 227 from Arizona. He proposed the symbol V^t for this gene. Resistance to the P-2 isolate in the New Mexico Acala lines 8076, 9575, 8861, 6532 and 1479 was also dominant in F_1 progeny; F_2 and F_3 progeny were not analysed.

Wilhelm *et al.* (1972) found that the greater resistance of *G. barbadense* Seabrook Sea Island 12-B-2 to mixed strains of *V. dahliae* was transferred as a single dominant gene when this cultivar was crossed with *G. hirsutum* Rex, Hartsville or the Alabama line H257. These latter cultivars and line already have substantial resistance to both Fusarium wilt and Verticillium wilt as young seedlings, presumably because of previous introgression of one or more genes from *G. barbadense* (Smith and Dick, 1960; Wiles, 1963). The high level of resistance in *G. barbadense* is probably due to two or more genes. When highly susceptible cultivars of *G. hirsutum* are crossed with resistant *G. barbadense*, resistance is incompletely dominant in F_1 hybrids, and segregation in F_2 and F_3 progeny is not distinct, indicating multiple genes and quantitative inheritance (Wiles, 1963; Bell and Presley, 1969; Wilhelm *et al.*, 1974a).

Resistance of *G. hirsutum* ssp. *mexicanum* var. *nervosum* is also incompletely dominant in F_1 progeny. Early investigations indicated that this resistance was due to a single dominant gene (Wilhelm *et al.*, 1974b). However, more detailed analyses of segregation in F_3 and F_4 generations, with defined conditions, show that the resistance to race 1 is polygenic (Makhbubov, 1990).

Host–Parasite Relationships

Symptom production and yield losses in cotton are due to water deficits caused by plugging of xylem vessels, especially in leaves, and to premature senescence of leaves leading to defoliation and loss of photosynthetic area (DeVay, 1989). Damage from the P-2 strain is due primarily to water deficits alone, whereas both factors contribute to losses from the P-1 strain. P-1 isolates cause a large increase in ethylene and abscisic acid production in leaves, which apparently contributes to both the plugging of vessels and defoliation. P-1 isolates also cause as much as ten-fold increases in ammonium ion concentrations in leaves, which may also contribute to symptom production (Bell, 1991a). The severity of symptoms shown by the plant is proportional to the amount of colonization by either strain of *V. dahliae* in the vascular system. Consequently, the density of propagules in petioles or leaves is proportional to cultivar susceptibility. Resistance therefore involves the physical and chemical confinement of the growth and spread of the pathogen in xylem vessels.

Resistance appears to depend on the sequential, co-ordinated formation of physical barriers (gels and tyloses) and phytoalexins (antibiotics) in xylem vessels, and lignins and tannins in xylem walls and surrounding xylem ray cells (Bell and Mace, 1984; Bell *et al.*, 1986; Mace, 1989). This co-ordinated active defence response occurs about 24 h sooner in xylem vessels of *G. barbadense*, compared with those of susceptible or moderately resistant cultivars of *G. hirsutum*. This early 'recognition' of conidia of *V. dahliae*, which is also expressed to heat-killed conidia (Bell and Stipanovic, 1978), apparently allows containment of the fungus in *G. barbadense*, whereas the lag of response in *G. hirsutum* allows the fungus to penetrate into adjoining xylem elements, or vessels, where defence responses must be reinitiated.

The major phytoalexins formed in vessels have been identified as hemigossypol, desoxyhemigossypol and the methyl ethers of these compounds. The methyl ethers occur in much higher concentrations in *G. barbadense* than in *G. hirsutum*. However, the methylated phytoalexins are less toxic than their unmethylated counterparts (Mace *et al.*, 1985). Thus the greater resistance of *G. barbadense* appears to be due to the speed of active defence responses rather than the quality of the response.

Resistance in Acala lines and cultivars is correlated with tannin and scopoletin concentrations in stems and leaves, both before and after infection (Stith, 1969; Bell, 1991b). Tannin oligomers in resistant cultivars are apparently smaller in molecular weight than those in susceptible cultivars because the tannins are more extractable with water and aqueous acetone. The smaller tannins extracted with water, compared with the larger ones extracted with 70% acetone, are more toxic to *V. dahliae*. Tannin concentrations in all except the lower leaves of resistant Acala cultivars grown outdoors are sufficiently high to suppress completely the growth of the fungus in culture. Scopoletin is derived from the same precursors as lignin, which has been implicated in the resistance of other plants to *V. dahliae*. Thus, differences in concentrations of scopoletin may reflect similar differences in concentrations of lignin, which acts as a mechanical barrier to confine the fungus in xylem vessels.

The P-1 and P-2 strains do not vary in their sensitivity to cotton phytoalexins or tannins at 25 °C (Mace *et al.*, 1985; Bell, 1991b). The most toxic phytoalexin to all strains is desoxyhemigossypol, which kills conidia at 10 ppm. Tannin concentrations greater than 0.2% kill the fungus. The differences in virulence among strains resides in their relative adaptation to high temperatures and ability to disturb hormonal and nitrogen metabolism in cotton plants. The P-1 strain grows considerably faster than the P-2 strain, especially in the yeast form, at 28–32 °C. This could give the P-1 strain a distinct advantage when soil temperatures first fall below 30 °C late in the growing season. The P-1 strain also differs from the P-2 strain in its ability to use nitrate and ammonium nitrogen and to cause accumulation of ammonia in diseased tissues (Bell, 1991a). These observations indicate that the P-1 strain may produce a unique phytotoxin.

References

ANON. (1949) Verticillium wilt of cotton. *Arizona Agricultural Experiment Station 60th Annual Report*, p. 58.

ASHWORTH, L.J. JR (1983) Aggressiveness of random and selected isolates of *Verticillium dahliae* from cotton and the quantitative relationship of internal inoculum to defoliation. *Phytopathology* 73, 1292–5.

ATLANOV, A.V. AND RAKHIMZHANOV, A.G. (1990) New sources and donors of wilt resistance genetically based on *Gossypium hirsutum* L. ssp. *Proceedings of the Fifth International* Verticillium *Symposium*, Leningrad, USSR, p. 68.

AUSHER, R., KRIKUN, J., FRANKEL, H., KATAN, J. AND RUBIN, D. (1973) Destroying *Verticillium* inoculum by flaming plant residues. *Phytoparasitica* 1, 131–2.

AZIMKHODZBAYEVA, N.N. AND RAMASANOVA, S.S. (1990) Application of organo-mineral fertilizers with lignin in Verticillium wilt control. *Proceedings of the Fifth International* Verticillium *Symposium*, Leningrad, USSR, p. 88.

BAKER, I.A. AND SAPPENFIELD, W.P. (1985) Comparative efficiencies of the 'sequential inoculation selection' (SIS) and the 'multi-adversity resistance' (MAR) systems of selection for multiple disease resistance in cotton. In: Brown, J.M. and Nelson,

T.C. (eds), *Proceedings of the Beltwide Cotton Production Research Conference*. National Cotton Council of America, Memphis, Tennessee, pp. 63–9.

BARNES, C.E. AND STATEN, G. (1961) The combining ability of some varieties and strains of *Gossypium hirsutum*. *New Mexico Agricultural Experimental Station Bulletin* 457.

BARROW, J.R. (1970a) Heterozygosity in inheritance of Verticillium wilt tolerance in cotton. *Phytopathology* 60, 301–3.

BARROW, J.R. (1970b) Critical requirements for genetic expression of Verticillium wilt tolerance in Acala cotton. *Phytopathology* 60, 559–60.

BARROW, J.R. (1973) Genetics of *Verticillium* tolerance in cotton. In: *Verticillium Wilt of Cotton*. Publication ARS-S-19, United States Department of Agriculture, Washington, DC, pp. 89–97.

BELL, A.A. (1973) Nature of disease resistance in Verticillium wilt of cotton. In: *Verticillium Wilt of Cotton*. Publication ARS-S-19, US Department of Agriculture, Washington D C pp. 47–62.

BELL, A.A. (1989) Role of nutrition in diseases of cotton. In: Engelhard, A.W. (ed.), *Soilborne Plant Pathogens: Management of Diseases with Macro- and Micro-elements*. American Phytopathological Society Press, St Paul, MN, pp. 167–204.

BELL, A.A. (1991a) Accumulation of ammonium ions in *Verticillium*-infected cotton and its relation to strain virulence, cultivar resistance, and symptoms. In: Brown, J.M. and Richter, D.A. (eds), *Proceedings of the Beltwide Cotton Production Research Conference*. National Cotton Council of America, Memphis, Tennessee, p. 186.

BELL, A.A. (1991b) Tannin concentrations in *Verticillium*-infected cotton: relationships to strain virulence and cultivar resistance. In: Brown, J.M. and Richter, D.A. (eds), *Proceedings of the Beltwide Cotton Production Research Conference*. National Cotton Council of America, Memphis, Tennessee.

BELL, A.A. (1992) Biology and ecology of *Verticillium dahliae*. In: Lyda, S.D. (ed.), *Comparative Pathology of Sclerotial-Forming Plant Pathogens: a* Phymatotrichum omnivorum *Symposium*. Texas A&M University Press, College Station, in press.

BELL, A.A. AND MACE, M.E. (1984) Physiology of Verticillium wilt of cotton. In: Brown, J.M. (ed.), *Proceedings of the Beltwide Cotton Production Research Conference*. National Cotton Council of America, Memphis, Tennessce, pp. 43–7.

BELL, A.A. AND PRESLEY, J.T. (1969) Temperature effects upon resistance and phytoalexin synthesis in cotton inoculated with *Verticillium albo-atrum*. *Phytopathology* 59, 1141–6.

BELL, A.A. AND STIPANOVIC, R.D. (1978) Biochemistry of disease and pest resistance in cotton. *Mycopathologia* 65, 91–106.

BELL, A.A. AND WHEELER, M.H. (1986) Biosynthesis and functions of fungal melanins. *Annual Review of Phytopathology* 24, 411–51.

BELL, A.A., MACE, M.E. AND STIPANOVIC, R.D. (1986) The biochemistry of cotton (*Gossypium*) resistance to pathogens. In: Green, M.A. and Hedin, P.A. (eds), *Natural Resistance of Plants to Pests: Roles of Allelochemicals*. Symposium Series 296, American Chemical Society, Washington, DC, pp. 36–54.

BEN-YEPHET, Y., LETHAM, D. AND EVANS, G. (1981) Toxicity of 1,2-dibromoethane and 1,3-dichloropropene to microsclerotia of *Verticillium dahliae*. *Pesticide Science* 12, 170–4.

BEN-YEPHET, Y., FRANK, Z.R., MALERO-VERA, J.M. AND DEVAY, J.E. (1989) Effect of crop rotation and metham-sodium on *Verticillium dahliae*. In: Tjamos, E.C. and Beckman, C.H. (eds), *Vascular Wilt Diseases of Plants*. NATO ASI Series H: Cell Biology, Volume 28, Springer-Verlag, New York, pp. 543–55.

BEWLEY, W.F. (1922) Sleepy disease of the tomato. *Annals of Applied Biology* 9, 116–33.

BIRD, L.S. (1982) The MAR (multi-adversity resistance) system. *Plant Disease* 66, 172–6.

BLANCO-LOPEZ, M.A., ALCAZAR, J.B., MALERO-VERA, J.M. AND JIMENEZ-DIAZ, R.M. (1989) Current status of Verticillium wilt of cotton in southern Spain: pathogen variation and population in soil. In: Tjamos, E.C. and Beckman, C.H. (eds), *Vascular Wilt Diseases of Plants*. NATO ASI Series H: Cell Biology, Volume 28, Springer-Verlag, New York, pp. 123–32.

BLANK, L.M. AND LEYENDECKER, P.J., JR (1951) Verticillium wilt spread to disease-free soil by infected cotton stalks. *Plant Disease Reporter* 35, 10–11.

BOZA BARDUCCI, T. AND GARCIA RADA, G. (1942) *El* Verticillium-*Wilt del Algodon-ero*. Molina Agricultural Experiment Station Bulletin No. 23.

BRINKERHOFF, L.A., VERHALEN, L.M. AND JOHNSON, W.M. (1977) Synthesis of resistance to four major Upland cotton diseases. In: *Proceedings of the Beltwide Cotton Production Research Conference*. National Cotton Council, Memphis, Tennessee, p. 232.

BROWN, F.H. AND WILES, A.B. (1970) Reaction of certain cultivars and weeds to a pathogenic isolate of *Verticillium albo-atrum* from cotton. *Plant Disease Reporter* 54, 508–12.

BUCHENAUER, H. AND ERWIN, D.C. (1971) Control of Verticillium wilt of cotton by spraying foliage with benomyl and thiabendazole solubilized with hydrochloric acid. *Phytopathology* 61, 433–4.

BUGBEE, W.M. AND PRESLEY, J.T. (1967) A rapid inoculation technique to evaluate the resistance of cotton to *Verticillium albo-atrum*. *Phytopathology* 57, 1264.

BUTLER, E.J. (1933) Cotton diseases. *Empire Cotton Growing Review* 10, 91–9.

BUTTERFIELD, E.J., DEVAY, J.E. AND GARBER, R.H. (1978) The influence of several crop sequences on the incidence of Verticillium wilt of cotton and on the population of *Verticillium dahliae* in field soil. *Phytopathology* 68, 1217–20.

CARPENTER, C.W. (1914) The Verticillium wilt problem. *Phytopathology* 4, 393.

CARPENTER, C.W. (1918) Wilt diseases of okra and the Verticillium wilt problem. *Journal of Agricultural Research* 12, 529–46.

CONWAY, K.E. AND MARTIN, M.J. (1983) The potential of soil solarization to control *Verticillium dahliae* in Oklahoma. *Proceedings of the Oklahoma Academy of Science* 63, 25–7.

COOPER, H.B., JR, DOBBS, J., LEHMAN, M., WILTON, A.C., TURNER, J.H. and SCHALLER, C.W. (1967) Breeding and evaluation of cotton for Verticillium wilt tolerance in Tulare County. In: *Cotton Disease Research in the San Joaquin Valley, California: Research Report for 1966–67*. University of California, Davis, pp. 106–53.

COTTON, J.R. (1965) *Breeding Cotton for Tolerance to Verticillium Wilt*. Publication ARS 34-80, United States Department of Agriculture, Washington, DC.

DEVAY, J.E. (1989) Physiological and biochemical mechanisms in host resistance and susceptibility to wilt pathogens. In: Tjamos, E.C. and Beckman, C.H. (eds), *Vascular Wilt Diseases of Plants*. NATO ASI Series H: Cell Biology, Volume 28, Springer-Verlag, New York, pp. 197–217.

DeVay, J.E. and Pullman, G.S. (1984) Epidemiology and ecology of diseases caused by *Verticillium* species, with emphasis on Verticillium wilt of cotton. *Phytopathologia Mediterranea* 23, 95–108.

Devey, M.E. and Roose, M.L. (1987) Genetic analysis of Verticillium wilt tolerance in cotton using pedigree data from three crosses. *Theoretical and Applied Genetics* 74, 162–7.

Devey, M.E. and Rosielle, A.A. (1986) Relationship between field and greenhouse ratings for tolerance to Verticillium wilt on cotton. *Crop Science* 26, 1–4.

Domsch, K.H., Gams, W. and Anderson, T.-H. (1980) *Verticillium* Nees ex Link 1824. In: *Compendium of Soil Fungi*. Academic Press, New York, pp. 828–45.

Duffield, P.C., Ortega, C.B. and Carillo, A.F. (1953) *Enfermedades del Algodonero en la Comarca Lagunera, Mexico*. Folleto de Divulgacion No. 12, Programa de Agricultura Cooperativo de La Secretaria de Agricultura y Ganaderia de Mexico, D. F. y La Fundacion Rockefeller.

Egamberdiyev, A.E. (1990) Amphydiploids of cotton as a new source of wilt resistance. *Proceedings of the Fifth International* Verticillium *Symposium*, Leningrad, USSR, p. 69.

El-Zik, K.M. (1985) Integrated control of Verticillium wilt of cotton. *Plant Disease* 69, 1025–32.

Erwin, D.C. (1981) Chemical control. In: Mace, M.E., Bell, A.A. and Beckman, C.H. (eds), *Fungal Wilt Diseases of Plants*. Academic Press, New York, pp. 563–600.

Erwin, D.C., Moje, W. and Malca, I. (1965) An assay of the severity of Verticillium wilt on cotton plants inoculated by stem puncture. *Phytopathology* 55, 663–5.

Erwin, D.C., Khan, R.A. and Buchenauer, H. (1974) Effect of oil emulsions on the uptake of benomyl and thiabendazole in relation to control of Verticillium wilt of cotton. *Phytopathology* 64, 485–9.

Erwin, D.C., Tsai, S.D. and Khan, R.A. (1979) Growth retardants mitigate Verticillium wilt and influence yield of cotton. *Phytopathology* 69, 283–7.

Evans, G. (1967) Verticillium wilt of cotton – the situation in the Namoi Valley. *Agricultural Gazette of New South Wales* 78, 581–3.

Evans, G. (1971) Influence of weed hosts on the ecology of *Verticillium dahliae* in newly cultivated areas of the Namoi Valley, New South Wales. *Annals of Applied Biology* 67, 169–75.

Evans, G., Wilhelm, S. and Snyder, W.C. (1966) Dissemination of the Verticillium wilt fungus with cotton seed. *Phytopathology* 56, 460–6.

Ezekiel, W.N. and Dunlap, A.A. (1940) Cotton diseases in Texas in 1939. *Plant Disease Reporter* 24, 434–9.

Fedorinchik, N.S. (1964) Biological method of controlling plant diseases. *Trudy Vsesoyuznogo Nauchno-Issledovatel'skogo Instituta Zashchity Rastenii*, No. 23, pp. 201–10.

Globus, G.A. and Muromtsev, G.S. (1990) The use of *Gliocladium roseum* as antagonist for defence of cotton from fitopathogene fungi. In: *Proceedings of the Fifth International* Verticillium *Symposium*, Leningrad, USSR, p. 90.

Grishechkina, L.D. (1990) The effect of some agrotechnical methods on quantity of causative agent of Verticillium wilt of cotton. In: *Proceedings of the Fifth International* Verticillium *Symposium*, Leningrad, USSR, p. 26.

Grishechkina, L.D. and Sidorova, S.F. (1984) Effect of mineral compounds on leaf infection by cotton wilt pathogen. *Mikologiya Fitopatologiya* 18, 66–70.

HASTIE, A.C. AND HEALE, J.B. (1984) Genetics of *Verticillium*. *Phytopathologie Mediterranea* 22, 130–62.

HEALE, J.B. (1988) *Verticillium* spp., the cause of vascular wilts in many species. *Advances in Plant Pathology* 6, 291–312.

HERBERT, F.W. AND HUBBARD, J.W. (1932) Verticillium wilt (Hadromycosis) of cotton in the San Joaquin Valley of California. *Circular* 211, United States Department of Agriculture, Washington, DC.

HILLOCKS, R.J. (1991) Screening for resistance to Verticillium wilt in Zimbabwe. *Tropical Agriculture (Trinidad)* 68, 144–8.

HINKLE, D.A. AND FULTON, N.D. (1963) Yield and Verticillium wilt incidence in cotton as affected by crop rotations. *Arkansas Experimental Station Bulletin* 675.

HUISMAN, O.C. AND GERIK, J.S. (1989) Dynamics of colonization of plant roots by *Verticillium dahliae* and other fungi. In: Tjamos, E.C. and Beckman, C.H. (eds), *Vascular Wilt Disease of Plants*. NATO ASI Series H: Cell Biology, Volume 28, Springer-Verlag, New York, pp. 1–17.

ISAAC, I. (1967) Speciation in *Verticillium*. *Annual Review of Phytopathology* 5, 201–22.

JOAQUIM, T.R. AND ROWE, R.C. (1990) Reassessment of vegetative compatibility relationships among strains of *Verticillium dahliae* using nitrate-nonutilizing mutants. *Phytopathology* 80, 1160–6.

JOHNSON, W.M., JOHNSON, E.K. AND BRINKERHOFF, L.A. (1980) Symptomatology and formation of microsclerotia in weeds inoculated with *Verticillium dahliae* from cotton. *Phytopathology* 70, 31–5.

KALANDAROV, S. (1990) New sources of cotton resistance to *Verticillium dahliae* Kleb. In: *Proceedings of the Fifth International* Verticillium *Symposium*, Leningrad, USSR, p. 105.

KASYANENKO, A.G., ALYAMOV, A.A. AND GORKOVTSEVA, E.A. (1976) *Gossypium arboreum* and *Gossypium herbaceum* as possible sources of wilt-resistant genes. In: *Proceedings of the Second International* Verticillium *Symposium*, University of California, Berkeley, p. 26.

KEINATH, A.P., FRAVEL, D.R. AND PAPAVIZAS, G.C. (1990) Evaluation of potential antagonists for biocontrol of *Verticillium dahliae*. In: *Proceedings of the Fifth International* Verticillium *Symposium*, Leningrad, USSR, p. 82.

KRAVTSOVA, T.I. (1990) The area of distribution of *Verticillium* wilt. In: *Proceedings of the Fifth International* Verticillium *Symposium*, Leningrad, USSR, p. 29.

KRIKUN, J. AND SUSNOSKI, M. (1971) Introduction and establishment of *Verticillium dahliae* in a newly-developed arid zone. In: *Proceedings of the First International* Verticillium *Symposium*, Wye College, Ashford, Kent, UK, p. 21.

LEYENDECKER, P.J., JR (1950) *Effects of Certain Cultural Practices on Verticillium Wilt of Cotton in New Mexico*. New Mexico Agricultural Experiment Station Bulletin 356.

LEYENDECKER, P.J., JR, STATEN, G., HOOVER, M., LYTTON, L.R., STOVALL, J.T. AND LEDING, A.R. (1947) Mesilla Valley, New Mex., survey for Verticillium wilt of cotton. *Plant Disease Reporter* 31, 483–4.

LI, J.-Y. AND SHEN, C.-Y. (1987) Cotton production and disease control in China. In: Brown, J.M. and Nelson, T.C. (eds), *Proceedings of the Beltwide Cotton Production Research Conference*. National Cotton Council of America, Memphis, Tennessee, pp. 54–6.

LIU, J.-Y. (1985) The identification of cotton resistance to *Verticillium dahliae* in seedling stage. *Journal of the Nanjing Agricultural University*, No. 2, 59–65.

LLOSA, P.T. (1938) Nuevo metoda para aislar hongos del tipo *Verticillium* o *Fusarium* de plantas atacadas por el wilt del algodonero. *Estacion Experimental Agricola de la Molina Boletin* No. 13.

MACE, M.E. (1989) Secondary metabolites produced in resistant and susceptible host plants in response to fungal vascular infection. In: Tjamos, E.C. and Beckman, C.H. (eds), *Vascular Wilt Diseases of Plants*. NATO ASI Series H: Cellular Biology, Volume 28, Springer-Verlag, New York, pp. 163–74.

MACE, M.E., STIPANOVIC, R.D. AND BELL, A.A. (1985) Toxicity and role of terpenoid phytoalexins in Verticillium wilt resistance in cotton. *Physiological Plant Pathology* 26, 209–18.

MAKHBUBOV, M.B. (1990) Inheritance of wilt resistance in intraspecific hybridization (*Gossypium* L.) In: *Proceedings of the Fifth International* Verticillium *Symposium*, Leningrad, USSR, p. 106.

MAROIS, J.J., FRAVEL, D.R. AND PAPAVIZAS, G.C. (1984) Ability of *Talaromyces flavus* to occupy the rhizosphere and its interaction with *Verticillium dahliae*. *Soil Biology and Biochemistry* 16, 387–90.

MARUPOV, A. (1990) Effect of predecessors and green manure cultures on manifestation of Verticillium wilt of cotton, infection level and biogenousness of soil. In: *Proceedings of the Fifth International* Verticillium *Symposium*, Leningrad, USSR, p. 94.

MATHRE, D.E., ERWIN, D.C., PAULUS, A.O. AND RAVENSCROFT, A.V. (1966) Comparison of the virulence of isolates of *Verticillium albo-atrum* from several of the cotton growing regions in the United States, Mexico, and Peru. *Plant Disease Reporter* 50, 930–3.

MELERO-VARA, J.M., BLANCO-LOPEZ, M.A., BEJARANO-ALCAZAR, J. AND JIMENEZ-DIAZ, R.M. (1990) Use of soil solarization to control cotton wilt induced by defoliating pathotypes of *V. dahliae* in southern Spain. In: *Proceedings of the Fifth International* Verticillium *Symposium*, Leningrad, USSR, p. 14.

MILES, L.E. (1934) Verticillium wilt of cotton in Greece. *Phytopathology* 24, 558–9.

MILES, L.E. AND PERSONS, T.D. (1932) Verticillium wilt of cotton in Mississippi. *Phytopathology* 22, 767–73.

MINTON, E.B. (1972) Effects of weed control in grain sorghum on subsequent incidence of Verticillium wilt in cotton. *Phytopathology* 62, 582–3.

MINTON, E.B. (1973) Chemical control. In: *Verticillium Wilt of Cotton*. Publication ARS-S-19, United States Department of Agriculture, Washington, DC, pp. 105–9.

MINTON, E.B. (1978) Neutralizers and pesticides and their sequence of application to acid-delinted cottonseed: effects on germination, stand, and Verticillium wilt of cotton. *Crop Science* 18, 831–5.

MINTON, E.B., BRASHEARS, A.D., KIRK, I.W. AND HUDSPETH, E.B., JR (1972) Effects of row and plant spacings on Verticillium wilt of cotton. *Crop Science* 12, 764–7.

MOSHIR-ABADI, H. (1981) *Verticillium dahliae* Kleb. in Iran. In: *Proceedings of the Third International* Verticillium *Symposium*, Bari, Italy, p. 46.

MUKHAMEZHANOV, M.V. (1966) Present and prospective scientific research of cotton wilt and its control. In: Mukhamedzhanov, M.V. (ed.), *Cotton Wilt* (Translated from Russian). United States Department of Agriculture and the National Science Foundation, Washington, DC, pp. 1–22.

MUNNECKE, D.E., KOLBEZEN, M.J. AND BRICKER, J.L. (1982) Effects of moisture, chloropicrin, and methyl bromide singly and in mixtures on sclerotia of *Sclerotium rolfsii* and *Verticillium albo-atrum. Phytopathology* 72, 1235–8.

OINGII, L. AND CHIYI, S. (1990) Species, physiological form and vegetative compatibility group of cotton *Verticillium* wilt pathogen in China. In: *Proceedings of the Fifth International* Verticillium *Symposium*, Leningrad, USSR, p. 17.

PALOMO GIL, A. AND HERNANDES, J.A. (1976) Effects of time and doses of application of the plant growth regulator cycocel on yield and quality of cotton fibers in soils infested with *Verticillium dahliae* in the Laguna Region, Mexico. In: *Proceedings of the Second International* Verticillium *Symposium*, University of California, Berkeley, p. 36.

PALOMO GIL, A. AND QUIRARTE, R. H. (1976) Effects of high plant populations and the number of auxiliary irrigation cycles on the yield and quality of fibers of two cotton varieties in soils infested with *Verticillium dahliae* Kleb. In: *Proceedings of the Second International* Verticillium *Symposium*, University of California, Berkeley, pp. 36–7.

PORTENKO, L.G. AND KAS'YANENKO, A.G. (1988) Genetics of the cotton wilt pathogen *Verticillium* Kleb X. Mutations of *Verticillium dahliae* Kleb. for virulence and aggressiveness. *Soviet Genetics* 23, 1312–16.

PRESLEY, J.T. (1950) Verticillium wilt of cotton with particular emphasis on variation of the causal organism. *Phytopathology* 40, 497–511.

PUHALLA, J.E. (1973) Differences in sensitivity of *Verticillium* species to ultraviolet irradiation. *Phytopathology* 63, 1488–92.

PUHALLA, J.E. (1976) Glycerol as a selective agent for auxotrophs of *Verticillium dahliae. Journal of General Microbiology* 94, 409–12.

PUHALLA, J.E. (1980) Spontaneous auxotrophic mutations in *Verticillium. Botany Gazette* 141, 94–100.

PUHALLA, J.E. AND BELL, A.A. (1981) Genetics and biochemistry of wilt pathogens. In: Mace, M.E., Bell, A.A. and Beckman, C.H. (eds), *Fungal Wilt Diseases of Plants*. Academic Press, New York, pp. 145–92.

PUHALLA, J.E. AND HUMMEL, M. (1983) Vegetative compatibility groups within *Verticillium dahliae. Phytopathology* 73, 1305–8.

PULLMAN, G.S. AND DeVAY, J.E. (1982) Effect of soil flooding and paddy rice culture on the survival of *Verticillium dahliae* and incidence of Verticillium wilt of cotton. *Phytopathology* 72, 1285–9.

PULLMAN, G.S., DeVAY, J.E., GARBER, R.H. AND WEINHOLD, A.R. (1981a) Soil solarization: effects on Verticillium wilt of cotton and soilborne populations of *Verticillium dahliae, Pythium* spp., *Rhizoctonia solani*, and *Thielaviopsis basicola. Phytopathology* 71, 954–9.

PULLMAN, G.S., DeVAY, J.E. AND GARBER. R.H. (1981b) Soil solarization and thermal death: a logarithmic relationship between time and temperature for four soilborne plant pathogens. *Phytopathology* 71, 959–64.

ROBERTS, C.L. AND STATEN, G. (1972) Heritability of Verticillium wilt tolerance in crosses of American Upland cotton. *Crop Science* 12, 63–6.

RUDOLPH, B.A. AND HARRISON, G.J. (1939) Attempts to control Verticillium wilt of cotton and breeding for resistance. *Phytopathology* 29, 753.

SACKSTON, W.E. (1983) Epidemiology and control of seed-borne *Verticillium* spp. causing vascular wilt. *Seed Science and Technology* 11, 731–47.

SAPPENFIELD, W.P., BALDWIN, C.H., WRATHER, J.A. AND BUGBEE, W.M. (1980) Breeding multiple disease resistant cottons for the North Delta. In: Brown, J.M. (ed.), *Proceedings of the Beltwide Cotton Production Research Conference.* National Cotton Council of America, Memphis, Tennessee, pp. 280–3

SCHNATHORST, W.C. (1969) A severe form of *Verticillium albo-atrum* in *Gossypium barbadense* in Peru. *Plant Disease Reporter* 53, 145–50.

SCHNATHORST, W.C. (1975) New approaches for obtaining field tolerance in *Gossypium hirsutum* to severe strains of *Verticillium dahliae*: a pathologist's perspective. In: *Proceedings of the Beltwide Cotton Production Research Conference.* National Cotton Council, Memphis, Tennessee, p. 148.

SCHNATHORST, W.C. (1981) Life cycle and epidemiology of *Verticillium.* In: Mace, M.E., Bell, A.A. and Beckman, C.H. (eds), *Fungal Wilt Diseases of Plants.* Academic Press, New York, pp. 81–111.

SCHNATHORST, W.C. AND MATHRE, D.E. (1966) Host range and differentiation of a severe form of *Verticillium albo-atrum* in cotton. *Phytopathology* 56, 1155–61.

SHADMANOV, R.K. AND SARANSKAYA, L.B. (1990) Improvement of wilt tolerance of cotton. In: *Proceedings of the Fifth International* Verticillium *Symposium*, Leningrad, USSR, p. 78.

SHAPOVALOV, M. AND RUDOLPH, B.A. (1930) *Verticillium* hadromycosis (wilt) of cotton in California. *Plant Disease Reporter* 14, 9–10.

SHEN, C.-Y. (1985) Integrated management of Fusarium and Verticillium wilts of cotton in China. *Crop Protection* 4, 337–45.

SHERBAKOFF, C.D. (1928) Wilt caused by *Verticillium albo-atrum. Plant Disease Reporter Supplement* 61, 283–4.

SHERBAKOFF, C.D. (1929) Verticillium wilt of cotton. *Phytopathology* 19, 94.

SHERBAKOFF, C.D. (1949) Breeding for resistance to Fusarium and Verticillium wilts. *Botanical Review* 15, 377–422.

SMITH, A.L. AND DICK, J.B. (1960) Inheritance of resistance to Fusarium wilt in Upland and Sea island cottons as complicated by nematodes under field conditions. *Phytopathology* 50, 44–8.

SMITH, H.C. (1965) The morphology of *Verticillium albo-atrum, V. dahliae,* and *V. tricorpus. New Zealand Journal of Agricultural Research* 8, 450–78.

STAFFELDT, E.E. (1959) Elimination of *Verticillium albo-atrum* by composting cotton gin wastes. *Plant Disease Reporter* 43, 1150–2.

STAFFELDT, E.E. AND FRYXELL, P.A. (1955) A measurement of disease reaction of cotton to Verticillium wilt. *Plant Disease Reporter* 39, 690–2.

STATEN, G. (1970) Breeding Acala 1517 cottons, 1926–1970. *New Mexico State University College of Agriculture and Home Economics Memoir Series* Number 4.

STERNE, R.E., MCCARVER, T.H. AND COURTNEY, M.L. (1974) Survival of plant pathogens in composted cotton gin trash. *Arkansas Farm Research* 28, 9.

STITH, L.S. (1969) Another look at Vert. In: *Proceedings of the Beltwide Cotton Production Research Conference.* National Cotton Council, Memphis, Tennessee, pp. 33–5.

TAUBENHAUS, J.J. (1936) Verticillium wilt of cotton. In: *Texas Agricultural Experiment Station 49th Annual Report*, Texas A&M University, College Station, p. 111.

TOLMSOFF, W.J. (1983) Heteroploidy as a mechanism of variability among fungi. *Annual Review of Phytopathology* 21, 317–40.

TOLMSOFF, W.J. AND HOWELL, C.R. (1984) Genetics and epigenetics as sources of variability in *Verticillium dahliae* and *V. albo-atrum*. In: Brown, J.M. (ed.), *Proceedings of the Beltwide Cotton Production Research Conference*. National Cotton Council of America, Memphis, Tennessee, pp. 37–43.

TYPAS, M.A. AND HEALE, J.B. (1976) Acriflavine-induced hyaline variants of *Verticillium albo-atrum* and *V. dahliae*. *Transactions of the British Mycological Society* 66, 15–25.

VERHALEN, L.M., BRINKERHOFF, L.A., FUN, K.-C. AND MORRISON, W.C. (1971) A quantitative genetic study of Verticillium wilt resistance among selected lines of Upland cotton. *Crop Science* 11, 407–12.

WADDLE, B.A. AND FULTON, N.D. (1955) Results of 1955 Verticillium wilt evaluation tests, Osceola, Arkansas. In: *Proceedings of the 16th Cotton Disease Council*. National Cotton Council, Memphis, Tennessee, pp. 8–10.

WAKER, H.J. AND ONKEN, A.B. (1969) *Fertilizing Irrigated Cotton, Southern High Plains of Texas*. Publication MP-913, Texas Agricultural Experiment Station, College Station.

WANJURA, D.F. AND BARKER, G.L. (1987) Cotton yield decline analysis for the Southern Great Plains. In: Brown, J.M. and Nelson, T.C. (eds), *Proceedings of the Beltwide Cotton Production Research Conference*. National Cotton Council of America, Memphis, Tennessee, pp. 559–63.

WEIR, B.L., DeVAY, J., STAPLETON, J., WAKEMAN, J. AND GARBER, D. (1987) Effect of solarization on potassium disease complex of cotton. In: Brown, J.M. and Nelson, T.C. (eds), *Proceedings of the Beltwide Cotton Production Research Conference*. National Cotton Council of America, Memphis, Tennessee, pp. 51–2.

WILES, A.B. (1963) Comparative reactions of certain cottons to Fusarium and Verticillium wilts. *Phytopathology* 53, 586–8.

WILES, A.B. (1973) Methods of inoculation and techniques for evaluation. In: *Verticillium Wilt of Cotton*. Publication ARS-S-19, United States Department of Agriculture, Washington, DC, pp. 63–8.

WILHELM, S. (1965) Analysis of biological balance in natural soils. In: Baker, K.F. and Snyder, W.C. (eds), *Ecology of Soil-borne Plant Pathogens: Prelude to Biological Control*. University of California Press, Berkeley, pp. 509–18.

WILHELM, S. (1981) Sources and genetics of host resistance in field and fruit crops. In: Mace, M.E., Bell, A.A. and Beckman, C.H. (eds), *Fungal Wilt Diseases of Plants*. Academic Press, New York, pp. 299–376.

WILHELM, S. AND SAGEN, J.E. (1985) Phenotype modification in cotton for control of Verticillium wilt through dense plant population culture. *Plant Disease* 69, 283–8.

WILHELM, S., EVANS, G., SNYDER, W.C., GEORGE, A., MATHRE, D., GARBER, R.H. AND HALL, D. (1966) Cultural control of *Verticillium* in cotton – a three-point approach. *California Agriculture* 20, 2–4.

WILHELM, S., SAGEN, J.E. AND TIETZ, H. (1972) Resistance to Verticillium wilt transferred from *Gossypium barbadense* to Upland cotton phenotype. *Phytopathology* 62, 798–9.

WILHELM, S., SAGEN, J.E. AND TIETZ, H. (1974a) Resistance to Verticillium wilt in cotton: sources, techniques of identification, inheritance trends, and the resistance potential of multiline cultivars. *Phytopathology* 64, 924–31.

WILHELM, S., SAGEN, J.E. AND TIETZ, H. (1974b) *Gossypium hirsutum* subsp. *mexicanum* var. *nervosum*, Leningrad strain – a source of resistance to Verticillium wilt. *Phytopathology* 64, 931–9.

WILSON, K.S. AND PORTER, C.L. (1958) The pathogenicity of *Verticillium albo-atrum* as affected by muck soil antagonists. *Applied Microbiology* 6, 155–9.

YOUNG, V.H., FULTON, N.D. AND WADDLE, B.A. (1959) Factors affecting the incidence and severity of Verticillium wilt disease of cotton. *Arkansas Agricultural Experimental Station Bulletin* 612.

YUNUSOV, M.R., IBRAGIMOV, F.U., POPOV, V.I. AND TKACHENKO, M.P. (1973) Effectiveness of new preparations. *Khlopkovodstvo* 4, 24–6.

4 | Fusarium Wilt

R.J. HILLOCKS

Fusarium wilt is very similar in aetiology to Verticillium wilt (described in Chapter 3) and causes similar symptoms. Verticillium wilt tends to be found in cotton-growing areas which experience cool weather towards the end of the growing season. Fusarium wilt is rarely a problem in areas with mean daily temperatures below 24 °C. It is not unknown, however, for both diseases to occur in the same field. Some other distinguishing features of the two vascular wilt diseases affecting cotton are described in Table 4.1.

History and Distribution

Fusarium wilt of cotton was first described by Atkinson (1892) in the USA. It was diagnosed first in Alabama and soon afterwards in several other cotton-growing states, becoming widespread throughout the cotton belt, especially on the sandy, acid soils of the south-east. During the 1950s and 1960s, the area of major cotton production began to move westwards and California became the main producer. The disease was at first absent or unimportant in these western states but, by the early 1960s, it had appeared in west Texas and lower California and also in Mexico (Blank, 1962). The disease was first noted in California in 1959 (Garber and Paxman, 1963), where the number of infested sites remained relatively few until the mid 1970s, since when their number has increased substantially.

The earliest report of the disease outside the USA came from Egypt in 1902 (Fahmy, 1927), where it spread rapidly with the release of the Sakel cultivar during the 1920s. It was also reported from East Africa about the same time. However, the reports from Tanzania in 1904 and Uganda in 1931 were unconfirmed (Ebbels, 1975) and seem likely to have been incorrect, in view of the rapid spread of the disease in Tanzania after it was positively identified

127

Table 4.1. Some distinguishing features of Fusarium and Verticillium wilts.

Fusarium wilt	Verticillium wilt
Favoured by high temperatures, i.e. daily mean above 23 °C	Favoured by cool temperatures, i.e. daily mean below 23 °C
Most favourable soil type varies with race of the pathogen. Race 1 is favoured by light soils of neutral to acid pH	Favoured by heavy soils of neutral to alkaline pH
Symptoms can appear at any stage of crop development, but usually begin 4 weeks after planting. Plants become more susceptible at flowering	The disease may be found in the crop from about 6 weeks after planting but disease incidence usually increases towards the end of the cropping season as temperatures begin to fall
Symptoms are similar to Verticillium wilt but chlorosis tends to be in patches without reddening. Part of the leaf may be chlorotic and flaccid with the rest appearing healthy. In advanced stages of disease extensive defoliation occurs. Vascular browning is pronounced, extending into the outer stele tissue	Symptoms begin with marginal chlorosis and, sometimes, reddening of the leaf. Chlorosis develops between the veins. The rate and extent of defoliation is dependent on the strain of the pathogen. Vascular browning is less pronounced than with Fusarium wilt

there in 1954 (Peat *et al.*, 1955). Within ten years of its discovery in Tanzania, Fusarium wilt had become sufficiently serious to justify the inclusion of selection for resistance in the country's cotton-breeding programme. By the mid-1970s the disease had spread more than 300 km around the shore of Lake Victoria, although most of the affected farms were within 8 km of the lake. Since then, the number of outbreaks occurring 50 or more kilometres from the lake has increased considerably and most of the western cotton-growing area is now affected (Hillocks, 1984; Kibani, 1987).

The first record of the disease in India dates from 1908 (Kulkarni, 1934), and subsequent investigations showed it to be serious in many Indian cotton areas. In Sudan, a wilt disease had been known for many years before it was confirmed as Fusarium wilt in 1960 (El-Nur and Fattah, 1970). It remains of limited distribution in Sudan but continues to cause problems in certain parts of the Gezira (Yassin and Daffalla, 1982).

The disease has appeared more recently in Israel (Dishon and Nevo, 1970) and in Zimbabwe (R. J. Hillocks, unpublished). The disease is recorded in a list of plant diseases occurring in Zimbabwe (Rothwell, 1983), but reports earlier than this are unsubstantiated. The causal organism was confirmed in 1987 by pathogenicity tests with isolates from two sites, the Cotton Research Institute in Kadoma and the Sanyati cotton estate, where the disease had probably been present for several years but was presumed to be Verticillium wilt (distinguishing features of the two diseases are described in Table 4.1). The failure of the disease to spread from Sanyati is probably due to the existence in Zimbabwe of an efficient seed certification scheme and because the participating farms happen to have been located far from the infested site.

Fusarium wilt now occurs in all the main cotton-growing areas of the world (see Fig. 4.1). It is one of the main cotton diseases in China (Cook, 1981) and parts of the former Soviet Union (e.g. Menlikiev, 1962), but is notably absent from west Africa, Turkey and Australia (Snyder and Smith, 1981).

Symptoms

Symptoms may appear at any stage of crop development, depending on inoculum density, temperature and host susceptibility. At high inoculum density or when infection initiates from the seed, plants may be killed at the seedling state. More typically, first symptoms become apparent in the field between one and two months after planting (see Plate 3A), often around the onset of flowering, due to changes in host physiology associated with the change from vegetative to reproductive phases of development.

In the seedling, first symptoms appear on the cotyledons as vein darkening, followed by peripheral chlorosis. The cotyledons become progressively more chlorotic and then necrotic before being shed. In older plants, the first external evidence of infection is yellowing at the margin of one or more of the lower leaves, but not necessarily the bottom-most leaf (see Plate 3B). As the

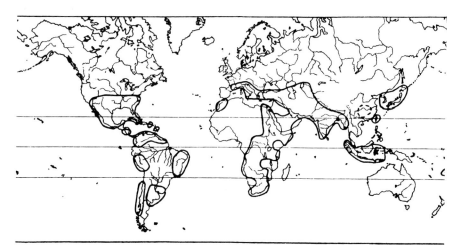

Fig. 4.1. Distribution of Fusarium wilt caused by *Fusarium oxysporum* f.sp. *vasinfectum* (CMI Distribution Map of Plant Diseases No. 362) (CAB International, UK).

disease progresses within the plant, more leaves develop chlorosis, which characteristically appears in patches between the main veins, the rest of the leaf remaining green. Gradually the chlorotic patches become necrotic and the leaves are easily detached from the stem. Infected plants are stunted compared with surrounding healthy plants, and appear wilted during the hottest part of the day. Under optimal conditions for disease development, all the leaves on affected plants succumb and are shed before the stem dries out and dies. Sometimes only one side of a plant shows symptoms or, if conditions become too cool for the disease, the plant may show partial recovery, with disease-free monopodia growing from the lower nodes. Such plants usually survive to produce at least a few mature bolls. From the time that symptoms become visible externally, the infected host also develops a brown discoloration of the vascular tissue, similar to that seen in plants infected with Verticillium wilt, which can be observed by peeling away the bark tissue.

Provided conditions are conducive to disease development, plants which show symptoms in the first two months after sowing usually die before setting any bolls. Plants which develop symptoms well after the onset of flowering often survive to produce a few bolls but tend to senesce sooner than non-infected plants.

Physiology of Symptom Expression

VASCULAR BROWNING

The accumulation of brown metabolites in the stele tissues is the most visible evidence of vascular infection. Vascular browning can be detected with the naked eye from a few days after vascular invasion (see Plate 2D). The oxidation of phenols and possibly their dehydrogenation by peroxidases give rise to flavonoids, tannins, lignin and quinones (Pegg, 1981), which are responsible for the discoloration. Some of these compounds may act as phytoalexins.

TOXINS

The cotton wilt *Fusarium* produces the toxin fusaric acid (5-*n* butyl picolinic acid) in culture (e.g. Charudatan, 1970) and also within the host (Lakshminarayan and Subramanian, 1955). However, the exact role of the toxin in the wilt syndrome remains unresolved, particularly in view of the fact that non-pathogenic strains of the fungus also produce fusaric acid.

WILTING

The most distinctive feature of a vascular wilt disease is the loss of turgor arising from the failure of the infected xylem to meet the water requirements of the plant. Wilting results from the combined effects of fungal metabolites and secondary host metabolites produced in response to infection. Toxins, phenolic compounds, pectolytic enzymes and growth-regulating compounds all contribute to the alteration of the physiology of the host, giving rise to the external symptoms of wilt (Dimond, 1970).

Losses

Estimates of losses caused by Fusarium wilt on a national scale are available only for the USA, where they were estimated for the 1989 season at 0.2% of lint yield (Blasingame, 1990). Although in most countries where the disease occurs losses on a national scale are not great, losses on individual farms may be considerable, especially in developing countries, where holdings are often very small. Also, the disease has an indirect effect in discouraging farmers from growing the crop in an area where the disease becomes prevalent.

Information on the effect of the disease on yield is conspicuously lacking. This may be due partly to the fact that, under favourable conditions for the disease, plants are killed at an early stage in crop development, allowing the surrounding healthy plants to compensate for the loss by increased growth. This is particularly true of the indeterminate African Upland cottons, where stand losses of up to 30% may not have a significant effect on yield. However, Ware and Young (1934) suggested that, in the USA, a threshold value of about 10% wilt in the crop was required before yield might be adversely affected. Chester (1946) estimated that, for every 5% increment in wilt up to 60%, there is approximately 3% loss of yield.

The extent to which yield is affected by the disease depends on the stage of crop development when the plants begin to show symptoms. Although plants which are infected late in the season survive to produce seed cotton, such infected plants produce significantly less seed cotton than non-infected plants (Ebbels, 1975). The disease is, therefore, probably most damaging when it occurs too late to kill the plant and allow surrounding plants to compensate, but sufficiently early in crop development to prevent infected plants reaching their full yield potential.

Causal Organism

Taxonomy

The causal organism is recognized as *Fusarium oxysporum* Schlecht f.sp. *vasinfectum* Atk. Sny. & Hans. The species *F. oxysporum* is very variable, containing a large number of saprophytic and pathogenic forms which have certain morphological features in common. The parasitic forms were grouped by Snyder and Hansen (1940) into *formae speciales* based on their selective pathogenicity. Forms which cannot be separated morphologically can be distinguished by their pathogenicity towards particular hosts.

MORPHOLOGY

The perfect state is not known for *F. oxysporum*, which produces two types of conidia in culture and within the host. Microconidia, borne on simple phialides, arising from a short conidiophore (Fig. 4.2a), are one-celled, or occasionally one-septate, and 5–20 × 2.2–3.5 μm in size. Macroconidia are borne on a more elaborately branched conidiophore or sporodochium. They are buff to salmon orange in the mass, fusiform–falcate in shape (Fig. 4.2b), curved inwards at both ends with a pedicellate base, usually three- to five-septate and 27–48 × 2.5–4.5 μm in size. Chlamydospores are readily formed in old cultures and in senescent host tissues. Their formation is both intercalary and terminal on the vegetative hyphae (Fig. 4.2c) and they are 7–13 μm in diameter (Booth and Waterson, 1964; Booth, 1971). The main morphological features which distinguish *F. oxysporum* from most other fusaria are the short microconidiophore, together with the formation of chlamydospores (Booth, 1977).

(a) (c)

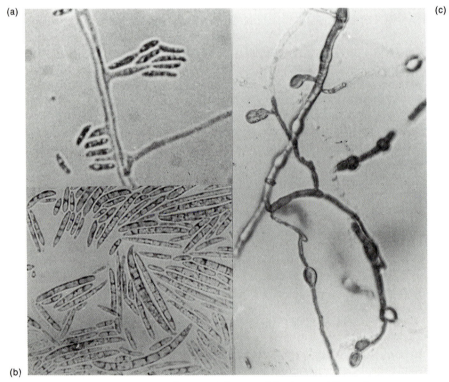

(b)

Fig. 4.2. *Fusarium oxysporum* f.sp. *vasinfectum*: (a) microconidia and conidiophore; (b) macroconidia; (c) hyphae bearing terminal and intercalary chlamydospores (× 500).

Requirements for growth

The requirements of the pathogen for maximum growth and sporulation were studied by El-Abyad and Saleh (1971). Optimal temperature for spore germination and growth through soil was 25 °C, but maximum sporulation occurred at 30 °C. Spore production and germination were greatest at 100% relative humidity, with no germination taking place below 80% RH. Mycelial growth in soil was greatest at 40% moisture-holding capacity and pH 5.6–7.2. Several sugars provided a good carbon source for mycelial growth and pigment production. Asparagine was a better source of nitrogen than inorganic nitrogen salts.

Isolation and growth in culture

The pathogen grows rapidly from the cut ends of infected, surface-sterilized stem segments incubated on PDA (see Plate 3C) and can be subcultured after 4–5 days at 28 °C. The preferred medium for morphological growth rate studies is 2% potato sucrose agar (PSA) at pH 6.5 on which growth is moderately fast (4.5 cm after 4 days at 25 °C) (Booth, 1977). The surface mycelium is white, appearing pink when macroconidia are produced, and sparse to floccose with a strong violet pigment in the medium. Conidial production is encouraged by exposure to black light and chlamydospore production can be stimulated by culturing the fungus on tap-water agar.

The fungus may be selectively isolated from infested soil, using a medium containing PCNB and streptomycin sulphate (Nash and Snyder, 1962). Pathogenicity tests can then be used to eliminate saprophytic isolates.

The pathogen can be stored for short periods on PSA slopes in the refrigerator but has a marked tendency to lose pathogenicity. However, storage over a period of several years can be achieved in sterile soil, without loss of pathogenicity (Windels *et al.*, 1988).

Pathogenic variation

When identifying the wilt fusaria it is important to use as criteria both cultural form and pathogenicity, for these fungi have a tendency to produce cultural and pathogenic variants. However, variation in the cotton wilt pathogen at the subspecies level is considered sufficiently consistent to distinguish races. Armstrong and Armstrong (1958, 1960) were able to subdivide *formae speciales* 'vasinfectum' into four races on the basis of their pathogenicity towards a set of differential hosts. The differentials consisted of a number of *Gossypium* species, *Nicotiana tabacum* cv. Gold Dollar and *Glycine max* (Table 4.2). Races 1 and 2 apparently originated in the USA, while race 3 is found in Egypt and race 4 in India. A fifth race was later identified in Sudan (Ibrahim, 1960) and race 6 was added by Armstrong and Armstrong (1978), after examining isolates of the pathogen from Brazil.

Table 4.2. Races of *Fusarium oxysporum* f.sp. *vasinfectum* and response to inoculation of six differential hosts.

Race no.	Origin	Differential hosts						
		G.a	*G.b.*[1]	*G.b.*[2]	*G.h.*	*G.m.*	*N.t.*	*L.l.*
Race 1	USA	R	S	S	S	R*	R*	S
Race 2	USA	R	S	S	S	S	S	S
Race 3	Egypt	S	R	S	R	–	R	–
Race 4	India	S	R	R	R	–	R	–
Race 5	Sudan	S	S	S	R	–	–	–
Race 6	Brazil	R	S	S	S	R	R	R

G.a.   *Gossypium arboreum* var. Rozi
G.b.[1] *Gossypium barbadense* cv. Ashmouni
G.b.[2] *Gossypium barbadense* cv. Sakel
G.h.   *Gossypium hirsutum* cv. Acala 44
G.m.   *Glycine max* cv. Yelredo
N.t.   *Nicotiana tabacum* cv. Gold Dollar
L.l.   *Lupinus luteus* cv. Weiko 3
R = resistant; S = susceptible; R* = not absolutely resistant to all isolates of race 1.
Source: Ebbels (1976) and Armstrong and Armstrong (1978).

     With the exception of Tanzania, where the pathogen appears to be race 1, isolates from other countries where the disease occurs have not been subjected to race testing with the full range of differential hosts. Results published by Charudatan (1969) indicate that the Russian isolate is similar to race 4 and an isolate from Italy belongs to race 1 or 2.

     DNA hybridization techniques have not yet been used to study the relationships between isolates of the wilt pathogen from different parts of the world. Some work has been done using vegetative compatibility studies, but the published work from Israel compared only isolates from within the country (race 3) (Katan and Katan, 1988). All isolates belonged to the same vegetative compatibility group (VCG) and did not form heterokaryons with any of the non-pathogenic fusaria isolated from the cotton rhizosphere. Assigbetse *et al.* (1991) have presented data to show that isolates from six geographical areas could be grouped into six VCGs. Furthermore, the strains corresponded to the six races of the pathogen (provided by the American Type Culture Collection) and were incompatible with each other.

HOST RANGE

Classification of *F. oxysporum* into *formae speciales* on the basis of host-specific pathogenicity must also be used with care. Investigations on the host range of a large number of wilt fusaria have revealed that they have a much wider host range than was allowed for in the system of Snyder and Hansen (Armstrong *et al.*, 1940). A concept of primary and secondary hosts was then developed by Armstrong and Armstrong (1968). Cotton is the primary host for *F. oxysporum* f.sp. *vasinfectum*, but there are numerous secondary hosts on which the fungus can multiply but produces only mild symptoms or no symptoms at all.

A plant species may be a secondary host for a particular race of the pathogen but a non-host for other races. For instance, natural infections by race 1 of the cotton wilt pathogen have been recorded on *Hibiscus esculentus*, but races 3 and 4 did not reproduce on this host (Grover and Singh, 1970). Natural infections by race 1 have also been recorded on *Hordeum vulgare*, *Cyperus esculentus*, *Malva parviflora* (Smith and Snyder, 1975), *Nicotiana tabacum* and *Glycine max* (Armstrong and Armstrong, 1958).

A number of other plants have been infected by inoculation. Charudatan and Kalyanasundaram (1966) successfully inoculated *Pennisetum typhoides*, *Eleusine coracana* and *Sorghum bicolor*, but the fungus was confined to the roots with no symptoms visible above ground. Wood and Ebbels (1972) found that the pathogens could reproduce on a wide range of common weeds associated with cotton cultivation in Tanzania.

Although not all of the species which are susceptible to inoculation would become infected under field conditions, the cotton wilt pathogen is clearly able to multiply on the roots of a number of crop and weed species. In one study, the pathogen multiplied more rapidly on a secondary host than on cotton (Smith and Snyder, 1975).

Seed Infection

In contrast to the situation with Verticillium wilt, a considerable body of evidence has accumulated to indicate that the Fusarium wilt pathogen is carried within the seed. Since seed transmission of the disease was first demonstrated by Elliot (1923), internal infection of cotton seed has been reported from East Africa (e.g. Perry, 1962), India (Kulkarni, 1934), Russia (Gubanov and Sabirov, 1972) and West Africa (Lagière, 1952).

Elliot (1923) obtained 6% infested seed from an infected crop. Kulkarni (1934) reported that almost 10% of seeds from wilted plants were infected. Perry (1962) found that only two out of every thousand seeds taken from an infected crop carried the fungus. However, when he tested seed taken from plants in which vascular browning extended into the bolls, 75 out of 165 were infected.

Seed infection is particularly important to the cotton industry in countries like Tanzania where planting seed is obtained from selected areas within the normal commercial crop. A thorough investigation of seed infection in relation to wilt symptoms in resistant and susceptible cultivars was undertaken in Tanzania (Hillocks, 1983). Most infected seed came from plants which developed symptoms late in the season because they were more likely to survive to maturity than plants infected before flowering. Infection levels in seed from these plants ranged from 2% in a resistant variety to 21% in a susceptible variety. However, some infected seed was obtained from plants not showing symptoms at harvest. This was particularly true of the resistant variety Auburn 56, suggesting that there may be considerable potential for seed transmission of the disease in 'tolerant' varieties. When the harvest as a whole was considered, estimated infection levels were between 0.01 and 1.72%. But even an infection level as low as 0.01% represents over 1500 infected seeds for each hectare harvested, assuming seed cotton yields similar to those in the experiments described by Hillocks (1983).

Survival of the Pathogen

F. oxysporum f.sp. *vasinfectum*, in common with other vascular wilt fusaria, can be considered as a soil-borne pathogen, but conforms to Garret's (1956) description of a soil-invader rather than a true soil-inhabitor. It is character-ized by an expanding parasitic phase within the host vascular tissue and a declining saprophytic phase after the host's death. The fungus grows well on artificial media in pure culture but is rapidly overgrown by many other soil-borne fungi in mixed cultures (Sabet and Khan, 1969). Although it has a limited capacity for saprophytic survival in the soil, the pathogen is nevertheless able to survive for long periods in the absence of cotton (e.g. Armstrong and Armstrong, 1948; Smith and Snyder, 1975). As is the case with other wilt fusaria (e.g. Nash et al., 1961), chlamydospores probably play an important role in survival of the fungus in the field. However, although the pathogen can remain viable in sterile soil in the form of chlamydospores for several years, it is not clear if long-term survival in field soil in the absence of the primary host is due entirely to the presence of dormant chlamydospores, or requires invasion of the roots of weed species and non-host crops. For instance, Smith and Snyder (1975) established that soil populations of the cotton wilt *Fusarium* could be maintained or even increased in fields cropped to barley.

Infection Cycle

The pathogen is introduced into previously disease-free soil through the planting of infected seed or is carried in from neighbouring infested areas on farm implements and the feet of farm workers. Infection of the plant then arises directly from the seed or from chlamydospores in the soil which are simulated to germinate by root exudates from the developing plant.

The germ-tube is able to grow through the soil for a short distance from the chlamydospore until it makes contact with the host root system. Rao and Rao (1966a) found that infection took place in the hypocotyl just below the soil surface, but Khadr and Snyder (1967) observed that the fungus penetrated both the taproot and lateral roots, directly behind the root-tip. After penetration, the fungus grows through the cortex to the endodermis, where a proportion of the hyphae grow through this layer both intercellularly and intracellularly. Khadr and Snyder (1967) found no evidence that the pathogen was inhibited by constitutive tannins present in the endodermal layers. In fact, the pathogen was capable of using tannin as a soluble carbon source. Conidia are rapidly produced when the fungus reaches the vascular tissue and are distributed systematically within the plant in the transpiration stream (Beckman *et al.*, 1976).

Conidia are carried passively as far as the first vessel end-wall, where they become trapped and must then germinate and grow through into the next vessel. Although most of the spores become initially trapped at about 2.5 cm above the point where the vessel was invaded, Beckman *et al.* (1976) were able, after only 24 hours, to detect a few infected vessels 10 cm above the inoculation point. In compatible host–pathogen interactions, the fungus becomes distributed throughout the vascular system in 10–15 days (Fig. 4.3 and Plate 3C) before beginning to invade the cortex. As the infected host begins to die and dry out, chlamydospores are formed within the invaded tissues and these are returned to the soil in falling leaf litter and other crop residues to complete the disease cycle. The epidemiology of Fusarium wilt therefore follows the

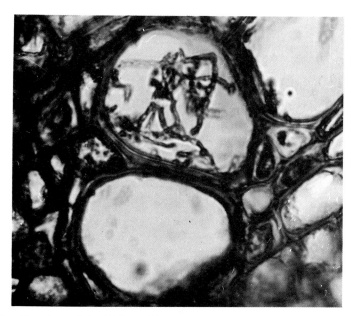

Fig. 4.3. T.S. xylem of plant infected by Fusarium wilt, showing mycelium within the vessel lumen 15 days after stem inoculation (× 460).

'simple interest' model of Van der Plank (1963), having no secondary cycle in which the conidia participate.

Factors Affecting Infection and Symptom Development

INOCULUM POTENTIAL

The relationship between populations of the pathogen in the soil and the amount of wilt in a crop is influenced by a number of factors, particularly climatic conditions, host resistance and the presence of parasitic nematodes.

In pot experiments in which inoculum was added to soil either as sand–oatmeal cultures (Rao and Rao, 1966b) or as conidia (Arjunarao, 1971b), disease incidence increased with increasing inoculum dose up to an optimum level, but then decreased with further increases in inoculum. In the experiments described by Arjunarao (1971b), spore germination in distilled water was 87% at 0.02×10^6 conidia ml^{-1}, 32% at 0.31×10^6 conidia ml^{-1} and 0% at 1.89×10^6 conidia ml^{-1}. The minimum spore concentration required to produce wilt symptoms within 30 days was 750 spores g^{-1} soil. Maximum disease incidence occurred at 15 000 spores g^{-1}, declining by 50% at 30 000 spores g^{-1}. However, this type of high inoculum inhibition may not occur in natural field situations. Inhibition in the case of the conidial inoculum may have been due to staling compounds accumulated in the culture medium. In the case of the sand–oatmeal inoculum, inhibition may have been caused by an increase in antagonistic micro-organisms, as these experiments were conducted with non-sterile soil.

Somewhat different results were reported by Garber *et al.* (1979), who found that no visible wilt symptoms were produced on plants grown in soil containing 5000 propagules g^{-1}. However, typical disease symptoms appeared on plants grown in soil containing 77 000 propagules g^{-1} of the pathogen. It appears that much higher inoculum levels are required to produce disease symptoms in pot experiments than in conditions of natural infection in the field. In California, for instance, Smith *et al.* (1970) reported that wilt was severe on land with pathogen populations of 800 propagules g^{-1} of soil and moderately severe with 400 propagules g^{-1}. However, in the field, most of the inoculum would have consisted of chlamydospores, whereas in the pot experiments of Garber *et al.* (1979) inoculum was added in the form of spore suspensions.

TEMPERATURE

Fusarium wilt is a disease of warm conditions. Optimum temperature for disease development is between 30 and 32°C, with little disease occurring if temperatures remain below 23 °C (Young, 1928). Cotton plants often recover from the disease and regrow if the temperature falls below the optimum for disease development after initial infection.

Soil type

The type of soil which favours the disease in cotton varies greatly with pathogenic race. In the USA, where race 1 and race 2 occur, and also in Tanzania, where the pathogen is similar to race 1, the disease is favoured by sandy soils of pH 5–6.5.

In Tanzania, the disease has not been observed on alkaline soils (Ebbels, 1976). According to the findings of Smith *et al.* (1970), field population levels of race 1 are negatively correlated with clay content of the soil. Snyder and Smith (1981) mention that red lateritic clay soils are usually suppressive to Fusarium wilts. In contrast, race 3 in Egypt (Fahmy, 1927) and race 4 in India (Subramanian, 1950) are more prevalent on clay soils.

Soil moisture

Tharp and Young (1939) showed that the disease was encouraged by moist conditions with an optimum of 85% saturation. However, El Abyad and Saleh (1971) reported that growth of the pathogen is retarded at moisture levels above 40% water holding capacity. Such apparent contradictions may be due to the two studies having used different races of the pathogen and/or different soil types. Also, as Ebbels (1975) has suggested, maximum host susceptibility and optimal growth of the pathogen in soil may not occur at the same soil moisture levels. The cotton plant may be less able to resist root invasion when soil moisture levels approach saturation. Furthermore, environmental conditions which most favour disease development after invasion of the vascular tissue are probably not the same as those which favour growth of the pathogen in soil and infection of the root. At least for race 1, it seems that infection is encouraged by warm moist conditions, but it is often observed that symptom expression is most severe in hot, dry conditions.

Fertilizer

In general, nitrogen fertilizers, particularly ammonium nitrogen, tend to increase wilt incidence, while potassium fertilizers tend to reduce wilt incidence. On the sandy alluvial soils in Arkansas, where Fusarium wilt and potassium deficiency often occur together, the use of potassium fertilizers to control the disease was recommended at an early stage in the history of the disease (Ware and Young, 1934).

Phosphate levels appear to have little effect on the disease. However, results obtained in the field depend on the initial nutrient status of the soil and host susceptibility is increased as much by general nutrient imbalance as by additions of one particular element. For instance, experiments conducted in Tanzania indicated a tendency for increased wilt incidence on sandy, infertile soils when nitrogen and phosphorus were applied in the absence of potassium, but fertilizer had little effect when potassium was included as a balanced application (Ebbels, 1976). Similar results were reported by Mohamed and Darrag (1964). The effect of fertilizer on disease incidence seems to depend to some extent on cultivar susceptibility. Sharoubeem (1966) found that potassium deficiency did not affect the resistance of cv. Ashmouni, but in the wilt-susceptible cv. Karnak either low or high potassium levels reduced wilt incidence, while moderate levels increased it. El-Nur and Fattah (1970) found that nitrogen fertilizer increased wilt incidence in susceptible varieties of *G. barbadense* but not in the resistant varieties. Similar results have been obtained in sand culture, where deficiency of nitrogen increased disease incidence of cv. Karnak but the effect on disease incidence of adding nitrogen to the nutrient solution depended on the concentration. Where nitrogen was added at up to 100 ppm, disease incidence was less than in the deficient treatment, but at 100–300 ppm, although host vigour was enhanced, wilt incidence was increased. At 500–1000 ppm of nitrogen, host vigour was decreased, but, in this case also, disease incidence increased (Naim *et al.*, 1966).

Nakamura *et al.* (1975) obtained a significant effect of fertilizer on wilt incidence in a wilt-susceptible variety only in autoclaved soil. There was a linear negative relationship between disease incidence and concentration of potassium, but a linear positive relationship between disease incidence and phosphorus. With nitrogen, wilt incidence initially decreased with increasing levels, but when large quantities were applied wilt incidence increased.

HERBICIDES

There appears to be little information available concerning the effect of herbicide application on disease incidence or host susceptibility to infection. However, some studies have been conducted concerning the direct effect of herbicides on the pathogen. Rodriguez-Kabana and Curl (1970) reported that atrazine at 80 μg ml^{-1} retarded fungal growth in liquid culture, but did so only for the first six days of incubation. Prometryne also has an inhibitory effect on the pathogen in sterile soil by reducing the capacity of microconidia to germinate (Chopra *et al.*, 1970). Trifluralin, on the other hand, was found to enhance chlamydospore production at concentrations of 0.6 to 40 μg g^{-1} of soil (Tang *et al.*, 1970).

Interactions with Other Soil-borne Fungi

Sabet and Khan (1969b) isolated six root-infecting fungi from the rhizosphere or roots of cotton plants in Egypt. When these fungi, in various combinations, were added to non-sterile soil together with the wilt pathogen, most had the effect of decreasing wilt incidence. This effect was most marked when the pathogen was mixed with *Sclerotium rolfsii*. *Rhizoctonia solani*, however, increased wilt incidence.

The relationship between *R. solani* and *F. oxysporum* f.sp. *vasinfectum* was further investigated by Rizk and Mohamed (1986), using two cotton varieties. The wilt-resistant variety of Giza 66 was found to be more susceptible to *R. solani* than the variety Karnak, which was highly wilt-susceptible. But, when inoculations were made using both pathogens, wilt incidence was increased in Karnak but unchanged in Giza.

In California, the phenomenon of sudden wilt was investigated by Schnathorst (1964) and found to be due to an interaction between the wilt pathogen and *Thielaviopsis basicola*.

Interactions with Root-knot Nematode

EFFECT ON DISEASE INCIDENCE

It has been demonstrated many times since the observations by Atkinson (1892) that Fusarium wilt incidence in the field is higher in the presence of root-knot nematode (*Meloidogyne* spp.) than in its absence, although both are important pathogens in their own right (e.g. Smith 1941; Cooper and Brodie, 1963; Jorgenson *et al.*, 1978). The nematode also increases infection by the wilt pathogen in pot experiments (Martin *et al.*, 1956; Minton and Minton, 1966; Yang *et al.*, 1976; Garber *et al.*, 1979). Although most of the work on the Fusarium wilt/root-knot disease complex has been conducted in the USA, a similar interaction has been reported from East Africa (Wickens and Logan, 1960) and in Egypt (Salem, 1980).

Much of the evidence for increased wilt incidence in the field in the presence of root-knot nematode has come from experiments in which soil fumigation has been used to reduce nematode populations, resulting in decreased wilt incidence (Smith and Dick, 1960; Brown, 1968; Jorgenson *et al.*, 1978; Hyer *et al.*, 1979). In these experiments little account has been taken of the effect of the fumigant on other members of the soil microflora, including the wilt fungus itself.

In pot experiments there is often little evidence of infection by the wilt pathogen in the absence of the nematode. Garber *et al.* (1979) found that no wilt developed with 650 *Fusarium* propagules g^{-1} of soil in the absence of the nematode but only 50 nematodes were required to produce symptoms with the same fungal population. Symptom expression was similar with 647 000 propagules g^{-1} of the fungus alone, as with 650 propagules g^{-1} and 3000 nematodes per plant.

Starr *et al.* (1989) investigated the effect of the two pathogens on plant mortality and yield, using three levels of fungal inoculum and five levels of nematode inoculum. There was no effect of *Fusarium* on yield. Most of the yield loss was due to the nematode. With respect to plant mortality, no interaction between the two pathogens was observed at high levels of *Fusarium* and low nematode levels, but there was a significant interaction at

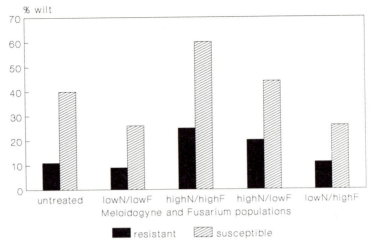

Fig. 4.4. Increase in percentage of *Fusarium*-infected plants during the first 17 weeks after planting with a wilt-resistant and wilt-susceptible cultivar, in plots where levels of the wilt pathogen (F) and root-knot nematode (N) were increased by the addition of inoculum or decreased by the application of chemicals. Ukirigura Agricultural Research Institute, Tanzania.

intermediate *Fusarium* populations with a high nematode population. Similar work was conducted in Tanzania (R. J. Hillocks and J. Bridge, unpublished), in which field populations of both organisms were altered by the application of fumigant and nematicide and the addition of inoculum. Varying the nematode population had a greater effect on wilt incidence than varying the *Fusarium* population (Fig. 4.4). The nematode also increased the number of plants which developed wilt symptoms late in the season. These plants survived to produce some seed cotton, which could be expected to contain a high proportion of internally infected seed (Hillocks, 1983). This is particularly significant in countries such as Tanzania where seed for planting the following season's crop is derived from the commercial crop. Under such circumstances, the presence of the nematode might be expected to increase the risk of infested seed entering the harvest.

There has been some controversy over whether or not root-knot nematode is able to break the resistance to Fusarium wilt of a cultivar. Although there may be differences between *Gossypium* species and cultivars in this respect, the evidence available for Upland cotton suggests that the nematode increases wilt incidence in both wilt-resistant and susceptible cultivars but that disease levels remain lower in the resistant cultivar (Wickens and Logan, 1960; Brown, 1964). However, strains of *M. incognita* vary in their ability to cause increased wilt incidence. Martin *et al.* (1956) found that one strain increased wilt incidence in a wilt-resistant cultivar to the level found in the susceptible cultivar. Other isolates caused smaller increases in susceptibility, so that there continued to be less wilt in the resistant cultivar than in the susceptible one. The magnitude of nematode-induced increases in wilt incidence might also be expected to vary between different cultivars with similar levels of wilt resistance, due for instance to differences in the genetics of wilt resistance or differences in susceptibility to root-knot (Bergeson, 1972).

MECHANISM OF NEMATODE-ENHANCED INFECTION

It was originally believed that increased wilt incidence caused by root-knot nematode was due to root damage which provided entry sites for the fungus. While root-wounding may be the cause of increased disease in some nematode–fungus interactions, there is evidence of a more complex relationship between *M. incognita* and *F. oxysporum* f.sp. *vasinfectum*.

Although there are exceptions, it is generally sedentary endoparasitic nematodes such as *Meloidogyne*, *Heterodera* and *Rotylenchulus* which cause increases in wilt incidence, and yet these nematodes do not cause extensive damage to the root cortex on penetration. Furthermore, *Belonolaimus* spp., migratory endoparasitic nematodes which cause little damage to the cortex when feeding, have also been associated with increased wilt incidence in cotton (Yang *et al.*, 1976). In contrast, *Hoplolaimus galeatus*, an ectoparasitic nematode which causes severe damage to cotton roots, does not increase wilt incidence (Yang *et al.*, 1976). Moreover, Perry (1963) was unable to find evidence that the wilt pathogen preferentially invaded cotton roots in areas penetrated by *M. incognita*. Also, when plants were inoculated with a strain of the nematode which was able to invade but unable to develop to maturity, there was no increase in wilt incidence.

The greatest damage to the root cortex caused by the sedentary endoparasites occurs with extrusion of the egg sack at maturity. The predisposing effect of *M. incognita* to Fusarium wilt reaches a maximum at three to four weeks after inoculation, coinciding with egg-sac development. It seems likely therefore, that damage to the cortex at this stage in the nematode's life cycle is at least partly responsible for enhanced invasion by soil-borne pathogens. However, in the case of Fusarium wilt of tomato and tobacco, mechanical wounding of the roots was not as effective as nematode infestation in causing disease increases (Jenkins and Coursen, 1957; Porter and Powell, 1967). Similar results were obtained by Carter (1981) for the interaction between root-knot nematode and *Rhizoctonia solani* on cotton seedlings.

It is possible that the nematode causes increased penetration by the fungus, mainly at the time of egg-sac development, but that fungal growth after penetration is also enhanced due to the high concentrations of nitrogen and carbon compounds found at the nematode feeding sites within the hypertrophied xylem tissue (see Owens and Specht, 1966; Melendez and Powell, 1967).

In addition to these localized effects on fungal penetration and growth in the roots, the nematode may also have a systemic effect on subsequent spread and proliferation of the fungus in the vascular tissue. Root invasion by the nematode causes an increase in susceptibility of the cotton plant to stem inoculation (Hillocks, 1986). This effect is associated with a reduced capacity for rapid vascular occlusion, which is an important component of host plant resistance to vascular invasion (Hillocks, 1985).

Association with Other Nematodes

In addition to *Meloidogyne* spp., some other nematode genera are reported to increase wilt incidence in cotton. The sting nematode, *Belonolaimus gracilis*, increased wilt incidence in the wilt-susceptible cv. Rowden and also in cv. Coker 100WR, which exhibits a degree of wilt resistance (Holdeman and Graham, 1955). The other nematode of major importance on cotton which enhances infection by *Fusarium* is the reniform nematode, *Rotylenchulus reniformis*. This nematode has been associated with increased wilt incidence in the USA (e.g. Neal, 1954) and in Egypt (Khadr *et al.*, 1972; El-Gindi *et al.*, 1974).

Control

ROTATION

Once established in the soil, it is almost impossible to eradicate *F. oxysporum*. The chlamydospores remain viable in the soil for several years and, furthermore, the pathogen is able to multiply on the roots of many weed and crop species (Wood and Ebbels, 1972; Smith and Snyder, 1975).

Graminaceous species have been considered as non-susceptible to Fusarium wilt, and yet several of them are able to sustain high populations of the fungus on their roots. In California, populations of the pathogen in field soils did not decline during the five years when the field was planted to barley and wheat. At some sites examined, the pathogen population was greater after a crop of barley than in those fields cropped continuously to cotton (Smith and Snyder, 1975).

In contrast to the results obtained in California, Goshaev (1971) reported that, in the former Soviet Union, planting barley reduces the incidence of wilt in the following season's cotton crop. Maize, lucerne, mustard and clover were also effective in this respect.

It seems difficult to draw any clear conclusions concerning the effectiveness of rotations in all situations. However, it is likely that the use of rotations involving a non-host of the wilt pathogen will reduce the rate at which the fungus builds up in the soil but that the most appropriate crop to use may vary with the particular race of the pathogen and soil type involved.

BIOLOGICAL CONTROL

Certain soils are known to be suppressive to Fusarium wilts and this has usually been related to the presence of antagonistic micro-organisms. Smith (1977) found that chlamydospore germination and hyphal growth were less in suppressive than non-suppressive soils. The suppressive soils contained a species of *Azotobacter* and populations of the bacterium around the fungus increased following spore germination. Similar results were obtained by Arjunarao (1971a,b), who found that soils which were free of the wilt pathogen contained a higher population of antagonistic actinomycetes than occurred in the soils infested with the pathogen.

Many other soil micro-organisms are antagonistic towards *F. oxysporum* f.sp. *vasinfectum* and in some cases these have been successfully used to reduce wilt incidence. In the former Soviet Union, *Bacterium agile* has given up to 53% reduction in wilt (Paletskaya *et al.*, 1972). In India, a strain of *Trichoderma harzianum* isolated from cotton plants in the field has shown some promise as a biocontrol agent, causing a decrease in wilt incidence in pot experiments when added to the soil (Sivan and Chet, 1986).

The phenomenon of cross-protection has also been investigated as a possible approach to biological control. Sabet *et al.* (1966) reported that *Cephalosporium maydis* interacts with the cotton wilt *Fusarium* to give an appreciable reduction in wilt incidence when mixed with the soil.

SOIL SOLARIZATION

Increases in soil temperature which occur under clear polythene sheeting have been shown to decrease the population of soil-borne pathogens (e.g. Katan, 1981), including *Fusarium* spp. (Smith *et al.*, 1980). Katan *et al.* (1983) showed that soil solarization decreased the incidence of wilt and increased cotton yields in Israel. The beneficial effect was apparent over three seasons. However, Ben-Yephet *et al.* (1980) found that *F. oxysporum* f.sp. *vasinfectum* was less affected by solarization than some other soil-borne pathogens such as *Verticillium dahliae*. Significant reductions in the viability of *Fusarium* populations required solarization to be combined with the application of metham-sodium.

CHEMICAL CONTROL

Fusarium soil populations can be decreased by soil fumigation, using, for instance, a mixture of chloropicrin and methyl bromide. But this type of treatment is usually confined to high-value crops grown on relatively small areas. In fields where wilt is associated with root-knot nematode, disease incidence can be decreased by controlling the nematode with nematicides (e.g. Brown, 1968; Jorgenson *et al.*, 1978). If nematode populations are sufficiently high and the crop potentially high-yielding, then nematicide application may give an economic return.

Some success in controlling seed-borne infection with seed-dressings has been reported from China (Li *et al.*, 1986) and from India (Sharma and Bedi, 1986). But the application of systemic fungicides to the seed or to the planting furrow has not been advocated for control of Fusarium wilt in cotton fields where the disease is already established. This is because the plant is susceptible to infection well beyond the seedling stage and the available fungicides are insufficiently persistent to provide the necessary protection.

RESISTANT CULTIVARS

The most effective approach to control has been the use of resistant cultivars. In the USA and several other countries, breeding programmes for wilt resistance have had some success in producing cultivars which suffer reduced stand losses when grown on wilt-infested land (see below). However, after almost a century of breeding for wilt resistance in the USA, it has not been possible to produce a commercially acceptable wilt-immune Upland cultivar.

Host–Plant Resistance

GENETICS AND SOURCES OF RESISTANCE

Variability for wilt resistance occurs in both Upland and *G. barbadense* cottons, but resistance is more complete in the latter. Fahmy (1927) reported that, in certain Egyptian Sea Island cultivars, resistance was determined by a single dominant major gene and one or more minor genes. Mohamed (1963) concluded that resistance in the *G. barbadense* cultivars Ashmouni and Menoufi was controlled by a single dominant gene. Resistance in American Sea Island cotton was considered to be controlled by two dominant genes with inter-locus additivity giving a high degree of resistance (Smith and Dick, 1960). The same workers also concluded from their field experiments that resistance in the Upland cultivar Cook 307-6 was of the major, dominant gene type, although several minor modifying genes were also present. It was possible to identify this dominant gene effect only after nematode populations had been decreased by soil fumigation.

The difficulty experienced in trying to produce highly wilt-resistant Upland cottons suggests that resistance in *G. hirsutum* may be more complex than the studies of Smith and Dick (1960) might suggest. This view is shared by Kappelman (1971a), whose pot experiments consistently showed that resistance was quantitatively inherited and controlled by several major genes and minor modifying genes.

The first wilt-resistant cottons were produced in the USA by mass selection on heavily infested land. The variety Rivers was selected in 1892 from a field of Georgia Sea Island. Further selection gave the variety Seabrook Sea Island, which remains a useful source of resistance to both Fusarium and Verticillium wilts (Wilhelm, 1981). Dillon was one of the earliest wilt-resistant Upland cultivars and it was followed in 1912 by Cook 307-6, which, although more resistant than previous Uplands, could not compare with the best wilt-susceptible varieties with respect to yield potential, lint quality and resistance to boll weevil (Ebbels, 1975). However, Cook 307-6 contributed to the pedigree of several later wilt-resistant varieties, including Empire, Delcot 277, Hartsville and McNair 220 (Wilhelm, 1981).

The Fusarium wilt–nematode evaluation programme at Talassee, Alabama, was initiated in 1952 to test for resistance to the wilt–nematode complex in cultivars derived from a number of breeding programmes across the cotton belt of the USA. Among the most wilt-resistant cultivars tested in the 1960s were Deltapine 16, McNair 511, Auburn 56 and Westburn 70 (Kappelman, 1980).

It is likely that many of the early wilt-resistant selections were made on land infested with root-knot nematode in addition to the wilt pathogen. As a result, cultivars such as Cook 307-6 shows some resistance to root-knot. Auburn 56 was developed from a cross between Cook 307-6 and Coker 100-Wilt (Bird, 1973) and McNair 511 was produced by direct selection from Auburn 56. With the release of Auburn 56, a variety became available to the growers for the first time that combined wilt resistance with good agronomic qualities and resistance to root-knot.

Although a few cultivars such as Clevewilt-6 may have been more root-knot resistant, there was no high-yielding, commercially acceptable variety available to the grower with better resistance to root-knot and Fusarium wilt than Auburn 56, even by the early 1970s, 20 years after its release. Varieties such as Auburn M and Dixie King-2 were performing well in regional wilt trials in the USA about this time but were still not sufficiently high-yielding on wilt-free land to gain widespread acceptance. Better yields were obtained by growing high-yielding varieties with only moderate wilt resistance, such as Coker 310 and Stoneville 603 (Kappelman, 1971b).

During the 1970s the Talassee programme placed more emphasis on improved agronomic qualities in wilt-resistant material, which resulted in the selection of, among others, Stoneville 603, Dixie King-3, McNair 235 and Deltapine 61, together with the Tamcot cultivars produced in the multi-adversity resistance (MAR) programme (Kappelman, 1980).

The first Upland cottons to be developed with resistance to the wilt–root-knot complex which improved upon that of Auburn 56 were the Bayou strains. These were derived from a cross between Deltapine 15 and Clevewilt-6 and had a greater capacity than Auburn 56 to reduce the build-up of *Meloidogyne* populations over the growing season (Jones and Birchfield, 1967).

Hyer *et al.* (1979) showed that, in fields infested with root-knot nematode and the wilt pathogen, a high degree of resistance to root-knot was more effective in reducing losses to the disease complex than was tolerance to root-knot combined with moderate wilt resistance. Also, Upland cultivars derived from crosses with the root-knot-resistant Mexico Wild by selection for resistance to the nematode but not to the wilt pathogen showed higher levels of resistance to the wilt/root-knot disease complex.

Results of this kind have given impetus to programmes to select for resistance to root-knot as the best means of producing cottons resistant to the complex. Auburn 623RNR, for example, was derived from a cross between Clevewilt-6 and Mexico Wild. It was one of the most wilt-resistant lines tested in field trials conducted over a six-year period (Kappelman, 1975b) and it also prevents the build-up of nematode populations over the growing season (Shepherd, 1982a). Several other Auburn lines have been registered as resistant to the wilt/root-knot complex. Auburn 634RNR and 612RNR were produced from a cross between Auburn 56 and Auburn 623, and Auburn 566RNR from a cross between Auburn 56 and Coker 201 (Shepherd, 1982a).

Resistance breeding outside the USA

Resistant varieties have been produced in other countries where the disease is a significant problem. The Egyptian long-staple varieties, Ashmouni, Giza 69 and Menoufi, are resistant and have also been used in breeding programmes in the former USSR (Wilhelm, 1981).

More recent varieties produced in Egypt, such as Giza-69 and Bahatim-185, also exhibit good wilt resistance (Abdel-Raheem *et al.*, 1974). Fusarium wilt is becoming increasingly important in the Sudan Gezira, where the main commercial variety, Barakat, is very susceptible. Potential cultivars for areas of the Gezira affected by wilt are therefore screened for wilt resistance (Yassin *et al.*, 1986).

The variety Reba W296 was selected in the 1950s in Central Africa from the progeny of a cross between Coker 100-Wilt and a selection from Allen (Lagière and Cognée, 1960). It carries resistance to wilt and bacterial blight. Reba B50 has superior wilt resistance and is grown in Paraguay (Kappelman, 1981) as well as in Africa. The Tanzanian cotton-breeding programme has produced high-yielding wilt-resistant commercial varieties, one of the most recent being UK 77 (Hillocks, 1984). Wilt-resistant varieties have also been produced in China (e.g. Li and Yang, 1989).

Breeding methods

Where variability for wilt resistance exists within the local cotton population, this can be exploited by selecting symptomless individuals on heavily infested land. This approach, however, has often proved unsuccessful due to the number of susceptible plants which escape infection. This can be overcome by using some form of inoculation which avoids disease escape. If there is insufficient variability for resistance in the local population, hybridization will be necessary between a suitable local cultivar and a wilt-resistant exotic.

Because of the polygenic inheritance of wilt resistance, as large a population as possible should be screened for resistance to increase the chance of identifying individuals with the desired combination of resistance genes. It is preferable, therefore, to postpone selection until the third generation after the initial cross and subsequent back-crosses. To reduce the time required to produce a new cultivar, selection can be practised in the F_2, provided seed is carried forward from a relatively large number of the more resistant individuals.

If the wilt-resistant parent differs widely from the local parent in important agronomic or quality characteristics, then back-crossing may be the best way of transferring resistance to the local cultivar. Back-crossing is not normally recommended for qualitatively inherited characters, but has been used in wilt-resistance breeding programmes (e.g. Shepherd, 1982a).

SCREENING TECHNIQUES

Once the necessary genes for resistance exist in the population to be screened, success in producing selections with superior resistance will depend on the effectiveness of the inoculation technique and grading scale used to evaluate the response. Mixing inoculum with soil before planting the seed (e.g. Perry, 1962; Rao and Rao, 1966b) gives the best comparison with screening in the field using natural infection. The method is not suitable for making single plant selections because some plants escape infection. Root-dip methods (e.g. Wiles, 1963; Miller and Cooper, 1967) tend to be very time-consuming as each plant has to be carefully uprooted and transplanted. Also, root damage may lead to severe symptom expression in resistant plants. The method does not totally eliminate the possibility of disease escape. As resistance is apparently expressed in the stem as well as in the root (Bugbee, 1970), methods of stem inoculation may be used and can give results which compare favourably with screening by natural infection (Bugbee and Sappenfield, 1972; Kappelman, 1975a, b). Stem inoculation has the advantage that it avoids disease escape, requires little inoculum and can be used in the greenhouse or in the field. Stem inoculation has been used in Tanzania to make single plant selections (Wickens, 1964), using a sewing-machine needle to make the punctures. Four-week old plants are inoculated by puncturing the hypocotyl after dipping the needle in a spore suspension containing 2×10^6 conidia ml^{-1}. Each plant is inoculated to minimize differences in inoculum dose. Symptoms can then be evaluated three weeks later, using a suitable scale of severity (Hillocks, 1984).

Once a selection reaches the more advanced stages of a wilt resistance breeding programme, when sufficient seed should be available for replicated trials, evaluation of wilt resistance must be carried out in the field under conditions of natural infection. If a suitable field is available which is already infested, the level of inoculum in the soil can be raised by ploughing in stem-inoculated cotton plants.

If a wilt-resistant cultivar is required for areas infected with root-knot nematode, it will probably be necessary to select for resistance to the wilt–nematode complex as a whole. As previously indicated, one approach is to select only for resistance to the nematode (see Chapter 10 for methods) but this would not be satisfactory for areas infested only with the wilt pathogen. Selection only for wilt resistance after inoculation with both pathogens should be used with care as it appears that cultivars selected in this way are not necessarily resistant to root-knot (Starr and Veech, 1986). It is therefore desirable to evaluate the material being screened for severity of root-galling and nematode egg production, in addition to scores for wilt severity.

Mechanisms of resistance

Preinvasive factors

There is some evidence of varietal differences in the composition of root exudates from cotton (Sulochana, 1962). Youseff and Heitefuss (1983) found that germination of conidia and chlamydospores was greater in medium containing exudates from a wilt-resistant cultivar than from a susceptible one. This effect may have been due to differences between the cultivars in amino acid and carbohydrate content of the exudate or due to the presence of antifungal compounds in greater concentration in exudate from the resistant cultivar. Hunter *et al.* (1978) have shown that terpenoid aldehyde compounds which exhibit antifungal activity are exuded and accumulate on the external surfaces of the root.

Prevascular factors

There have been very few studies on the early stages of infection of the cotton root by the Fusarium wilt pathogen. There appear to be no chemical or physical barriers to penetration of the root and growth of the fungus through the cortex to the vascular tissue. In the case of Verticillium wilt, the endodermis presents a barrier to the fungus due to the presence of flavonol compounds which are inhibitory to the fungus (Garber and Houston, 1966; Mace and Howell, 1974). Growth of the Fusarium wilt pathogen, however, does not appear to be inhibited at the endodermis and, indeed, the fungus is able to use tannin compounds as a source of carbon (Khadr and Snyder, 1967).

Vascular occlusion

Vascular occlusion in response to wounding or the presence in the xylem of foreign chemicals or micro-organisms is a general phenomenon in plants. Occlusion may be caused by the accumulation of polysaccharide gels around the perforation plate of the vessel end-wall (Vandermolen *et al.*, 1972) and/or by the formation in the vessels of ingrowths of the vascular parenchyma, called tyloses (e.g. Mace, 1978).

The gel seems to localize infection by vascular wilt pathogens in a wide range of hosts, including cotton (Vandermolen *et al.*, 1972). Upward movement of spores in the transpiration stream is initially halted at the vessel end-wall, and secondary conidia produced in the contiguous vessel then become trapped in the gel. This vascular gel does not appear to inhibit hyphal growth or sporulation and therefore is likely to provide only a temporary barrier to the upward spread of the fungus. Xylem vessels in cotton plants become occluded by gels within 24 hours after inoculation (Bugbee, 1970). More permanent occlusion may then be effected by the growth of tyloses. However, the evidence for the role of tyloses in cotton infected with vascular wilt diseases has been derived from work with Verticillium wilt (Mace, 1978), but they are produced as a general response to wounding or the presence of foreign matter in the vessels, including the presence of *Fusarium* conidia (see Fig. 4.5).

Vascular occlusion during the early stages of infection by *Fusarium* is more extensive in wilt-resistant than in wilt-susceptible cottons (Beckman, 1968; Bugbee, 1970). Also, protection from infection by a virulent isolate of the pathogen by previous inoculation with an avirulent isolate was shown to

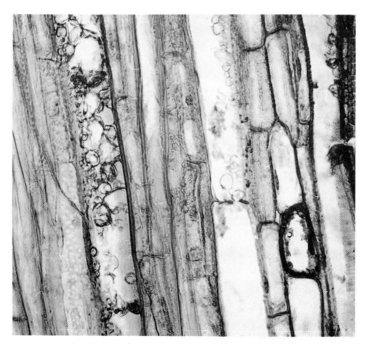

Fig. 4.5. L. S. vascular tissue of a wilt-resistant cotton cultivar, showing tyloses in the xylem lumen.

be due to the induction of vascular occlusion by the non-pathogen (Hillocks, 1986). Furthermore, there is evidence that occlusion in response to inoculation of the stem with the wilt pathogen is less effective in plants which have been attacked by root-knot nematode (Hillocks, 1985). These results support the conclusion of Harrison and Beckman (1982) that vascular occlusion is the primary mechanism of resistance to Fusarium wilt in cotton. The susceptible interaction between host and pathogen might then depend on a fungal product which rapidly degrades the gel (Beckman, 1968).

Phytoalexins

Gossypol and related terpenoid aldehydes (TAs) are the main antifungal phenolic compounds produced in the genus *Gossypium* (Bell *et al.*, 1975; Bell and Stipanovic, 1978). Terpenoid aldehydes are present in glands in the epidermal layers of the plant (Adams *et al.*, 1960) and also in the roots (Mace *et al.*, 1974). Although absent from the vascular tissue of healthy plants, TA compounds accumulate in the xylem of plants infected by Verticillium wilt (e.g. Bell, 1967, 1969). TAs are toxic to *Verticillium* conidia (Zaki *et al.*, 1972) and accumulate more rapidly during the initial stages of infection by *V. dahliae* in a wilt-resistant *G. barbadense* cultivar than in a wilt-susceptible Upland cultivar (Bell, 1969; Bell and Presley, 1969).

Although the role of TA phytoalexins in cultivar-specific resistance to Verticillium wilt has been established, their role in resistance to Fusarium wilt is more controversial. Bugbee (1970) reported that germination of *Fusarium* conidia was less than 1% in xylem extract from infected cotton plants, compared with 77% in extract from healthy plants. Four days after inoculation with the wilt pathogen, inhibition of conidial germination was apparent in the resistant cultivar (*G. barbadense*) but not in the susceptible one (*G. hirsutum*). Similar results were published by Kaufman *et al.* (1981), except that in this case the resistant cultivar was *G. hirsutum* (Acala SJ-2) and the susceptible cultivar was *G. barbadense* (Pima S-5). By 72 hours after inoculation, the amount of fungitoxic substance in the resistant plants was more than eight times greater than that in the susceptible plants. Harrison and Beckman (1982) also confirmed that the Fusarium wilt pathogen induced more rapid and more extensive accumulation of TAs in a resistant (*G. barbadense*) than in a susceptible *(G. hirsutum)* cultivar during the first two days after inoculation. However, while *Verticillium* hyphae taken from vessels containing high concentrations of TA failed to grow, the growth of similar *Fusarium* hyphae was uninhibited. Furthermore, Hillocks (1986) has shown that the cross-protective effect against a virulent isolate of the Fusarium wilt pathogen by previous inoculation with a non-virulent isolate was not related to the stimulation of TA synthesis.

Although the results described above appear contradictory, they are not directly comparable. The species of *Gossypium* used and the methods of measuring fungitoxicity were not consistent in the various reports discussed, and the race of the pathogen was not specified. Subramanian and Dolia (1974) have suggested that New World cottons are resistant to the Indian race of the Fusarium wilt pathogen because the race is highly sensitive to TA compounds, while American and Russian races of the pathogen have developed tolerance to them. There also appear to be qualitative differences between TA compounds produced by different *Gossypium* species (Bell *et al.*, 1975) and some TA compounds are more toxic than others. Therefore TA compounds may be a more important factor in resistance to Fusarium wilt in some species than in others. Further work is required on the role of phytoalexins in Fusarium wilt resistance, using particular races of the pathogen and comparing resistant and susceptible cultivars within the same *Gossypium* species.

References

ABDEL-RAHEEM, A., HAGGAG, M.E.A. AND ABOU-DAOUD, M.S. (1974) The reaction of some Egyptian cotton varieties and their crosses to Fusarium wilt. *Zeitschrift für Pflanzenschutz* 81, 516–21.

ADAMS, R., GEISSMAN, T.A. AND EDWARDS, J.D. (1960) Gossypol, a pigment of cotton seed. *Chemical Review* 60, 555–74.

ARJUNARAO, V. (1971a) Biological control of cotton wilt. 1. Soil fungistasis and antibiosis in cotton fields. *Proceedings of the Indian Academy of Sciences Part B* 73, 265–72.

ARJUNARAO, V. (1971b) On the pathogenicity of *Fusarium vasinfectum* Atk. *Proceedings of the Indian Academy of Sciences Part B* 73, 273–84.

ARMSTRONG, G.M. AND ARMSTRONG, J.K. (1948) Nonsusceptible hosts as carriers of wilt fusaria. *Phytopathology* 38, 806–26.

ARMSTRONG, G.M. AND ARMSTRONG, J.K. (1958) A race of the cotton wilt fusaria causing wilt of Yelredo soybean and flue-cured tobacco. *Plant Disease Reporter* 42, 147–51.

ARMSTRONG, G.H. AND ARMSTRONG, J.K. (1960) American, Egyptian and Indian cotton wilt fusaria: their pathogenicity and relationship to other fusaria. *US Department of Agriculture Technical Bulletin* No. 1219, 1–19.

ARMSTRONG, G.M. AND ARMSTRONG, J.K. (1968) Formae speciales and races of *Fusarium oxysporum* causing tracheomycosis in the syndrome of disease. *Phytopathology* 58, 1242–6.

ARMSTRONG, G.M. AND ARMSTRONG, J.K. (1978) A new race of the cotton wilt *Fusarium* from Brazil. *Plant Disease Reporter* 62, 421–3.

ARMSTRONG, G.M., MACLACHLAN, J.D. AND WEINDLING, R. (1940) Variation in pathogenicity and cultural characteristics of the cotton wilt organism. *Phytopathology* 30, 515–20.

ASSIGBETSE, K., DOSSA, C., PANDO-BAHUON, A. AND BOISSON, C. (1991) Vegetative compatibility in two formae speciales of *Fusarium oxysporum*. *Abstracts from the Conference on Vascular Pathogens*, British Society for Plant Pathology, held at the University of Wales, Swansea, p. 27.

ATKINSON, G.F. (1892) Some diseases of cotton. 3. Frenching. *Bulletin of Alabama Agricultural Experimental Station*, 41, 19–29.

BECKMAN, C. H. (1968) An evaluation of possible resistance mechanisms in broccoli, cotton and tomato vascular infection by *Fusarium oxysporum*. *Phytopathology* 58, 429–33.

BECKMAN, C.H., VANDERMOLEN, G.E. AND MUELLER, W.C. (1976) Vascular structure and distribution of vascular pathogens in cotton. *Physiological Plant Pathology* 9, 87–94.

BELL, A.A. (1967) Formation of gossypol in infected or chemically irritated tissues of *Gossypium* spp. *Phytopathology* 57, 759–64.

BELL, A.A. (1969) Phytoalexin production and Verticillium wilt resistance in cotton. *Phytopathology* 59, 1119–27.

BELL, A.A. AND PRESLEY, J.T. (1969) Heat-inhibited or heat-killed conidia of *Verticillium albo-atrum* induce disease resistance and phytoalexin synthesis in cotton. *Phytopathology* 59, 1147–51.

BELL, A.A. AND STIPANOVIC, R.D. (1978) Biochemistry of disease and pest resistance in cotton. *Mycopathologia* 65, 91–106.

BELL, A.A., STIPANOVIC, R.D., HOWELL, C.R. AND FRYXELL, P.A. (1975) Antimicrobial terpenoids of *Gossypium*: hemigossypol, 6-methoxyhemigossypol and 6-deoxy-hemigossypol. *Phytochemistry* 14, 225–31.

BERGESON, G.B. (1972) Concepts of nematode–fungus associations in plant disease complexes: a review. *Experimental Parasitology* 32, 301–14.

BIRD, L.S. (1973) Cotton. In: Nelson, R.R. (ed.), *Breeding Plants for Disease Resistance*. Pennsylvania State University Press, University Park and London, pp. 181–99.

BLANK, L.M. (1962) Fusarium wilt of cotton moves west. *Plant Disease Reporter* 46, 396.

BLASINGAME, D. (1990) Disease loss estimate committee report. *Proceedings of the Beltwide Cotton Production Research Conference*, National Cotton Council, Memphis, Tennessee, p. 4.

BOOTH, C. (1971) *The Genus Fusarium*. Commonwealth Agricultural Bureaux, Farnham Royal, Bucks, UK, 237 pp.

BOOTH, C. (1977) *Fusarium: a Laboratory Guide to the Identification of the Major Species*. Commonwealth Mycological Institute, Kew, UK. 58pp.

BOOTH, C. AND WATERSON, J.M. (1964) *Fusarium oxysporum* f.sp. *vasinfectum*. *CMI Descriptions of Pathogenic Fungi and Bacteria* No. 28, Commonwealth Agricultural Bureaux, Farnham Royal, Bucks, UK.

BROWN, A.G.P. (1964) Field trials with three Fusarium wilt resistant cotton selections in Tanganyika. *Empire Cotton Growing Review* 41, 194–7.

BROWN, A.G P. (1968) Effects of fumigation to control nematodes on Fusarium wilt of cotton. *Empire Cotton Growing Review* 35, 128–32.

BUGBEE, W.M. (1970) Vascular response of cotton to infection by *Fusarium oxysporum* f.sp. *vasinfectum*. *Phytopathology* 60, 121–3.

BUGBEE, W.M. AND SAPPENFIELD, W.P. (1972) Greenhouse evaluation of *Verticillium*, *Fusarium* and root-knot nematode on cotton. *Crop Science* 12, 112–14.

CARTER, W.W. (1981) The effect of *Meloidogyne incognita* and tissue wounding on severity of seedling disease of cotton caused by *Rhizoctonia solani*. *Journal of Nematology* 13, 374–6.

CHARUDATAN, R. (1969) Studies on strains of *Fusarium vasinfectum* Atk. 1. On their morphology and pathogenicity on cotton. *Proceedings of the Indian Academy of Sciences* 70, 139–56.

CHARUDATAN, R. (1970) Studies on strains of *Fusarium vasinfectum* Atk. 2. *In vitro* production of toxin and enzymes and immunoserology. *Phytopathologische Zeitschrift* 67, 129–43.

CHARUDATAN, R. AND KALYANASUNDERAM, R. (1966) Carrier hosts of *Fusarium vasinfectum* Atk. *Phytopathologische Zeitschrift* 55, 239–41.

CHESTER, K.S. (1946) The loss from cotton wilt and the temperature of wilt development: a study of new uses for old data. *Plant Disease Reporter* 30, 253–60.

CHOPRA, B.K., CURL, E.A. AND RODRIGUEZ-KABANA, R. (1970) Influence of prometryn in soil on growth-related activities of *Fusarium oxysporum* f.sp. *vasinfectum*. *Phytopathology* 60, 717–22.

COOK, R.J. (1981) Fusarium diseases in the People's Republic of China. In: Nelson, P.E., Toussoun, T.A. and Cook, R.J. (eds), *Fusarium Diseases: Biology and Control*. Pennsylvania State University Press, University Park and London, pp. 53–5.

COOPER, W.E. AND BRODIE, B.B. (1963) A comparison of Fusarium wilt indices of cotton varieties with root-knot and sting nematodes as predisposing agents. *Phytopathology* 53, 1077–80.

DIMOND, A.E. (1970) Biophysics and biochemistry of the vascular wilt syndrome. *Annual Review of Phytopathology* 8, 301–22.

DISHON, I. AND NEVO, D. (1970). The appearance of Fusarium wilt in the Pima cotton cultivar. *Hassadeh* 56, 2281–3 (in Hebrew).

EBBELS, D.L. (1975) Fusarium wilt of cotton: a review, with special reference to Tanzania. *Cotton Growing Review* 52, 295–339.

EBBELS, D.L. (1976) Diseases of Upland cotton in Africa. *Review of Plant Pathology* 55, 747–63.

EL-ABYAD, M.S. and SALEH, Y.E. (1971) Studies with *Fusarium oxysporum* f.sp. *vasinfectum*, the cause of cotton wilt in Egypt: germination, sporulation and growth. *Transactions of the British Mycological Society* 57, 427–37.

EL-GINDI, D.M., OTEIFA, B.A. AND KHADR, A.S. (1974) Inter-relationships of *Rotylenchulus reniformis, Fusarium oxysporum* f.sp. *vasinfectum* and potassium nutrition of cotton (*Gossypium barbadense*). *Potash Review* 5, 1–5.

ELLIOT, J.A. (1923) Cotton wilt: a seed borne disease. *Journal of Agricultural Research* 23, 387–93.

EL-NUR, E. AND FATTAH, A.A. (1970) Fusarium wilt of cotton. *Technical Bulletin* No. 2, Sudan Gezira Research Station. 16pp.

FAHMY, T. (1927) The Fusarium wilt disease of cotton and its control. *Phytopathology* 17, 749–67.

GARBER, R.H. AND HOUSTON, B.R. (1966) Penetration and development of *Verticillium albo-atrum* in the cotton plant. *Phytopathology* 56, 1121–6.

GARBER, R.H. AND PAXMAN, G.A. (1963) Fusarium wilt of cotton in California. *Plant Disease Reporter* 47, 398–400.

GARBER, R.H., JORGENSON, E.C., SMITH, S. AND HYER, A.H. (1979) Interaction of population levels of *Fusarium oxysporum* f.sp. *vasinfectum* and *Meloidogyne incognita* on cotton. *Journal of Nematology* 11, 133–7.

GARRET, S.D. (1956) *Biology of Root Fungi*. Cambridge University Press, Cambridge. 293pp.

GOSHAEV, D. (1971) The role of preceding crops of cotton in the suppression of Fusarium wilt. *Review of Plant Pathology* 51, 2513.

GROVER, R.K. AND SINGH, G. (1970) Pathology of wilt of okra (*Abelmoschus esculentus*) caused by *Fusarium oxysporum* f.sp. *vasinfectum* (Atk.) Snyder and Hansen, its host range and histopathology. *Indian Journal of Agricultural Science* 40, 989–96.

GUBANOV, R.K. AND SABIROV, B.G. (1972) Transmission of Fusarium wilt infection by cotton seed. *Review of Plant Pathology* 51, 3376.

HARRISON, N.A. AND BECKMAN, C.H. (1982) Time/space relationship of colonisation and host response in wilt resistant and wilt susceptible cotton cultivars inoculated with *Verticillium dahliae* and *Fusarium oxysporum*. *Physiological Plant Pathology* 21, 193–207.

HILLOCKS, R.J. (1983) Infection of cotton seed by *Fusarium oxysporum* f.sp. *vasinfectum* in cotton varieties resistant or susceptible to Fusarium wilt. *Tropical Agriculture* 60, 141–3.

HILLOCKS, R.J. (1984) Production of cotton plants resistant to Fusarium wilt with special reference to Tanzania. *Tropical Pest Management* 30, 234–46.

HILLOCKS, R.J. (1985) The effect of root-knot nematode on resistance to Fusarium wilt in the stems of cotton plants. *Annals of Applied Biology* 107, 213–18.

HILLOCKS, R.J. (1985) Cross protection between strains of *Fusarium oxysporum* f.sp. *vasinfectum* and its effect on vascular resistance mechanisms. *Journal of Phytopathology* 117, 216–25.

HOLDEMAN, Q.L. AND GRAHAM, T.W. (1955) Effect of sting nematode on expression of Fusarium wilt in cotton. *Phytopathology* 44, 683–5.

HUNTER, R.E., HALLOIN, J.M., VEECH, J.A. AND CARTER, W.W. (1978) Exudation of terpenoids by cotton roots. *Plant and Soil* 50, 237–40.

HYER, A.H., JORGENSON, E.C., GARBER, R.H. AND SMITH, S. (1979) Resistance to root-knot nematode in control of root-knot nematode–Fusarium wilt disease complex in cotton. *Crop Science* 19, 898–901.

IBRAHIM, F.M. (1966) A new race of cotton wilt *Fusarium* in the Sudan Gezira. *Empire Cotton Growing Review* 43, 296–9.

JENKINS, W.R. AND COURSEN, B.W. (1957) The effect of root-knot nematode *Meloidogyne incognita acrita* and *M. hapla* on Fusarium wilt of tomato. *Plant Disease Reporter* 41, 182–6.

JONES, J.E. AND BIRCHFIELD, W. (1967) Resistance of the experimental cotton variety, Bayou and related strains to root-knot nematode and Fusarium wilt. *Phytopathology* 57, 1327–31.

JORGENSON, E.C., HYER, A.H., GARBER, R.H. AND SMITH, S. (1978) The influence of root-knot nematode control by fumigation on the *Fusarium* root-knot nematode complex of cotton in California. *Journal of Nematology* 10, 228–31.

KAPPELMAN, A.J. (1971a) Inheritance of resistance to Fusarium wilt in cotton. *Crop Science* 11, 672–4.

KAPPELMAN, A.J. (1971b) Fusarium wilt resistance in commercial cotton varieties. *Plant Disease Reporter* 55, 896–9.

KAPPELMAN, A.J. (1975a) Fusarium wilt resistance in cotton (*Gossypium hirsutum*). *Plant Disease Reporter* 10, 803–5.

KAPPELMAN, A.J. (1975b) Correlation of Fusarium wilt of cotton in field and glasshouse. *Crop Science* 15, 270–2.

KAPPELMAN, A.J. (1980) The Fusarium wilt–nematode evaluation programme at Talassee, Alabama – progress through the years. *Proceedings of the Beltwide Cotton Production Research Conference*, Memphis, Tennessee, pp. 302–3.

KAPPELMAN, A.J. (1981) Fusarium wilt resistance of two cotton cultivars from Paraguay. *Plant Disease Reporter* 65, 344–5.

KATAN, J. (1981) Solar heating (solarization) of soil from for control of soilborne pests. *Annual Review of Phytopathology* 19, 211–36.

KATAN, J., FISHER, G. AND GRINSTEIN, A. (1983) Short and long-term effects of soil solarization and crop sequence on Fusarium wilt and yield of cotton in Israel. *Phytopathology* 73, 1215–19.

KATAN, T. AND KATAN, J. (1988) Vegetative-compatibility grouping of *Fusarium oxysporum* f.sp. *vasinfectum* from the tissue and rhizosphere of cotton plants. *Phytopathology* 78, 852–5.

KAUFMAN, Z., NETZER, D. AND BARASH, I. (1981) The apparent involvement of phytoalexins in the resistant response of cotton plants to *Fusarium oxysporum* f.sp. *vasinfectum*. *Phytopathologische Zeitschrift* 102, 178–82.

KHADR, A.S. AND SNYDER, W.C. (1967) Histology of early stages of penetration in Fusarium wilt of cotton. *Phytopathology* 57, 99 (Abstr.).

KHADR, A.S., SALEM, W.C. AND OTEIFA, B.A. (1972) Varietal susceptibility and significance of the reniform nematode, *Rotylenchulus reniformis* in Fusarium wilt of cotton. *Plant Disease Reporter* 56, 1040–2.

KIBANI, J.H.M. (1987) Fusarium wilt disease of cotton in Western Tanzania. *Tanzania Agricultural Research Organization Newsletter* 2, 9.

KULKARNI, G.S. (1934) Studies in the wilt disease of cotton in the Bombay Presidency. *Indian Journal of Agricultural Science* 4, 976–1045.

LAGIÈRE, R. (1952) Possibilities of transmission of Fusarium wilt of cotton by the seed. *Coton et Fibres Tropicales* 15, 146–8.

LAGIÈRE, R. AND COGNE, M. (1960) Sélection pour résistance à la fusariose (*Fusarium oxysporum* f.sp. *vasinfectum*). *Coton et Fibres Tropicales* 15, 188–91.

LAKSHMINARAYAN, K. AND SUBRAMANIAN, D. (1955) Is fusaric acid a vivotoxin? *Nature* 176, 697–8.

LI, C.B., GUO, J.C., LI, Y. K. AND SONG, X.Y. (1986) A study on the methods of seed disinfection against Fusarium wilt pathogens in cotton. *China Cottons* 6, 36–7.

LI, G.X. AND YANG, F.X. (1989) Regional Fusarium and Verticillium wilt resistant cotton variety tests over 16 years in China. *China Cottons* 5, 26–7.

MACE, M.E. (1978) Contributions of tyloses and terpenoid aldehyde phytoalexins to Verticillium wilt resistance in cotton. *Physiological Plant Pathology* 12, 1–11.

MACE, M.E. AND HOWELL, C.R. (1974) Histochemistry and identification of condensed tannin precursors in roots of cotton seedlings. *Canadian Journal of Botany* 52, 2423–6.

MACE, M.E., BELL, A.A. AND STIPANOVIC, R.D. (1974) Histochemistry and isolation of gossypol and related terpenoids in roots of cotton seedlings. *Phytopathology* 64, 1297–302.

MARTIN, W.J., NEWSOM, L.D. AND JONES, J.E. (1956) Relationship of nematodes to the development of Fusarium wilt in cotton. *Phytopathology* 46, 285–9.

MELENDEZ, P.L. AND POWELL, N.T. (1967) Histological aspects of the Fusarium wilt–root knot complex in flue-cured tobacco. *Phytopathology* 57, 286–91.

MENLIKIEV, N.Y. (1962) Fusarium wilt of fine-staple cotton and a study of *Fusarium oxysporum* f.sp. *vasinfectum* strains as the causal agent of the disease in conditions of the Vakash Valley. *Review of Applied Mycology* 43, Abstract No. 3381.

MILLER, D.A. AND COOPER, W.E. (1967) Greenhouse technique for studying Fusarium wilt in cotton. *Crop Science* 7, 75–6.

MINTON, N.A. AND MINTON, E.B. (1966) Effects of root-knot and sting nematodes on expression of Fusarium wilt of cotton in three soils. *Phytopathology* 56, 319–22.

MOHAMED, H.A. (1963) Inheritance of resistance to Fusarium wilt in some Egyptian cottons. *Empire Cotton Growing Review* 40, 292–5.

MOHAMED, H.A. AND DARRAG, J.E. (1964) Fusarium wilt of cotton in the United Arab Republic. 4. Effect of nitrogen fertilizer. *Plant Disease Reporter* 48, 950–3.

NAIM, M.S., SHAROUBEEM, H.H. AND HABIB, A.A. (1966) The relation of different nitrogen levels to the incidence of vascular wilt and growth vigour of Egyptian cotton. *Phytopathologische Zeitschrift* 55, 257–64.

NAKAMURA, K., NISHIMURA, A., FERREIRA, M.E., BANZATTO, D.A., AND KRONKA, S.N. (1975) The effect of fertilizer on Fusarium wilt of cotton. *Review of Plant Pathology* 55, 240.

NASH, S. M. AND SNYDER, W.C. (1962) Quantitative estimates by plate counts of the bean root-rot *Fusarium* in field soils. *Phytopathology,* 52, 567–72.

NASH, S.M., CHRISTOU, T. AND SNYDER, W.C. (1961) Existence of *Fusarium solani* f. *phaseoli* as chlamydospores. *Phytopathology* 51, 308–12.

NEAL, D.C. (1954) The reniform nematode and its relationship to the incidence of Fusarium wilt of cotton at Baton Rouge, Louisiana. *Phytopathology* 44, 447–50.

OWENS, R.C. AND SPECHT, H.M. (1966) Biochemical alterations induced in host tissue by root-knot nematodes. *Contributions to the Boyce Thompson Institute* 23, 181–198.

PALETSKAYA, L.N., KISELEVA, N.T., ZURAVLEVA, V.P., SARAYEVA, A.N. AND GORINA, E.I. (1972) An experiment with the use of antagonistic bacteria against Fusarium wilt of cotton. *Review of Plant Pathology* 51, 4039.

PEAT, J.E., MUNRO, J.M. AND ARNOLD, M.H. (1955) Lake Province, Tanganyika Territory. *Progress Reports from Experimental Stations* 1953–54. Empire Cotton Growing Corporation, p. 20.

PEGG, G.F. (1981) Biochemistry and physiology of pathogenesis. In: Mace, M.E., Bell, A.A. and Beckman, C. H. (eds), *Fungal Wilt Diseases of Plants.* Academic Press, New York and London, pp. 193–253.

PERRY, D.A. (1962) Fusarium wilt of cotton in the Lake Province of Tanganyika. *Empire Cotton Growing Review* 39, 14–16.

PERRY, D.A. (1963) Interaction of root-knot and Fusarium wilt of cotton. *Empire Cotton Growing Review* 40, 41–7.

PORTER, D.M. AND POWELL, N.T. (1967) Influence of certain *Meloidogyne* spp. on Fusarium wilt development in flue-cured tobacco. *Phytopathology* 57, 282–5.

RAO, M.V. AND RAO, A. S. (1966a) Influence of the inoculum potential of *Fusarium oxysporum* f.sp. *vasinfectum* on its development in cotton roots. *Phytopathologische Zeitschrift* 56, 393–7.

RAO, M.V. AND RAO, A.S. (1966b) Fusarium wilt of cotton in relation to inoculum potential. *Transactions of the British Mycological Society* 49, 403–9.

RISK, R.H. AND MOHAMED, H.A. (1986) Influence of *Rhizoctonia solani* on reaction of cotton plants to Fusarium wilt. *Agricultural Research Review* 61, 19–24.

RODRIGUEZ–KABANA, R. AND CURL, E.A. (1970) Effects of atrazine on growth of *Fusarium oxysporum* f.sp. *vasinfectum. Phytopathology* 60, 65–9.

ROTHWELL, A. (1983) A revised list of plant diseases occurring in Zimbabwe. *Kikia* 12, 233–351.

SABET, K.A. AND KHAN, I.D. (1969) Inhibition and stimulation of *Fusarium oxysporum* f.sp. *vasinfectum* in combination with other root-infecting fungi. *Cotton Growing Review* 46, 210–22.

SABET, K.A., SAMRA, A.S. AND MANSOOR, I.S. (1966) Interaction between *Fusarium oxysporum* f.sp. *vasinfectum* and *Cephalosoprium maydis* on cotton and maize. *Annals of Applied Biology* 58, 93–101.

SALEM, A.R.M. (1980) Effect of *Meloidogyne incognita* and *Fusarium oxysporum* f.sp. *vasinfectum* on cotton. *Egyptian Journal of Phytopathology* 12, 27–30.

SCHNATHORST, W.C. (1964) A fungal complex associated with the sudden wilt syndrome in California cotton. *Plant Disease Reporter* 48, 90–2.

SHARMA, J.P. AND BEDI, P.S. (1986) Seed treatment for control of Fusarium wilt of cotton in the Punjab. *Pesticides* 20, 19–28.

SHAROUBEEM, H.H. (1966) Effect of different levels of potassium on growth-vigour of cotton plants in relation to *Fusarium* spp. associated with the vascular wilt disease. *Review of Applied Mycology* 45, 545.

SHEPHERD, R.L. (1982a) Registration of three germplasm lines of cotton. *Crop Science* 22, 692.

SHEPHERD, R.L. (1982b) Genetic resistance and its residual effects for control of root-knot nematode–Fusarium wilt complex in cotton. *Crop Science* 22, 1151–5.

SIVAN, A. AND CHET, I. (1986) Biological control of *Fusarium* species in cotton, wheat and muskmelon by *Trichoderma harzianum*. *Journal of Phytopathology* 116, 39–47.

SMITH, A.L. (1941) The reaction of cotton varieties to Fusarium wilt and root-knot nematode. *Phytopathology* 38, 943–7.

SMITH, A.L. AND DICK, J.B. (1960) Inheritance of resistance to Fusarium wilt in Upland and Sea Island cotton as complicated by nematodes under field conditions. *Phytopathology* 50, 44–8.

SMITH, S.N. (1977) Comparison of germination of pathogenic *Fusarium oxysporum* chlamydospores in host rhizosphere soil conducive and suppressive to wilts. *Phytopathology* 67, 476–81.

SMITH, S.N. AND SNYDER, W.C. (1975) Persistence of *Fusarium oxysporum* in fields in the absence of cotton. *Phytopathology* 65, 190–6.

SMITH, S.N., SNYDER, W.C. AND MOYNIHAN, F. (1970) Populations of *Fusarium oxysporum* f.sp. *vasinfectum* in field soils in relation to cotton wilt. *Proceedings of the Beltwide Cotton Production Research Conference*. Memphis, Tennessee, pp. 69–70.

SMITH, S.N., PULLMAN, G.S. AND GARBER, R.H. (1980) Effect of soil-solarization on soil-borne populations of *Fusarium* spp. *Proceedings of the Beltwide Cotton Production Research Conference*, Memphis, Tennessee, p. 18.

SNYDER, W.C. AND HANSEN, H.N. (1940) The species concept in *Fusarium*. *American Journal of Botany* 27, 64–7.

SNYDER, W.C. AND SMITH, S.N. (1981) Current status. In: Mace, M.E., Bell, A.A. and Beckman, C.H. (eds), *Fungal Wilt Disease of Plants*. Academic Press, New York and London, pp. 25–50.

STARR, J.L. AND VEECH, J.A. (1986) Susceptibility to root-knot nematode in cotton lines resistant to the Fusarium wilt–root-knot nematode complex. *Crop Science* 26, 543–6.

STARR, J.L., JEGER, M.J. MARTYN, R.D. AND SCHILLING, K. (1989) Effects of *Meloidogyne incognita* and *Fusarium oxysporum* f.sp. *vasinfectum* on plant mortality and yield of cotton. *Phytopathology* 79, 640–6.

SUBRAMANIAN, C.V. (1950) Soil conditions and wilt diseases in plants with special reference to *Fusarium vasinfectum* on cotton. *Proceedings of the Indian Academy of Sciences Section B* 31, 67–102.

SUBRAMANIAN, D. AND DOLIA, C.B. (1974) Physiology of wilt resistance. *Indian Journal of Genetics and Plant Breeding* 34, 251–3.

SULOCHANA, C.B. (1962) Amino acids in root exudates from cotton. *Plant and Soil* 16, 312–26.

TANG, A., CURL, E.A. AND RODRIGUEZ-CABANA, R. (1970). Effect of trifluralin on inoculum density and spore germination of *Fusarium oxysporum* f.sp. *vasinfectum* in soil. *Phytopathology* 60, 1082–6.

THARP, W.H. AND YOUNG, V.H. (1939) Relation of soil moisture to Fusarium wilt of cotton. *Journal of Agricultural Research* 58, 47–61.

VANDERMOLEN, G.E., BECKMAN, C.H. AND RODEHORST, E. (1972) Vascular gelation: a general response phenomenon following infection. *Physiological Plant Pathology* 11, 95–100.

VAN DER PLANK, J.E. (1963) *Plant Disease: Epidemics and Control*. Academic Press, New York and London.

WARE, J.O. AND YOUNG, W.H. (1934) Control of cotton wilt and rust. *Arkansas Agricultural Experimental Station Bulletin* No. 308, p. 23.

WICKENS, G.M. (1964) Methods for detection and selection of heritable resistance to Fusarium wilt of cotton. *Empire Cotton Growing Review* 41, 172–93.

WICKENS, G.M. AND LOGAN, C. (1960) Fusarium wilt and root-knot of cotton in Uganda. *Empire Cotton Growing Review* 37, 15–17.

WILES, A.B. (1963) Comparative reactions of certain cottons to Fusarium and Verticillium wilts. *Phytopathology* 53, 586.

WILHELM, S. (1981) Sources and genetics of host resistance in field and fruit crops. In: Mace, M.E., Bell, A.A. and, Beckman, C.H. (eds), *Fungal Wilt Diseases of Plants*. Academic Press, New York and London, pp. 300–76.

WINDELS, C.F., BURNES, P.M. AND KOMMEDAHL, T. (1988) Five-year preservation of *Fusarium* species on silica gel and soil. *Phytopathology* 78, 107–9.

WOOD, C.M. AND EBBELS, D.L. (1972) Host range and survival of *Fusarium oxysporum* f.sp. *vasinfectum* in N.W. Tanzania. *Cotton Growing Review* 49, 79–82.

YANG, H., Powell, N.T. AND BARKER, K.R. (1976) Interaction of concomitant species of nematodes and *Fusarium oxysporum* f.sp. *vasinfectum* on cotton. *Journal of Nematology* 8, 74–80.

YASSIN, A.M. AND DAFFALLA, G.A. (1982) A preliminary note on the present status of cotton with syndrome in the Sudan. *Coton et Fibres Tropicales* 4, 379–83.

YASSIN, A.M., KHALIFA, H. AND ABBAS, I.M. (1986) A quick method for effective screening of cotton cultivars against pathogenic wilt: A useful tool for the breeder. *Tropical Pest Management* 32, 115–17.

YOUNG, V.H. (1928) Cotton wilt studies. 1. Relation of soil temperature to the development of cotton wilt. *Arkansas Agricultural Experimental Station Bulletin* No. 226.

YOUSEFF, B.A AND HEITEFUSS, R. (1983) Side effects of herbicides on cotton wilt caused by *Fusarium oxysporum* f.sp. *vasinfectum*. 3. Microbial studies. *Zeitschrift für Pflanzenkrankheiten und Pflanzenschutz* 90, 160–72.

ZAKI, A.J., KEEN, N.Y. AND ERWIN, D.C. (1972) Implication of vergossin and hemigossypol in the resistance of cotton to *Verticillium albo-atrum*. *Phytopathology* 62, 1402–6.

5 Fungal Diseases of the Root and Stem

C.M. KENERLEY* AND M.J. JEGER†

Introduction

This chapter covers published work on pathogens affecting the root and stem of cotton, but excludes the seedling disease complex (Chapter 1) and the vascular wilts (Chapters 3 and 4) covered elsewhere in this book. Thus the pathogens are those which can infect and rot the roots and lower stems of cotton once stands are established. Seedlings emerge some 5–15 days after planting, under suitable conditions, and the basic plant architecture has been established by 45–55 days (El-Zik, 1985). The dynamics of seedling diseases are seen in seed rots, pre-emergence damping-off, post-emergence damping-off and subsequent root damage until the stand is considered complete (Bird, 1974). Thus there is some ambiguity in classifying pathogens according to whether their effects are manifested prior to stand establishment (and thus seedling diseases) or as diseases of older plants (and thus considered here).

In practice, despite this ambiguity, there are two major diseases considered in this chapter, Phymatotrichum root rot and charcoal rot. As will be seen, these diseases have very different characteristics, the former being a disease with a very localized distribution in south-western USA and northern Mexico (Percy, 1983) but with serious and damaging consequences, and the latter having a worldwide distribution (Dhingra and Sinclair, 1978) but with some ambiguity over its importance – Garrett (1956) went so far as to call the causal agent on 'impostor pathogen' in terms of its unwarranted implication with some diseases. Other pathogens in this chapter are those implicated in the seedling disease complex, *Thielaviopsis basicola*, *Rhizoctonia solani*, *Fusarium* spp. and *Pythium* spp., which have some residual effect on infected

* Department of Plant Pathology and Microbiology, Texas A&M University, College Station, Texas 77843-2132, USA; and † Natural Resources Institute, Central Avenue, Chatham Maritime, Kent ME4 4TB, UK.

plants surviving at stand establishment, and plurivorous pathogens, including *Sclerotium rolfsii* and *Cylindrocladium crotalariae*, which have been recorded as affecting cotton. Finally, reports of minor or localized foliar pathogens on stems and their effects are noted.

Phymatotrichum Root Rot

HISTORY AND DISTRIBUTION

Pammel (1888, 1889) was the first to report that a fungus was the causal agent of the 'root rot and dying of cotton' in Texas in the 1880s and that the disease was more severe in 'black cretaceous' soils. The fungus was identified as *Ozonium auricomum* Lk. (Pammel, 1988) and subsequently renamed *Ozonium omnivorum* (Shear, 1907). The pathogen was reported on alfalfa in Texas in 1892 (Curtis, 1892), and by 1906 the appearance of spore mats on the soil surface had been correlated with the fungus attacking the roots of cotton in Arizona (Streets and Bloss, 1973). In 1916 Duggar proposed the name *Phymatotrichum omnivorum* (Shear) Duggar based on the conidial morphology. Continuous research into the epidemiology, aetiology and control of this devastating pathogen has occurred since the 1920s in Texas and Arizona. However, the pathogen is still a serious and recalcitrant problem in its indigenous range, despite the voluminous number of research papers and dedicated research efforts.

The pathogen is indigenous to the south-western United States and Mexico, with isolated regions occurring in Arkansas, California, Louisiana, Nevada, Oklahoma and Utah (Streets and Bloss, 1973; Lyda, 1978). Ezekiel (1945) suggested that the northward distribution of Phymatotrichum root rot was limited by low air temperatures. Percy (1983), using the edaphic factors of base exchange capacity, Ca and Na content, pH and clay content and a non-edaphic factor of mean air temperature, developed a map of the potential distribution of *P. omnivorum* in North America. A comparison between the generated map and the reported distribution demonstrated that the two were co-terminous. An examination of conducive soils on other continents revealed that, even though Africa and Australia have compatible soil types, the pathogen has never been reported from these areas (Percy, 1983). Even within a major land resource area, there are similar soils containing sites where cotton annually succumbs to the pathogen physically adjacent to sites with no apparent disease. Smith and Hallmark (1987) examined soils with these characteristics in the Blackland Prairies and Coast Prairie of Texas for chemical and physical properties that could be related to disease incidence. They found that, within a land resource area, extractable Mg^{2+}, Zn^{2+} and Cu^{2+} were greater in some cases in non-root rot sites. The authors suggested that the factors found to be different between the root rot and non-root rot sites were related to plant stress, but further research was required to establish these relationships. Mueller *et al.* (1983) also attempted to correlate soil chemical factors (sodium, potassium, calcium and magnesium contents) with incidence of *P. omnivorum* in Arizona. They examined infested and adjacent non-infested soils in 13 fields and found no differences in cation contents of the soils in most comparisons.

Symptoms

The initial symptom on cotton plants is a bronzing or slight yellowing of the leaves and an increase in leaf temperature (Streets and Bloss, 1973; Watkins, 1981). Leaves become flaccid, develop visible wilt and within two to three days permanently wilt and die. Leaves desiccate, but generally remain attached to the plant (Plate 3D, E). Olsen *et al.* (1983) examined the mechanism of wilt development and demonstrated that rates of recovery from wilting and root area were not significantly different between infected and non-infected plants. However, with wilting, resistance to water flow in both roots and lower stems significantly increased in diseased plants, indicating that wilting was probably a result of increased resistance to water flow in roots.

During initial wilting, root decay is usually confined to the lower taproot, with laterals appearing healthy. Within a few days of wilting, the entire root system will rapidly decay (Plate 3F). Discoloration of xylem elements in the roots and lower stems can often be found (Brinkerhoff and Streets, 1946; Olsen *et al.*, 1983; Rush *et al.*, 1985). Generally, discoloured sunken lesions can be found near the soil line where the fungus assumes a floccose growth habit. Cotton seedlings can be colonized as soon as contact is made between the root and the fungus, and the level of colonization may be extensive without any above-ground symptoms being expressed (Rush *et al.*, 1984a; Kenerley and Jeger, 1990). Root cortex sloughing appears to be required before any foliar symptoms are observed (Rush *et al.*, 1984a). During periods of low soil moisture availability (drier than −22 bars), atypical symptoms have been reported (Rush *et al.*, 1985). These included gradual wilting, followed by leaf necrosis and defoliation of mature leaves, but young leaves near the apical meristem remained alive. Roots displayed sunken lesions well below the soil line, and the fungus could not be recovered above the coalesced lesions. When soil moisture was replenished, the fungus assumed acropetal growth, girdling plants near the soil surface.

Losses

Yearly variation of losses to *P. omnivorum* depends primarily upon the environmental conditions of soil moisture, soil temperature and rainfall. Average loss of raw fibre yield has been estimated to be 3.5% and 2.2% in Texas and Arizona, respectively (Streets and Bloss, 1973). Losses are most severe in the alkaline, calcareous soils of Texas (Lyda, 1978). An intensive countrywide survey of Hill County, Texas (Blackland Prairie, Grand Prairie and East Cross Timber Land Resource Areas in north-central Texas), was conducted in 1979 and 1981, using aerial photography (S.D. Lyda, unpublished data). Approximately 25% of the cotton-producing areas (6429 ha in 1979; 3339 ha in 1981) was infested with Phymatorichum root rot in both years of the survey. Estimated lint loss was 12.4% and 8.3% in 1979 and 1981, respectively. Mulrean *et al.* (1984) combined field surveys with aerial photography and image analysis to estimate losses based on loss per hectare, hectares affected and fibre quality and marketability in Upland and Pima cotton. Yields were reduced by 10 and 13% for *Phymatotrichum*-infested Upland and Pima cottons, respectively, compared with healthy plants. Yield loss components consisted of fewer bolls set and less lint per mature boll for Upland cotton. In Pima cotton, losses were associated with lower total boll set. Seed quality was affected in both varieties, whereas fibre quality was affected only in Upland cotton. Based on these two studies, previous yield loss estimates are conservative, as a greater acreage is probably infected and, in addition, fibre quality may be reduced.

CAUSAL ORGANISM

Taxonomy

The teleomorphic stage of the fungus has not been conclusively demonstrated although reports to the contrary have been published. The anamorphic stage of the fungus was assigned to the genus *Phymatotrichum* (erected by Bonorden in 1851) by Duggar, with a revised description of the species (Duggar, 1916). Hennebert (1973), in his revision of Botrytis-like genera, placed *Phymatotrichum* in synonymy with *Botrytis* and recognized the new genus *Phymatotrichopsis* with the type species *P. omnivora*. Although this taxonomic reassessment has been adopted in some taxonomic keys, the majority of references have retained the species designation of *Phymatotrichum omnivorum*. As this is the more familiar epithet, we will continue its use in this chapter.

Morphology

As no teleomorphic stage is recognized, the primary morphological features that characterize this fungus are the formation of blastoconidia, sclerotia, and hyphal strands with acicular, cruciform branches on the strands. Conidial formation *in vitro* is very rare, but spore mats are occasionally produced on the soil surface in moist, shaded areas of alfalfa or cotton fields (Streets and Bloss, 1973). Initially white in mass, the mats become cinnamon–buff within several days as the conidia mature. The single-celled blastoconidia are formed on spheroidal to ellipsoidal conidiophores, which may be simple or branched (Fig. 5.1). Blastoconidia may also rise from undifferentiated hyphae (Duggar, 1916).

In sterile soil cultures and along root surfaces *in situ*, yellowish to dark

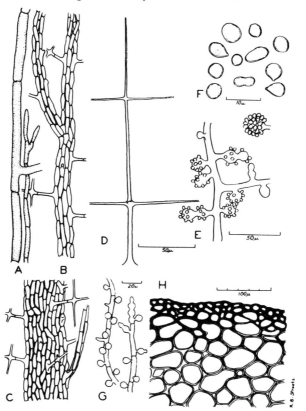

Fig. 5.1. *Phymatotrichum omnivorum*. A, Single hypha; B, strand formed by branching hyphae surrounding a large central hypha; C, mature strand with cruciform hyphae; D, cruciform hypha with acicular points; E, conidia on conidiophores from a sporemat; F, conidia freed from conidiophore; G, conidiophores before conidiation; H, cross-section of a sclerotium showing thick-celled rind cells and parenchyma (Streets and Bloss, 1973; with permission of the American Phytopathological Society).

brown strands of intertwined hyphae may be found. Mature strands are composed of a central hypha (20–40 µm in diameter) with two to eight layers of surrounding hyphae (branch hyphae) of lesser diameter (Rogers and Watkins, 1938; Neal *et al.*, 1934; Alderman and Stowell, 1986). During development, the central and surrounding hyphae enlarge, with the central hypha undergoing greater wall thickening (Alderman and Stowell, 1986). As differentiation occurs, narrow, acicular hyphae with cruciform branching, become oriented perpendicular to the strand (Rogers and Watkins, 1938; Alderman and Stowell, 1986). On cotton roots, the strands form a continuous, interconnected network that can extensively envelop the root. Alderman and Stowell (1986) also demonstrated that hyphae from separate sclerotia can undergo anastomosis, forming a common strand.

The sclerotia are the long-term overseasoning propagules that allow *P. omnivorum* to persist in soil for as long as 12 years. The sclerotia appear as swellings along the strands and are generally produced in clusters (Dana, 1931; King *et al.*, 1931; White *et al.*, 1987). They are elongate to spheroidal, often flattened around the strand connection, taking the shape of the channels in which they are formed and having a diameter of 1–2 mm (King and Loomis, 1929b; Neal *et al.*, 1934; Lyda, 1978). Immature sclerotia are first white, becoming tan to dark brown with age (Streets and Bloss, 1973). The development of the sclerotium is the result of the branching and division of the cells of the central hypha of the strand (Neal *et al.*, 1934). Mature sclerotia consist of a central core of thin-walled parenchymatous cells, two or three layers of thick-walled cells and an outer rind (Neal *et al.*, 1934; Streets and Bloss, 1973).

Physiology

The *in vitro* carbon and nitrogen nutrition of *P. omnivorum* has been examined by several researchers (Moore, 1937; Lyda, 1978; Keithly, 1985). Keithly (1985) reported that, in Murashige and Skoog (MS) liquid medium (Murashige and Skoog, 1962), *P. omnivorum* had an extended exponential growth phase and did not accumulate phenolic compounds during this phase. In the MS basal medium under conditions of stationary culture, *P. omnivorum* was found to grow on a variety of polysaccharide, disaccharide or simple sugar sources, with sucrose supporting the most rapid rate of growth. Fungal growth was further enhanced by the addition of thiamine, myo-inositol and EDTA to the basal medium with sucrose. Potassium nitrate was determined to be the nitrogen source supporting the best growth, ammonium chloride the least and the MS nitrate stock solution intermediate. Bloss and Wheeler (1975) found that strand formation *in vitro* was inhibited by increasing the levels of P and K in various media. The branch hyphae failed to aggregate around the central hyphae when P and K concentrations were increased.

The fungus has been found to tolerate high levels of CO_2 (Gunasekaran, 1973a; Lyda and Burnett, 1975; Hill and Lyda, 1976). On solid medium with glycogen as the carbon source, CO_2 levels of 0.03 and 0.5% favoured growth, with inhibition occurring at 5.0 and 50.0% levels (Gunasekaran, 1973a). However, the fungus was capable of sustaining growth at these high levels for the duration of the experiment (45 days). In a non-sterile soil system flushed with different gas mixtures (air, 0.5, 5.0 and 50% CO_2), the fungus was capable of growth at 0.5 and 5.0% CO_2, but not at the highest level (Lyda and Burnett, 1975). Lyda and Burnett (1971b) correlated the formation of sclerotia with CO_2 concentrations in sealed culture chambers containing sterile Houston black clay and the fungus. Sclerotial initiation coincided with elevated CO_2 levels, and levels of CO_2 declined during sclerotial maturation. The regulatory role of CO_2 (particularly in its dissolved forms of bicarbonate or carbonate) remains to be resolved (Lyda and Burnett, 1975).

A microanalysis of sclerotial tissue and mycelia revealed that during sclerotial development, fat, protein, glucans and hemicellulose were accumulated (Ergle and Blank, 1947). Mycelia contained higher concentrations of fat, protein and reducing sugars than sclerotia, but total carbohydrates were much lower. DNA isolated from sclerotia of *P. omnivorum* is also more highly methylated than DNA isolated from mycelia (Jupe *et al.*, 1986). These data are consistent with the role of sclerotia as overseasoning storage propagules and may indicate that DNA methylation is a property of inactive genes (Jupe *et al.*, 1986). The lipid composition was found to consist of 47.9 and 21.4% polar lipids in mycelia and sclerotia, respectively. Sterols occurred in higher concentrations in the sclerotia (10%) than in mycelia (3.6%). Monoglycerides were more prevalent in the sclerotia (17.5%) than in the mycelia (1.6%) (Gunasekaran and Weber, 1974). The most common unsaturated fatty acids found in mycelia and sclerotia were linoleic and oleic, with palmitic the dominant saturated fatty acid (Gunasekaran *et al.*, 1974). Additional details of the polar lipid components have been presented by Gunasekaran *et al.* (1974). Other studies have demonstrated the production of glycogen synthetase, glycogen phosphorylase, phosphorylase kinase, and cAMP at different levels during the growth of the fungus (Gunasekaran, 1982; Sambandam and Gunasekaran, 1987). Phosphorylase and phosphorylase kinase activity peaked at 30 days and subsequently declined, whereas glycogen synthase reached a maximum at 20 days and then gradually declined (Sambandam and Gunasekaran, 1987). Activity of cAMP phosphodiesterase reached a maximum between 21 and 28 days, indicating that phosphorylation is active in the early stages of growth in *P. omnivorum* (Gunasekaran, 1982). Low levels of inducible cellulases in culture filtrates have also been reported (Gunasekaran, 1980).

Isolation and growth in culture

Gunasekaran (1973b) demonstrated that the optimum temperature for growth on agar media was 28 °C. Sclerotial germination and strand formation in non-sterile Houston black clay was optimal between 27 and 32 °C (White, 1988). An optimum pH of 5.0 was determined on solid media (over seven days' incubation) and in liquid broth (during the first four weeks of growth) (Gunasekaran, 1973b). Lyda (1976) showed that respiration rate was similar over a pH range of 3–7. However, this fungus is generally confined to soils with a pH >7.0 (Lyda, 1978). Soil water potentials of −0.02 to −1.9 MPa supported sclerotial germination and strand formation from the germinating sclerotia. However, sclerotia either failed to germinate or germinated by producing only small-diameter hyphae if water potentials were above or below this range (White, 1988).

The fungus can be isolated from plants just beginning to show symptoms by removing small sections of the taproot, surface-disinfecting in sodium hypochlorite and then plating on to water agar, supplemented with antibacterial agents (e.g. streptomycin sulphate). This technique is even more effective when strands can be found on the taproot. Sclerotia can be recovered from soil by wet-sieving soil through a screen (425 μm, USA Standard Testing Sieve no. 40). Sclerotia can be separated from other debris by several washes and decanting the liquid; then, a 2.5 M sucrose solution is added, in which the sclerotia will float. These sclerotia can be surface-disinfected and transferred to water agar and observed for the characteristic hyphal growth.

Inoculum preparation

Sclerotia are the most common forms of inoculum desired and can be produced in large quantities using a modified soil-culture procedure of Dunlap (White *et al.*, 1987). Houston black clay (air-dried, 250 g) is added to an Erlenmeyer flask (500 ml) and moistened with water (110 ml). Soil is autoclaved and then sorghum seeds (imbibed with water) are added to the surface. The flask is autoclaved again, inoculated with a large disc (2 × 2) of agar containing the fungus and incubated at 28 °C (Lyda and Burnett, 1971a). Mature sclerotia can be obtained within six to eight weeks. The fungus can be maintained as mycelia by serial transfer on to potato dextrose agar or as sclerotia in the soil flasks.

Host range

P. omnivorum has an extensive host range that includes some 2000 species of dicotyledonous cultivated and wild plants (Taubenhaus *et al.*, 1929; Blank, 1953; Streets and Bloss, 1973). Monocotyledonous plants were originally considered resistant or immune to the pathogen (Ezekiel *et al.*, 1932; Taubenhaus and Ezekiel, 1932). When planted in alternate rows with dicotyledonous plants, monocotyledonous plants were not killed as compared with the dicotyledonous plants even though some lesions were visible on maize roots (Taubenhaus and Ezekiel, 1932). A comparison of plant sap from the two plant divisions showed that sap from monocots inhibited the growth of the fungus *in vitro*, but dicot sap did not (Ezekiel *et al.*, 1932). However, Rogers (1942) found fungal colonization of some monocots, such as day lily and blue stem grass. Others have reported lesions on roots of maize grown in soil infested with sclerotia of *P. omnivorum* (Lyda, 1978). Rush *et al.* (1984c) demonstrated that, at the point of contact between hyphae of *P. omnivorum* and roots of sorghum, small burgundy-coloured lesions appeared, but the pathogen did not continue to develop. There was no reduction in shoot or root weight of sorghum plants grown in soil infested with the pathogen compared with controls, but the pathogen was capable of producing sclerotia. The number of sclerotia formed was significantly reduced compared with the number produced when cotton was used as the host. Thus, it would appear that the pathogen is capable of surviving saprophytically in the presence of monocots and reproducing by sclerotial formation.

Variation

Variation within this species has been demonstrated for restriction fragment length polymorphisms, sclerotial production, vegetative growth, disease progression and susceptibility to mycoparasites. Riggs and Lyda (1989) examined 13 isolates from cotton and eight isolates from other hosts and found genetic differences within these groups based on phenogram analysis of restriction digests. Difference in sclerotial production among the isolates were found, but these differences were not correlated with the phenogram groupings. Growth rate was found to vary inversely with sclerotial production.

Ten isolates of *P. omnivorum* were found to vary in their sensitivity to infection by strains of the mycoparasite *Gliocladium roseum*. The relative ease of invasion and the length of infectiousness were found to vary among the *P. omnivorum* isolates (C.M. Kenerley, unpublished data).

DISEASE CYCLE

Sclerotia are the primary source of inoculum for initiating disease. Even though conidia can occasionally be found among diseased plants, they have not been shown to be infective propagules. Wheeler and Hine (1972) demonstrated that strands on dead roots are capable of surviving for a year. However, inoculation of taproots with viable strands of *P. omnivorum* in field, greenhouse or laboratory studies did not result in disease (sclerotia or infected roots placed adjacent to taproots did cause disease) (Alderman and Hine, 1982). Analysis of spatial patterns of disease progress have demonstrated that disease development in cotton is consistent with plant-to-plant spread by hyphal strands (Jeger *et al.*, 1987).

Contact with growing roots is made by strands from germinating sclerotia (White *et al.*, 1987) or adjacent infected plants (Jeger *et al.*, 1987). On young plants (< 25 days) Rush *et al.* (1984c) demonstrated that the initial reaction is a slight water-soaking and discoloration of the root at the point of contact. The fungus continued to proliferate and formed a hyphal mantle over the root, but no above-ground symptoms were observed. The root cortex and fungal mantle were then sloughed (19–25 days), leaving the root appearing healthy. However, during the next ten days, fungal strands were observed growing on the taproots. Strands and associated lesions may also be found on the lateral roots as strands grow from the taproot into the surrounding soil (Kenerley and Jeger, 1990).

Penetration of the root periderm cells and invasion of underlying cambium and xylem portions of the root have been reported (Peltier *et al.*, 1926; Watkins, 1938). Strands continue to grow towards the soil line and death is rapid as the transition zone is penetrated. Discoloration of the xylem elements in roots and lower stems of uninfected plants has been observed in several studies (Olsen *et al.*, 1983; Rush *et al.*, 1985), but the mechanism of increased root resistance to water flow in infected plants has not been determined

(Olsen *et al.*, 1983). A toxin has been purified from sclerotial exudates and has been shown to cause death in cotton suspension cells (King and Hope, 1932). At what stage in the pathogenesis of this fungus the toxin is involved remains to be established.

Several studies have demonstrated the occurrence of sclerotia among diseased plants in the field (Taubenhaus and Ezekiel, 1930; King and Hope, 1932; Lyda, 1984) or the formation of sclerotia in containers where cotton plants have been inoculated with the pathogen (Rush *et al.*, 1984c; White *et al.*, 1987; Kenerley and Jeger, 1990). Preliminary reports have indicated that sclerotia are most frequently formed in containers after plant death, in soils that are near saturation (White *et al.*, 1987; Kenerley and Jeger, 1990). Sclerotia can also form among the roots of plants that do not appear to be infected or to show any detrimental effect on root or shoot growth (Rush *et al.*, 1984c).

EPIDEMIOLOGY

Early workers developed maps illustrating the spatial patterns of disease increase in cotton by recording the location and distribution of cotton root rot in several fields (King and Loomis, 1929a; McNamara *et al.*, 1931; Rogers, 1942). McNamara and Hooton (1933) concluded that *P. omnivorum* spread radially and the rate of spread between rows was the same as within rows. However, these are annual rates of spread, as within-season assessments were not made frequently enough to analyse the spatial development. Recently, Jeger *et al.* (1987) and Koch *et al.* (1987) re-examined the spatial dynamics of Phymatotrichum root rot and showed clearly that, although both between and within-row spread occurred, within-row spread contributed most to disease development.

The expansion of runs (sequences of diseased plants) within rows was found to be a major component of disease increase. The number of runs increased rapidly during the first month of disease incidence, but then levelled off and decreased as the runs began to merge. The mean number of diseased plants within runs increased linearly throughout most of the assessment periods, but their distribution was highly skewed. Spatial autocorrelation analysis supported the view that plant-to-plant spread (mostly strand growth, but root contact may also play a role) within rows was more important than spread across rows. Regression coefficients of a linear regression model developed to predict the relative frequency of the transition from healthy to diseased plants indicated that contiguous (within-row) neighbours were more important than adjacent (across-row) ones. These results were consistent in terms of disease development (assessed as symptom expression) resulting from sclerotia and plant-to-plant spread by the fungus, but the evidence is not direct, as the movement of strands of the pathogen were not monitored *in situ* on a field scale.

Percy and Rush (1985) monitored growth of hyphae of *P. omnivorum* in a greenhouse study in which 12 cotton plants were evenly spaced approximately 5 cm apart and one end of a container infested with sclerotia at the time of planting. Initial soil temperature was 21.1 °C and was increased by 2.8 °C increments per week until a final temperature of 29.4 °C was reached. Observations were made with a boroscope placed at four intervals along the plant row. Fungal growth was fitted, using a non-linear Gompertz model; the fungus was capable of growing through 61 cm of non-sterile Houston black clay in less than 35 days. Thus, disease development in cotton can proceed by strand growth along rows of plants.

Koch *et al.* (1987) manipulated within-row plant density (1, 5, 9 and 13 plants m^{-1}) and between-row spacings (69 and 138 cm) to determine how these variables influenced disease progress in cotton. Between-row spacings had no effect on the rate of disease progress, initial disease incidence or area under the disease progress curve. Within-row density of one plant m^{-1} reduced initial disease incidence, area under the disease progress curve and final disease incidence. The plots with five plants m^{-1} reduced the initial disease incidence and delayed build-up of diseased plants compared with 9 or 13 plants m^{-1}, but not the rate of disease progress. A distance of 20–69 cm between plants was estimated as limiting the increase of Phymatotrichum root rot. This distance will vary depending upon edaphic factors, as McNamara *et al.* (1931) estimated that strands of *P. omnivorum* could extend 1.1 m beyond the last dead plant in a disease front.

As disease varies within fields and among years, Gerik *et al.* (1985) determined that large plots (33 m^2 or greater) should be used to give the best estimate of the true mean and to provide the lowest variance in field experiments on *P. omnivorum* in cotton. They (Gerik *et al.*, 1985) further suggested that plot size should be increased through the number of adjacent rows rather than row length.

The horizontal and vertical distribution of sclerotia of *P. omnivorum* can influence the time and magnitude of initial disease incidence (Rush *et al.*, 1984b). Lyda (1984) sampled fields in 22 counties in Texas (primarily Houston black clay) and found that 93% of the viable sclerotia occurred between 30 and 90 cm deep. Sclerotia were recovered in 20 out of 100 9 m^2 quadrants with 79% found in the 60–90 cm depth. The vertical and horizontal distribution showed clustered patterns. In a greenhouse study, when soil temperature was maintained at 27 °C, inoculum (sclerotia) placed at 5 cm below the soil surface resulted in 50% mortality within 40 days (Rush *et al.*, 1984b). However, placement of inoculum at 60 cm induced a similar disease level only after a 21-day delay. Viable strand lengths could be found in soil cores at depths of 15–30 and 30–60 cm, but not in the 0–15 or 60–90 cm depths in a Gila silt loam from cotton fields in Arizona in early summer (June) (Alderman and Hine, 1982). Strand lengths increased in samples taken in July, peaked in August and were recovered at all depths, but most frequently in the lower depths. The recovery of strands from the surfaces of infected roots indicated that strand viability increased with soil depth (Alderman and Hine, 1982). The influence of vertical stratification of strands on disease progress remains to be resolved.

Lyda and Burnett (1970) demonstrated in greenhouse tests the effect of various levels of sclerotial inoculum on development of disease. Maximum disease (100% of plants in test) occurred when 125–625 sclerotia kg^{-1} of dry soil were infested uniformly or centrally positioned in the soil. As few as 5 sclerotia kg^{-1} of dry soil resulted in greater than 50% disease regardless of their position. There appeared to be a delay in symptom development where sclerotia were centrally positioned compared with uniformly infested treatments. In addition to inoculum density, edaphic factors have been found to greatly influence disease development of *P. omnivorum* in cotton. Optimal disease development has been found to occur at 27 °C with a delay in plant death of almost four weeks at soil temperatures of 22 °C (Lyda and Burnet, 1971b). No plant death was recorded after 12 weeks where soils were maintained at 12 or 17 °C. Chavez *et al.* (1967) also demonstrated greatly reduced mortality in cotton plants grown in soil infested with *P. omnivorum* when soil temperature was maintained below 28 °C. Rush *et al.* (1984a) found that no foliar symptoms developed until the soil temperature at the depth of the inoculum was greater than 22 °C.

Soil moisture has also been found to influence symptom appearance and disease progress (Taubenhaus and Dana, 1928; Rush *et al.*, 1984b; Kenerley and Jeger, 1990). Kenerley and Jeger (1990) found that soil temperature and moisture in field sites were of greater importance in symptom development than the extent of prior fungal colonization on roots of cotton plants incubated in a greenhouse (three weeks before field planting). A reduction of soil water potential to approximately −25 bars delayed the appearance of the disease. Rush *et al.* (1984b) reported that the rate of disease progress was reduced in field sites with a reduction of water potential to −15 bars. However, an increase in soil water potential from <−20 to −0.7 bars resulted in an increased rate of disease development. Interactive effects of moisture and temperature on disease development have also been demonstrated. Jeger and Lyda (1986) analysed epidemics of Phymatotrichum root rot of cotton over a 14-year period and found that final incidence was directly related to cumulative precipitation (with one exception) and inversely related to air temperatures greater than 34 °C.

CONTROL

Cultural practice

There has been little success in controlling *P. omnivorum* by cultural practices (e.g. crop rotation, deep tillage, clean fallow) (King and Loomis, 1929a; Rogers, 1942; Streets and Bloss, 1973; Rush and Lyda, 1984). These approaches are generally not sufficient for reducing the number and viability of the sclerotia within heavy clay soils. Rush and Lyda (1984) compared deep chiselling and deep chiselling accompanied by anhydrous ammonia with conventional practices. Both treatments reduced onset and disease incidence compared with conventional agronomic practices, but the cost involved and the erratic results appear to prohibit this approach. Organic amendments have been used, but the inconsistent results have prevented this method from being adopted in most cases (Streets and Bloss, 1973; Matocha *et al.*, 1988).

Chemical treatments

A number of soil fumigants and fungicides have been tested for their efficacy against *P. omnivorum* with very few demonstrating consistent and adequate control. Lyda *et al.* (1967) demonstrated that 1,3-dichloropropene applied at 700 l ha^{-1} in shallow treatments and 45, 90 or 140 l ha^{-1} applied deep gave good control of the fungus. Unfortunately, the economics of applying these quantities to row crops dictates that other control measures be examined. Anhydrous ammonia has been shown to be highly lethal to sclerotia *in vitro* (Rush and Lyda, 1982; Riggs and Lyda, 1987). However, field tests of this compound with deep chiselling were not an improvement over deep chiselling alone (Rush and Lyda, 1984). This compound does not disperse well in soil and is highly adsorbed to clays with a conversion to NH_4 (a non-toxic form), rendering the compound ineffective.

The pathogen has been shown to be highly sensitive to the triazole fungicides (Whitson and Hine, 1986; Riggs and Lyda, 1988). However, these compounds have herbicidal activity in cotton if applied to the seed. Field evaluations in Arizona demonstrated that foliar or granular side-dressing applications of propiconazole, six to nine weeks after planting, provided statistically significant control of root rot. Olsen and George (1987) applied propiconazole and triadimenol in subsurface drip irrigation for two years and found a significant reduction in dead plants. Slow-release formulations have been developed which overcome the toxicity effects in cotton, but the materials were inconsistent in controlling the pathogen in the field. There is a continued interest in these compounds as they are highly efficacious in rather small quantities.

Resistance

Growing resistant cultivars is considered the most effective means of controlling diseases of cotton (El-Zik and Frisbie, 1985). Gains have been made in combining small incremental increases in apparent resistance to Phymatotrichum root rot in agronomically acceptable cottons (El-Zik and Frisbie, 1985) but this may itself be an agronomic trait rather than representing constitutive resistance. Bird (1982) has developed a holistic multi-adversity resistance (MAR) system for genetic improvement of cotton.

Biological control

Mycoparasites of the sclerotia have been identified (Kenerley and Stack, 1987; Kenerley *et al.*, 1987), but a delivery method that allows for placement of the biocontrol agent at the site of overseasoning sclerotia (30–60 cm in soil profile) or agents that can colonize the site of infection for a sufficient period of time to interact with the pathogen have not been realized.

Charcoal Rot

HISTORY AND DISTRIBUTION

The charcoal rot fungus was first described on cotton, cowpea, jute and peanut in north-central India in the early 1900s, producing a dry rot of the main roots and lower stems of these crops. Since then, reports of the pathogen on cotton and other crops in tropical and sub-tropical countries have proliferated (Holliday and Punithalingam, 1970; Watkins, 1981).

The genus *Macrophomina* was established by Petrak (1923). Ashby (1927) proposed the binomial *Macrophomina phaseoli* (Maubl.) Ashby, which was commonly used until the review by Goidanich (1947) which led to the designation *Macrophomina phaseolina* (Tassi) Goid, now generally accepted. Synonyms for the pathogen include *Sclerotium bataticola* Taub and *Rhizoctonia bataticola* (Taub) Butl. (Holliday and Punithalingam, 1970). *M. phaseolina* is a seed- and soil-borne pathogen with both a wide distribution and a wide host range (Dhingra and Sinclair, 1978). Charcoal rot has been reported on cotton in the USA, from the south-eastern states and Oklahoma and Texas (Watkins, 1981), but is of little economic significance there compared with the Indian subcontinent, east and central Africa, and elsewhere in the tropics and sub-tropics. A considerable amount of work was done on root rot disease of cotton in India in the 1930s, covering environmental and edaphic effects, crop rotation and extent of losses (see, for example, Vasudeva and Ashraf, 1939), but much of this work did not distinguish between the charcoal rot fungus and *Rhizoctonia solani*, a tendency which has continued to recent times in the literature from the subcontinent.

Macrophomina phaseolina is also found affecting a range of non-cultivated plant species. In the Arizona Desert, for example, Milhail *et al.* (1989) found that the presence of several indigenous plant species was inversely related to population densities of the fungus.

SYMPTOMS

The most frequent symptoms of charcoal rot across a wide range of crops are a dry or wet dark rot of the lower stem (Holliday and Punithalingam, 1970). In cotton there is a discoloration of the stele in the taproot and pith of the main stem (Lee *et al.*, 1986); in severe cases there is a dissolution of stem and root tissues except for liquefied strands (Ghaffar and Erwin, 1969). Sclerotia of *M. phaseolina* in infected tissues are sometimes reported (Lee *et al.*, 1986). In cotton plants apparently suffering from a wilt syndrome in the Sudan, *M. phaseolina* and other fungi were found in addition to *Fusarium oxysporum* f.sp. *vasinfectum* (Yassin and Dafalla, 1982).

The charcoal rot fungus has been reported to cause severe leaf spot and blight disease on cotton. Bharathudu and Rao (1976) reported small pink-coloured circular spots on cotton cotyledons and then on leaves, causing premature leaf fall. Srivastava and Mor (1976) reported a yellowing/browning of cotton leaves initiated from the apical ends, followed by the appearance of minute pycnidia; isolations and infectivity tests were made. Isolates of *M. phaseolina* can also cause a leaf blight of mungbean and are implicated in a head rot on sunflower (Taneja and Grover, 1982).

As will be discussed in later sections, with this particular fungus it is important to distinguish clearly between factors influencing root infection and factors influencing symptom development in a range of crops (Bruton *et al.*, 1987).

LOSSES

Losses due to soil-borne fungal diseases of cotton, in general, are poorly documented over large areas. In the principal cotton-producing area (27 000 ha) of Piura, in Peru, losses due to soil-borne diseases amounted to about $500 000 out of a total loss estimated at $4 000 000 (Delgado and Agurto, 1984). *M. phaseolina* was cited as one of the pathogens contributing to this loss.

Chauhan (1986) tested systemic and non-systemic fungicides against root rot in Haryana, India. Yield levels in treated plots ranged from 1400 to 1900 kg ha^{-1} compared with mean values in control non-treated plots of 1300 kg ha^{-1}, indicating a potential loss of about 30%, although it should be noted that disease was still present in the treated plots. There have been anecdotal reports of losses in cotton relating to incidence and soil conditions (Thakar, 1984). In studies aimed at monitoring the survival of *M. phaseolina* in groundnut fields, it was found that severity was directly proportional to population densities of sclerotia and that yield was inversely proportional to severity within a growing season (Short *et al.*, 1980); similar detailed studies on yield losses have not been made with cotton.

CAUSAL ORGANISM

Taxonomy and morphology

Only the anamorphic stage is known. As noted above, the designation *Macrophomina phaseolina* (Tassi) Goid is now generally accepted. The following description of morphology is taken directly from Holliday and Punithalingam (1970) (and also see Fig. 5.2):

Fig. 5.2. *Macrophomina phaseolina.* A, Vertical section of pycnidium; B, part of pycnidial wall and conidiophores; C, conidiophores and young conidia; D, conidia; E, sclerotium; F section of sclerotium. (Holliday and Punithalingham, 1970.)

Sclerotia within roots, stems, leaves and fruits, black, smooth, hard, 100 μm – 1 mm diameter (in culture 50–300 μm). Pycnidia dark brown, solitary or gregarious on leaves and stems, immersed becoming erumpment, 100–200 μm diameter, opening by apical ostioles; pycnidial wall multicellular with heavily pigmented thick-walled cells on the outermost side. Conidiophores (phialides) hyaline, short, obpyriform to cylindrical, 5–13 × 4–6 μm. Conidia hyaline, ellipsoid to obvoid, 14–30 × 5–10 μm.

Cultural conditions and inoculum preparation

Pycnidia production in culture is obtained on propylene oxide-sterilized tissues (Goth and Ostazeski, 1965). Non-sporulating isolates from different hosts produced pycnidia on autoclaved leaf tissues of pearl millet and wheat under defined light conditions on water agar within two to seven days (Chidambaram and Mathur, 1975). In darkness, only sclerotia developed. Between five and 16 out of 50 isolates of *M. phaseolina*, originating from soil or *Phaseolus* seed, produced conidia on dried hypocotyl segments for different combinations of temperature and light (again found necessary); fluctuating temperature between 18 and 42 °C gave the best result (Watanabe, 1972). Knox-Davies (1965, 1966) produced pycnidia on groundnut meal irradiated with UV light and on filter paper treated with vegetable oil on peptone or asparagine agar.

Selective media have been developed and used for assaying *M. phaseolina* populations (Meyer *et al.*, 1973); such media typically give higher population estimates than those based on sclerotial flotation methods. Estimates of sample size necessary to optimize recovery of *M. phaseolina* from soil and sampling strategies have been determined (Campbell and Nelson, 1986; Mihail and Alcorn, 1987). Mycelial inoculum persisted in soil for up to seven days in the absence of a host. Cloud and Rupe (1991) compared three selective media for enumerating sclerotia. Khan (1972) reported that 'peak vigour and inoculum potential' were achieved with ten-day-old inoculum from culture and gave the highest percentage of disease in the shortest period of time.

Host range

The charcoal rot fungus is plurivorous, with a host range of over 290 plant species reported by Dhingra and Sinclair (1978).

Variation

There is a wide variation in isolate characteristics with some largely vegetative and others forming abundant sclerotia in culture. It has not been demonstrated that these differences are associated with host specializations. Isolates from a single host are normally found to attack a wide range of other hosts (Holliday and Punithalingam, 1970). Vilela and Delgado (1987) tested 18 isolates causing charcoal rot in cotton, growing them in culture and testing for pathogenicity. Significant differences in many characters were found but all isolates were highly pathogenic to cotton in northern Peru. Pearson and Schwerk (1986) found that isolates of *M. phaseolina* from maize grew on a medium with potassium chlorate, but that isolates from soyabean did not. They postulated that these differences might reflect metabolic activities associated with host specialization. Lee *et al.* (1986) tested isolates of *M. phaseolina* from bean, cotton and groundnut for pathogenicity on cotton. There was some variation in resistance but all isolates infected cotton. Diourte (1987) inoculated sorghum, groundnut, bean and cotton with isolates from each of these hosts. There was a general trend to host preference for the same-host isolate.

Disease cycle

The main form of transmission is from plant debris in the soil. Sclerotia are probably the main source of infection, although this can also occur from conidia (Thirumalachar, 1971). Mechanical wounds are not necessary for infection to occur (Lee *et al.*, 1986). The level of root infection of cotton was related to sclerotial populations at different depths but only low correlations were found (Hussain and Ghaffar, 1975). Sclerotial populations on residue buried deep in the soil were relatively stable (Short *et al.*, 1980), although overall there were fluctuations within season and over seasons. Population densities were directly related to the number of consecutive years of growing hosts. Short *et al.* (1980) reported a two-fold increase over three years. Survival of sclerotia produced *in vivo* on soyabean was longer than when produced in culture.

Large sclerotia produce more germ-tubes than small sclerotia (Short and Wyllie, 1978). Size of sclerotium was directly related to the number of cells per sclerotium and depended upon available nutrients in the substrate. Sclerotia germinated within 2–3 mm of soyabean seed surfaces and each produced between one and seven germ-tubes.

Epidemiology

Macrophomina phaseolina is a root inhabitant *sensu* Garrett (1956), who referred to its imposter role in several diseases. There may be some justification in this view but, equally, on cereals and legumes its status as a serious pathogen has been demonstrated many times in the sub-tropics and tropics, where it cannot be assumed that crops are 'nutritionally-balanced' or growing 'in good physical conditions' (Watkins, 1981). Its status on cotton is likely to be similar.

Much remains to be done in the area of epidemiology, in terms of environmental and nutritional effects and the interactions with other pathogenic organisms. For example, in the latter case, it has been claimed that, in the presence of *M. phaseolina*, the Fusarium wilt fungus has poor pathogenic competitive ability (Khan, 1972).

Disease has been claimed to be most severe at high temperatures, between 35 and 39 °C (Holliday and Punithalingam, 1970), and, indeed, the avoidance of high temperatures through early sowing and harvesting has been recommended for cotton (Thirumalachar, 1971). However, it is important to distinguish the effects of temperature on root infection, colonization and subsequent symptom development if a proper disease chronology is to be constructed with an epidemiological perspective (Bruton *et al.*, 1987). This has yet to be done for charcoal rot in cotton. There is much ambiguity with other crops. For example, maximum colonization of bean or wheat stems occurred at 15–20 °C and decreased above this temperature; at 15 °C wheat was more colonized and the converse was found at higher temperatures (Dhingra and Chagas, 1981). When cotton roots were inoculated with *M. phaseolina* in a greenhouse at 40 °C, severity was not as great as seen at similar temperatures in the field (Ghaffar and Erwin, 1969). At a soil temperature of 20–40 °C and with soil moisture stress, the severity was much greater than without soil moisture stress. Thus, it has been claimed that soil moisture effects are more important than temperature effects, with generally an inverse relationship with symptom expression (Thakar, 1984). However, again, distinction is not usually made between root infection *per se* and symptom development. It remains highly improbable, in our view, that low levels of soil moisture lead to high levels of root infection. For example, by disentangling environmental effects at different stages of an epidemic, Bruton *et al.* (1987) showed clearly that vine decline in cantaloup was directly related to soil matric potential prior to flowering; wet conditions at the critical time for infection led to subsequent high levels of disease. In cotton, however, by plating out root segments on PDA, it was found that percentage infection varied inversely with soil moisture (Sheikh and Ghaffar, 1979). Further experimental investigation is required to corroborate their finding (or conversely) and to establish the range of soil moisture over which it applies.

The effect of fertilization and soil nutrient status seems clearer. Nitrogen inhibits saprophytic colonization of plant tissues (Dhingra and Chagas, 1981) and leads to decline in *M. phaseolina* populations (Filho and Dhingra, 1980a, b). The decline depends on both the form of nitrogen applied and the soil type. In sandy clay loams, *M. phaseolina* populations declined with nitrogen applied as sodium nitrate or ammonium sulphate, but in sandy loams the effect of NO_3 was less than that of NH_3 (Filho and Dhingra, 1980b). The same authors also looked at the effect of soils amended with straw or nitrogen-enriched straw at different C : N ratios. In the former case, populations of *M. phaseolina* subsequently increased, but this did not occur in the latter case, where populations declined more rapidly initially, especially at low C : N ratios (Filho and Dhingra, 1980a).

As stated in the previous section, population densities of *M. phaseolina* build up over seasons (Short *et al.*, 1980) and there is a relationship between population densities and root infection (Hussain and Ghaffar, 1975). Further quantification was provided by Sheikh and Ghaffar (1979), who found that at 25% water-holding capacity in a sandy loam, pH 8.3, a density of 40 sclerotia g^{-1} gave an incidence of 50% infection in cotton plots. Such quantification, especially over a wider range of inoculum densities and soil types, is essential for future epidemiological investigations.

Some studies have looked at the spatial aspects of *M. phaseolina* population dynamics (Mihail, 1989; Olanya and Campbell, 1989). The density of *M. phaseolina* propagules (measured as colony-forming units of microsclerotia) did not vary significantly with host density within a season in maize (Olanya and Campbell, 1989) but did significantly increase after two successive crops of the latex-producing plant *Euphorbia lathyris* (Mihail, 1989). A range of analyses indicated a random or possibly slightly aggregated pattern of propagules in maize and an aggregated one in *E. lathyris*. Olanya and Campbell (1989) considered the more random pattern in maize was due to a lack of reproductive activity in the fungus and to decay of propagules during root growth, whereas, in a very susceptible host such as *E. lathyris*, the authors suggested that populations increase as host debris, incorporated into the soil, decomposes.

CONTROL

Cultural

Early sowing and harvesting to avoid extreme temperatures has been recommended as a means of avoiding severe charcoal rot (Thirumalachar, 1971), and, similarly, frequent irrigation to avoid soil moisture stress. The effects of maintaining nutritional status of the soil have been noted (Filho and Dhingra, 1980a,b; Dhingra and Chagas, 1981). Crop rotations can also be expected to play a part in control. Francl *et al.* (1988) grew soyabean in rotation with other crops, including maize and cotton, and compared the build-up of *M. phaseolina* populations compared with soyabean grown alone. Paradoxically, from the perspective of this chapter, the greatest relative reduction was obtained with cotton. This is easily resolved, however, by turning the finding around; in a normal cropping sequence of cotton it would not be desirable to introduce soyabean, which would tend to increase *M. phaseolina*, compared with a cotton monoculture.

Intercropping has also been found to be effective by using *Phaseolus aconitifolius*, which, it is claimed, acts by reducing soil temperatures (Rajpurghit, 1983). The control obtained was equal to that obtained by a range of different chemical treatments.

Chemical treatment

Pentachloronitrobenzene (PCNB) and carbendazim have been used as both seed and soil treatments to control cotton root rot in India, where normally *M. phaseolina* (as *Rhizoctonia bataticola*) is not distinguished from *R. solani*. Reductions in root rot reported ranged from 20 to 40% (Mathur and Singh, 1973). Seed and soil application treatments with a range of fungicides reduced cotton mortality from about 52% in controls to 25–30% (soil application) and 43% (seed-dressing).

Chauhan (1986) evaluated seed treatments of seven fungicides and selected pairwise combinations. The percentage root rot at harvest ranged from 14 to 22% in treated plots compared with 28% in the controls. Benzimidazole-derived fungicides differed in their effects on these isolates of *M. phaseolina* (as *R. bataticola*). Benomyl and carbendazim were most inhibitory but a sesame isolate was least sensitive to all fungicides.

Mathur *et al.* (1971) reported adaptation of *M. phaseolina* (as *R. bataticola*) to PCNB and carboxin when sequentially cultured on increasing concentrations. Cross-resistance to both fungicides did not develop.

Resistance

Resistance to *M. phaseolina* has not been found in cotton (Holliday and Punithalingam, 1970). Sources of resistance were considered completely lacking in Pakistan (Akhtar, 1977). Tiwari and Shroff (1982) looked at the variation in root rot incidence (measured as mean percentage mortality) in 11 cotton entries. Unfortunately, again, no distinction was made between *M. phaseolina* (as *R. bataticola*) and *R. solani* in these screens.

More recently, Lee *et al.* (1986) reported variation in resistance levels in cotton infected with a range of *M. phaseolina* isolates. It is not clear whether this variation represents usable resistance in a breeding programme.

Biological control

The lack of sources of resistance has led to the search for antagonistic micro-organisms (Akhtar, 1977). Butt and Ghaffar (1972) found that the fungus *Stachybotrys atra* controlled *M. phaseolina*, as well as other pathogens, in culture plates but not in soil. Unfortunately, there were also phytotoxic effects on cotton. *Rhizobium* strains indigenous to Pakistan were found to inhibit *M. phaseolina* growth on culture plates. Seeds dipped in a bacterial cell suspension gave a significant reduction in root rot of mungbean, okra and sunflower in greenhouse pot experiments (Zaki and Ghaffar, 1972). The search for direct antagonistic effects on *M. phaseolina* has been limited to date. A more profitable venture might be the elucidation of tillage, rotational and nutritional effects on soil-borne pathogens, mediated through ecological interactions with a wide range of soil biota (De Vay *et al.*, 1989; Rickerl *et al.*, 1989).

Other Root and Stem Diseases

Several of the fungi contributing to the seedling disease complex have been implicated in disease of older plants. This is only to be expected as older surviving plants may still bear lesions of the original seedling disease. A good example is the soreshin phase of *Rhizoctonia solani* infection, with reddish brown, sunken lesions on roots at or below the soil surface (Walla and Bird, 1981). The practice commented on earlier in this chapter of not distinguishing between *R. solani* and *M. phaseolina* (as *R. bataticola*) in cotton root rot in India makes appraisal of the effect of *R. solani* beyond the seedling stage very difficult. It should be noted in passing that reported effects of *Rhizoctonia solani* on Verticillium wilt are contradictory in the extreme (cf. Khoury and Alcorn, 1973; Srinivasan and Kannan, 1975).

A range of *Fusarium* spp. is implicated in the cotton seedling disease complex – mainly *F. oxysporum* f.sp. *vasinfectum*, although *F. moniliforme* and *F. solani* are consistently isolated from diseased seedlings and roots of older diseased plants (Chapter 2). Species of *Pythium* are well-known cotton seed and seedling pathogens, mainly *P. ultimum* (DeVay, 1982). Some species can cause root necrosis and subsequent stunting of mature plants, especially *P. irregulare* (Roncadori and McCarter, 1972a,b; Roncadori *et al.*, 1974). *P. irregulare* is most active as a root rot pathogen in the early spring but the length of time that symptoms persist is dependent upon favourable environmental conditions after infection. Although stunted plants may outgrow the condition, the fruiting cycle may be delayed and yield reduced. There may be a significant interaction of Pythium root rot with the plant-parasitic nematode *Hoplolaimus columbus* Sher. (Roncadori *et al.*, 1974).

Thielaviopsis basicola, which is implicated with the cotton seedling disease complex, also causes black root rot, which can affect older plants (Watkins, 1981; Chapter 2). This pathogen is of particular interest in that fungicidal control is not currently possible, despite extensive evaluation of chemicals for control of seedling disease (Minton *et al.*, 1982). As the fungus has a wide host range, it is not surprising that it also affects a range of weed species commonly found in cotton fields (Klimova, 1979). The black root rot disease appears to be most important in eastern Europe, for example as in the last-cited study by Klimova (1979).

Cylindrocladium crotalariae causes Cylindrocladium black rot of groundnut (Rowe and Beute, 1973), a disease of relatively recent origin in the United States. Studies were made on the host range of this fungus, including tobacco, cotton, maize and small grains. It was found that, generally, cotton was not visibly affected when inoculated, except in a few plants, but that the fungus was reisolated in many cases and groundnut planted following cotton was consequently infected.

Diseases caused by *Sclerotium rolfsii* occur throughout the tropics and sub-tropics, affecting many dicotyledonous crops and several monocotyledonous species (Aycock, 1966; Punja, 1985). The fungus has been reported as affecting cotton, especially as a consequence of other pest management practices. The use of herbicide EPTC increased the severity of *S. rolfsii* in cotton (Peeples *et al.*, 1976). On the other hand, a range of other herbicides have been shown to reduce disease levels in cotton (Youssef *et al.*, 1985). The use of the nematicide aldicarb in groundnut and cotton increased yields and its use in groundnut following cotton, similarly treated, reduced the incidence of southern blight caused by *S. rolfsii*. Again, somewhat paradoxically, as in the case of *Macrophomina phaseolina* and the soyabean–cotton rotation, cotton–groundnut rotations were considered useful for reducing the incidence of southern blight. The presence of *S. rolfsii* (and *R. solani*) in the rhizosphere of cotton plants has been claimed to increase microflora counts, including total bacterial counts, actinomycetes and fungal plate counts (Neweigy *et al.*, 1983).

Several foliar pathogens cause lesions on stems of mature plants (Watkins, 1981). Bacterial blight, caused by *Xanthononas campestris* pv. *malvacearum*, can cause cankers on stems called blackarm, which, when severe, can girdle the stems. Similar cankers can be caused by Asochyta blight, caused by *Asochyta* spp.

Colletotrichum gossypii (although not listed as an accepted taxon by Sutton, 1991) causes anthracnose lesions on all plant parts, including seedling stems. In South America, notably in Brazil and Venezuela, a witches'-broom disease of cotton is present which results from the abnormal growth of otherwise quiescent axillary and terminal buds (Costa and Fraga, 1937; Malaguti, 1955). The causal agent is claimed to be a form of the anthracnose pathogen, *C. gossypii* var. *cephalosporioides*.

References

AKHTAR, C.M. (1977) Biological control of some plant diseases lacking genetic resistance of the host crops in Pakistan. *Annals of the New York Academy of Sciences* 287, 45–56.

ALDERMAN, S.C. AND HINE, R.B. (1982) Vertical distribution in soil of an induction of disease by strands of *Phymatotrichum omnivorum*. *Phytopathology* 72, 409–12.

ALDERMAN, S.C. AND STOWELL, L.J. (1986) Strand ontogeny in *Phymatotrichum omnivorum*. *Transactions of the British Mycological Society* 86, 207–11.

ASHBY, S.F. (1927) *Macrophomina phaseoli* (Maubl.) comb. nov.: the pycnidial stage of *Rhizoctonia bataticola* (Taub.) Butl. *Transactions of the British Mycological Society* 12, 141–7.

AYCOCK, R. (1966) Stem rot and other diseases caused by *Sclerotium rolfsii*, or the status of Rolf's disease after 70 years. *North Carolina Agricultural Experimental Station Technical Bulletin* No. 174.

BHARATHUDU, C. AND RAO, A.S. (1976) A new leaf spot and blight on *Gossypium barbadense* caused by *Macrophomina phaseolina*. *Indian Phytopathology* 29, 440.

BIRD, L.S. (1974) The dynamics of cotton seedling disease. In: *The Relation of Soil Micro-organisms to Soilborne Plant Pathogens, Southern Cooperative Series Bulletin* No. 183. Virginia Polytechnic Institute and State University, Blacksburg, Virginia, pp. 75–80.

BIRD, L.S. (1982) The MAR (multi-adversity resistance) system for genetic improvement of cotton. *Plant Disease* 66, 172–6.

BLANK, L.M. (1953) The rot that attacks 2000 species. In: Stefferud, A. (ed.), *Plant Diseases*. USDA Yearbook Agriculture, pp. 298–301.

BLOSS, H.E. AND WHEELER, J.E. (1975) Influence of nutrients and substrata on formation of strands and sclerotia by *Phymatotrichum omnivorum*. *Mycologia* 67, 303–10.

BRINKERHOFF, L.A. AND STREETS, R.B. (1946) Pathogenicity and pathological histology of *Phymatotrichum omnivorum* in a woody perennial pecan. *Arizona Experimental Station Bulletin* 111, 103–26.

BRUTON, B.D., JEGER, M.J. AND REUVENI, R. (1987) *Macrophomina phaseolina* infection and vine decline in cantaloupe in relation to planting date, soil environment and plant maturation. *Plant Disease* 71, 259–63.

BUTT, Z.L. AND GHAFFAR, A. (1972) Inhibition of fungi, actinomycetes and bacteria by *Stachybotrys atra*. *Mycopathologia et Mycologia Applicata* 47, 241–51.

CAMPBELL, C.L. AND NELSON, L.A. (1986) Evaluation of an assay for quantifying populations of sclerotia of *Macrophomina phaseolina* from soil. *Plant Disease* 70, 645–7.

CHAUHAN, M.S. (1986) Systemic and non-systemic fungicides against root rot of cotton in Haryana. *Indian Journal of Mycology and Plant Pathology* 16, 226–7.

CHAVEZ, H.B., MCINTOSH, T.H. AND BOYLE, A.M. (1967) Greenhouse infection of cotton by *Phymatotrichum omnivorum*. *Plant Disease Reporter* 51, 926–7.

CHIDAMBARAM, P. AND MATHUR, S.B. (1975) Production of pycnidia by *Macrophomina phaseolina*. *Transactions of the British Mycological Society* 64, 165–8.

CLOUD, G.L. AND RUPE, J.C. (1991) Comparison of three media for enumeration of sclerotia of *Macrophomina phaseolina*. *Plant Disease* 75, 771–2.

COSTA, A.S. AND FRAGA, C. (1937) Superbrotamento ou ramulose de algodoeiro. *Piracicaba, Brazil* 12, 249–59.

CURTIS, G.W. (1892) Alfalfa root rot. *Texas Agricultural Experiment Station Bulletin* 22, 211–15.

DANA, B.F. (1931) Soil cultures for the laboratory production of sclerotia in *Phymatotrichum omnivorum*. *Phytopathology* 21, 551–6.

DELGADO, M.A. AND AGURTO, V.R. (1984) Estimado del monto de danos producidos por patogenos radioulares y fibrovasculares mediante mustreos, ana lisis y evaluacioues periodicas de la poblacio de plantas de differentes zonas algodoneras de piura. *Fitopatologia* 19, 27–38.

DEVAY, J.E. (1982) Role of *Pythium* species in the seedling disease complex of cotton in California. *Phytopathology* 82, 151–4.

DEVAY, J.E., EL-ZIK, K.M., BOURLAND, F.M., GARBER, R.H., KAPPELMAN, A.J., LYDA, S.D., MINTON, E.B., ROBERTS, P.A. AND WALLACE, T.P. (1989) Strategies and tactics for managing plant pathogens and nematodes. In: Frisbie, R.E., El-Zik, K.M. and Wilson, L.T. (eds), *Integrated Pest Management Systems and Cotton Production*. John Wiley & Sons, New York, pp. 225–66.

DHINGRA, O.D. AND CHAGAS, D. (1981) Effect of soil temperature, moisture and nitrogen on competitive saprophytic ability of *Macrophomina phaseolina*. *Transactions of the British Mycological Society* 77, 15–20.

DHINGRA, O.D. AND SINCLAIR, J.B. (1978) *Biology and Pathology of* Macrophomina phaseolina. Imprensia Universidade Federal de Viscosa, Brazil. 166 pp.

DIOURTE, M. (1987) Pathogenic variation and morphological studies of *Macrophomina phaseolina* (Tossi) Goid. M.Sc. Thesis, Texas ACM University. 48 pp.

DUGGAR, B.M. (1916) The Texas root rot fungus and its conidial stage. *Annals of the Missouri Botanical Gardens* 3, 11–23.

EL-ZIK, K.M. (1985) Integrated control of Verticillium wilt of cotton. *Plant Disease* 69, 1025–32.

EL-ZIK, K.M. AND FRISBIE, R.E. (1985) Integrated crop management systems for pest control. In: Mandova, N.B. (ed.), *CRC Handbook of Natural Pesticides: Methods. Volume 1. Theory, Practice and Detection*. CRC Press, Inc., Boca Raton, Florida, pp. 21–122.

ERGLE, D.R. AND BLANK, L.M. (1947) A chemical study of the mycelium and sclerotia of *Phymatotrichum omnivorum*. *Phytopathology* 37, 153–61.

EZEKIEL, W.N. (1945) Effect of low temperatures on survival of *Phymatotrichum omnivorum*. *Phytopathology* 35, 296–301.

EZEKIEL, W.N., TAUBENHAUS, J.J. AND FUDGE, J.F. (1932) Growth of *Phymatotrichum omnivorum* in plant juices as correlated with resistance of plants to root rot. *Phytopathology* 22, 459–74.

FILHO, E.S. AND DHINGRA, O.D. (1980a) Population changes of *Macrophomina phaseolina* in amended soils. *Transactions of the British Mycological Society* 74, 471–81.

FILHO, E.S. AND DHINGRA, O.D. (1980b) Survival of *Macrophomina phaseolina* sclerotia in nitrogen amended soils. *Phytopathologische Zeitschrift* 97, 136–43.

FRANCL, L.J., WYLLIE, T.D. AND ROSENBROCK, S.M. (1988) Influence of crop rotation on population density of *Macrophomina phaseolina* in soil infested with *Heterodera glycines*. *Plant Disease* 72, 760–4.

GARRETT, S.D. (1956) *Biology of Root-infecting Fungi.* Cambridge University Press, Cambridge.

GERIK, T.J., RUSH, C.M. AND JEGER, M.J. (1985) Optimizing plot size for field studies of *Phymatotrichum* root rot of cotton. *Phytopathology* 75, 240–3.

GHAFFAR, A. AND ERWIN, D.C. (1969) Effect of soil water stress on root rot of cotton caused by *Macrophomina phaseoli. Phytopathology* 59, 795–7.

GOIDANICH, G. (1947) Revisione del genere *Macrophomina* Petrak. Species tipica: *Macrophomina phaseolina* (Tassi), G. Goid. n. comb. nec *M. phaseoli* (Maubl) Ashby. *Annali della Sperimentazinoe Agraria Rom* NS 1, 449–61.

GOTH, R.W. AND OSTAZESKI, S.A. (1965) Sporulation of *Macrophomina phaseoli* on propylene oxide-sterilized leaf tissues. *Phytopathology* 55, 1156.

GUNASEKARAN, M. (1973a) Physiological studies on *Phymatotrichum omnivorum*. IV. Effect of pH and the interaction of temperature, minerals and carbon source on growth *in vitro. Mycopathologia et Mycologia Applicata* 50, 249–53.

GUNASEKARAN, M. (1973b) Physiological studies on *Phymatotrichum omnivorum*. V. Effect of interaction of carbon dioxide, minerals and carbon source on growth *in vitro. Mycopathologia et Mycologia Applicata* 50, 313–21.

GUNASEKARAN, M. (1980) Physiological studies on *Phymatotrichum omnivorum*. XI. Cellulolytic enzymes. *Mycologia* 72, 759–66.

GUNASEKARAN, M. (1982) Physiological studies on *Phymatotrichum omnivorum*. XIII. cAMP phosphodiesterase. *FEMS Letters* 13, 83–6.

GUNASEKARAN, M. AND WEBER, D.J. (1974) Physiological studies on *Phymatotrichum omnivorum*. VI. Lipid composition of mycelium and sclerotia. *Mycopathologia et Mycologia Applicata* 52, 261–6.

GUNASEKARAN, M., HESS, W.M. AND WEBER, D.J. (1974) Lipids and ultrastructure of *Phymatotrichum omnivorum. Transactions of the British Mycological Society* 63, 519–25.

HENNEBERT, G.L. (1973) *Botrytis* and *Botrytis*-like genera. *Persoonia* 7, 183–204.

HILL, T.F. AND LYDA, S.D. (1976) Gas exchange by *Phymatotrichum omnivorum* in a closed, axenic system. *Mycopathologia* 59, 143–7.

HOLLIDAY, P. AND PUNITHALINGAM, E. (1970) *Macrophomina phaseolina. CMI Descriptions of Pathogenic Fungi and Bacteria* No. 275. Commonwealth Agricultural Bureaux, Farnham Royal, Bucks, UK.

HUSSAIN, T. AND GHAFFAR, A. (1975) Colonisation of *Macrophomina phaseolina* on cotton roots in relation to sclerotial population in soil. *Pakistan Journal of Botany* 7, 85–7.

JEGER, M.J. AND LYDA, S.D. (1986) Epidemics of Phymatotrichum root rot (*Phymatotrichum omnivorum*) in cotton: environmental correlates of final incidence and forecasting criteria. *Annals of Applied Biology* 109, 523–34.

JEGER, M.J., KENERLEY, C.M., GERIK, T.J. AND KOCH, D.O. (1987) Spatial dynamics of Phymatotrichum root rot in row crops in the Blackland region of north central Texas. *Phytopathology* 77, 1647–56.

JUPE, E.R., MAGILL, J.M. AND MAGILL, C.W. (1986) Stage-specific DNA methylation in a fungal plant pathogen. *Journal of Bacteriology* 165, 420–3.

KEITHLY, J.H. (1985) The physiology of parasitism by *Phymatotrichum omnivorum* (Shear) Duggar. Ph.D. Thesis, Texas A&M University. 83 pp.

KENERLEY, C.M. AND JEGER, M.J. (1990) Root colonization by *Phymatotrichum omnivorum* and symptom expression of Phymatotrichum root rot in cotton in relation to planting date, soil temperature and soil water potential. *Plant Pathology* 39, 489–500.

KENERLEY, C.M. AND STACK, J.P. (1987) Influence of assessment methods on selection of fungal antagonists of the sclerotium-forming fungus *Phymatotrichum omnivorum*. *Canadian Journal of Microbiology* 33, 632–5.

KENERLEY, C.M., JEGER, M.J., ZUBERER, D.A. AND JONES, R.W. (1987) Populations of fungi associated with sclerotia of *Phymatotrichum omnivorum* buried in Houston black clay. *Transactions of the British Mycological Society* 89, 437–45.

KHAN, I.D. (1972) Effect of peak vigour and inoculum potential on competitive pathogenic ability and interaction of four cotton root pathogens in soil. *Zeitschrift für Pflanzenkrankheiten und Pflanzenschutz* 79, 714–28.

KHOURY, F.Y. AND ALCORN, S.M. (1973) Influence of *Rhizoctonia solani* on the susceptibility of cotton plants to *Verticillium albo-atrum* and on root carbohydrates. *Phytopathology* 63, 352–8.

KING, C.J. AND HOPE, C. (1932) Distribution of the cotton root rot fungus in soil and plant tissues in relation to control by disinfectants. *Journal of Agricultural Research* 45, 725–40.

KING, C.J. AND LOOMIS, H.F. (1929a) Cotton root rot investigations in Arizona. *Journal of Agricultural Research* 39, 199–221.

KING, C.J. AND LOOMIS, H.F. (1929b) Further studies of cotton rot in Arizona, with a description of a sclerotium stage of the fungus. *Journal of Agricultural Research* 39, 641–76.

KING, C.J., LOOMIS, H.F. AND HOPE, C. (1931) Studies on sclerotia and mycelial strands of the cotton root-rot fungus. *Journal of Agricultural Research* 42, 827–40.

KLIMOVA, A.P. (1979) *Thielaviopsis basicola* (Berk. et Br.) Ferraris on weed plants of cotton fields in Turkmenistan. *Isvestya Akademii Nauk Turkmenskoi SSR Biologichestih Nauk* 5, 53–8.

KNOX-DAVIES, P.S. (1965) Pycnidium production by *Macrophomina phaseoli*. *South African Journal of Agricultural Science* 8, 205–18.

KNOX-DAVIES, P.S. (1966) Further studies on pycnidium producing by *Macrophomina phaseoli*. *South African Journal of Agricultural Science* 9, 595–600.

KOCH, D.O., JEGER, M.J., GERIK, T.J. AND KENERLEY, C.M. (1987) Effects of plant density on progress of Phymatotrichum root rot in cotton. *Phytopathology* 77, 1657–62.

LEE, C.C., BIRD, L.S., THAXTON, P.M. AND HOWELL, M.L. (1986) The association of *Macrophomina phaseolina* with cotton. *Acta Phytophylactica Sinica* 13, 169–73.

LYDA, S.D. (1976) Optimizing mycelial respiration of *Phymatotrichum omnivorum*. *Mycologia* 68, 1011–19.

LYDA, S.D. (1978) Ecology of *Phymatotrichum omnivorum*. *Annual Review of Phytopathology* 16, 193–209.

LYDA, S.D. (1984) Vertical and horizontal distribution of *Phymatotrichum* sclerotia in Texas soils. *Phytopathology* 74, 814 (Abstr.).

LYDA, S.D. AND BURNETT, E. (1970) Sclerotial inoculum density of *Phymatotrichum omnivorum* and development of Phymatotrichum root rot in cotton. *Phytopathology* 70, 729–31.

LYDA, S.D. AND BURNETT, E. (1971a) Influence of temperature on *Phymatotrichum* sclerotial formation and disease development. *Phytopathology* 61, 728–30.

LYDA, S.D. AND BURNETT, E. (1971b) Changes in carbon dioxide levels during sclerotial formation by *Phymatotrichum omnivorum*. *Phytopathology* 61, 858–61.

LYDA, S.D. AND BURNETT, E. (1975) The role of carbon dioxide in growth and survival of *Phymatotrichum omnivorum*. In: BRUEHL, G.W. (ed.) *Biology and Control of Soil Borne Plant Pathogens*. American Phytopathological Society, St Paul, pp. 63–8.

LYDA, S.D., ROBISON, G.D. AND LEMBRIGHT, H.W. (1967) Soil fumigation control of Phymatotrichum root rot in Nevada. *Plant Disease Reporter* 51, 331–3.

McNAMARA, H.C. and HOOTON, C.R. (1933) Sclerotia-forming habits of the cotton root-rot fungus in Texas black-land soils. *Journal of Agricultural Research* 46, 807–19.

McNAMARA, H.C., HOOTON, C.R. AND PORTER, D.D. (1931) Cycles of growth in cotton root rot at Greenville, Texas. *US Department of Agriculture Circular* 173. 17 pp.

MALAGUTI, G. (1955) La escobilla del algodon en Venezuela. *Agronomia Tropical (Maracay, Venezuela)* 5, 73–86.

MATHUR, R.L. AND SINGH, R.R. (1973) Control of root rot of cotton caused by *Rhizoctonia bataticola* (Taub) Butler with soil application of Brassicol. *Science and Culture* 39, 221–2.

MATHUR, R.L., MASIH, B. AND SANKHLA, H.C. (1971) Adaptability of *Alternaria burnsii* and *Rhizoctonia bataticola* to fungicides. *Indian Phytopathology* 24, 548–52.

MATOCHA, J.E., MOSTAGHIMI, S. AND CRENSHAW, C. (1988) Effect of soil amendments on cotton growth and incidence of *Phymatotrichum* root rot. *Proceedings of the Beltwide Cotton Production Research Conference* 48, 41–5.

MEYER, W.A., SINCLAIR, J.B. AND KHARE, M.N. (1973) Biology of *Macrophomina phaseoli* in soil studied with selective media. *Phytopathology* 63, 613–20.

MIHAIL, J.D. (1989) *Macrophomina phaseolina*: spatio-temporal dynamics of inoculum and of disease in a highly susceptible crop. *Phytopathology* 79, 848–55.

MIHAIL, J.D. AND ALCORN, S.M. (1987) *Macrophomina phaseolina*: spatial patterns in a cultivated soil and sampling strategies. *Phytopathology* 77, 1126–31.

MIHAIL, J.D., ORUM, T.V. AND ALCORN, S.M. (1989) *Macrophomina* in the Sonaran desert. *Canadian Journal of Botany* 67, 76–82.

MINTON, E.B., PAPAVIZAS, G.C. AND LEWIS, J.A. (1982) Effect of fungicide seed treatments and seed quality on seedling diseases and yield of cotton. *Plant Disease* 66, 832–5.

MOORE, E.J. (1937) Carbon and oxygen requirements of the cotton root-rot organism *Phymatotrichum omnivorum*, in culture. *Phytopathology* 27, 918–30.

MUELLER, J.P., HINE, R.B., PENNINGTON, D.A. AND INGLE, S.J. (1983) Relationship of soil cations to the distribution of *Phymatotrichum omnivorum*. *Phytopathology* 73, 1365–8.

MULREAN, E.N., HINE, R.B. AND MUELLER, J.P. (1984) Effect of Phymatotrichum root rot on yield and seed and lint quality in *Gossypium hirsutum* and *Gossypium barbadense*. *Plant Disease* 68, 381–3.

MURASHIGE, T. AND SKOOG, F. (1962) A revised medium for rapid growth and bioassays with tobacco tissue culture. *Physiologia Plantarum* 15, 473–97.

NEAL, D.C., WESTER, R.E. AND GUNN, K.C. (1934) Morphology and life history of the cotton root-rot fungus in Texas. *Journal of Agricultural Research* 49, 539–48.

NEWEIGY, N.A., EISA, N.A., ZIEDAN, M.I. AND FELAIFEL, M.S.A. (1983) Microflora of soil and rhizosphere of cotton plants Giza 70, as influenced by *Rhizoctonia solani* and *Sclerotium rolfsii* and fungicide application. *Annals of Agricultural Science (Faculty of Agriculture, Ain Shams University, Cairo, Egypt)* 28, 167–82.

OLANYA, O.M. AND CAMPBELL, C.L. (1989) Density and spatial pattern of propagules of *Macrophomina phaseolina* in corn rhizospheres. *Phytopathology* 79, 1119–23.

OLSEN, M.W. AND GEORGE, S. (1987) Applications of systemic fungicides through subsurface drip irrigation for control of *Phymatotrichum* root rot of cotton. *Phytopathology* 77, 1748.

OLSEN, M.W., MISAGHI, I.J., GOLDSTEIN, D. AND HINE, R.B. (1983) Water relations in cotton plants infected with *Phymatotrichum*. *Phytopathology* 73, 213–16.

PAMMEL, L.H. (1888) Root rot of cotton, or 'cotton blight'. *Texas Agricultural Experimental Station Bulletin* 4, 50–65.

PAMMEL, L.H. (1889) Cotton root rot. *Texas Agricultural Experimental Station Bulletin* 7, 61–91.

PEARSON, C.A.S. AND SCHWERK, F.W. (1986) Variable chlorate resistance in *Macrophomina phaseolina* from corn, soyabean and soil. *Phytopathology* 76, 646–9.

PEEPLES, J.L., CURL, E.A. AND RODRIGUEZ-KABANA, R. (1976) Effect of the herbicide EPTC on the biocontrol activity of *Trichoderma viride* against *Sclerotium rolfsii*. *Plant Disease Reporter* 60, 1050–4.

PELTIER, G.L., KING, C.J. AND SAMSON, R.W. (1926) Ozonium root rot. *US Department of Agriculture Bulletin* 1417. 28 pp.

PERCY, R.G. (1983) Potential range of *Phymatotrichum omnivorum* as determined by edaphic factors. *Plant Disease* 67, 981–3.

PERCY, R.G. AND RUSH, C.M. (1985) Evaluation of four upland cotton genotypes for a rate-limiting resistance to Phymatotrichum root rot. *Phytopathology* 75, 463–6.

PETRAK, F. (1923) Mykologische Notizen, VI. *Annales Mycologici* 21, 314–15.

PUNJA, Z.K. (1985) The biology, ecology and control of *Sclerotium rolfsii*. *Annual Review of Phytopathology* 23, 97–127.

RAJPURGHIT, T.S. (1983) Control of root rot of cotton caused by *Rhizoctonia bataticola* (Taub) Butler. *Madras Agricultural Journal* 70, 751–4.

RICKERL, D.H., CURL, E.A. AND TOUCHTON, J.T. (1989) Tillage and rotation effects on *Collembola* populations and *Rhizoctonia* infestation. *Soil and Tillage Research* 15, 41–9.

RIGGS, J.L. AND LYDA, S.D. (1987) Response of sclerotia of *Phymatotrichum omnivorum* to five soil fumigants. *Proceedings of the Beltwide Cotton Production Research Conference* 47, 46–7.

RIGGS, J.L. AND LYDA, S.D. (1988) Laboratory and field tests with triazole fungicides to control *Phymatotrichum* root rot of cotton. *Proceedings of the Beltwide Cotton Production Research Conference* 48, 45–8.

RIGGS, J.L. AND LYDA, S.D. (1989) Genetic and pathogenicity differences among several isolates of *Phymatotrichum omnivorum*. *Proceedings of the Beltwide Cotton Production Research Conference* 49, 23–7.

ROGERS, C.H. (1942) Cotton root rot studies with special reference to sclerotia, cover crops, rotations, tillage, seedling rates, soil fungicides, and effects on seed quality. *Texas Agricultural Experimental Station Bulletin* 614. 45 pp.

ROGERS, C.H. AND WATKINS, G.M. (1938) Strand formation in *Phymatotrichum omnivorum*. *American Journal of Botany* 25, 244–6.

RONCADORI, R.W. AND McCARTER, S.M. (1972a) Effect of Pythium root rot on vegetative growth, time of fruiting and yield of cotton. *Proceedings of the Beltwide Cotton Production Research Conference* 32, 24–5.

RONCADORI, R.W. AND McCARTER, S.M. (1972b) Effect of soil treatment, soil temperature and plant age of Pythium root rot of cotton. *Phytopathology* 62, 373–6.

RONCADORI, R.W., LEHMAN, P.S. AND McCARTER, S.M. (1974) Effect of *Phythium irregulare* on cotton growth and yield and joint action with other soil-borne pathogens. *Phytopathology* 64, 1303–6.

ROWE, R.C. AND BEUTE, M.K. (1973) Susceptibility of peanut rotational crops (tobacco, cotton and corn) to *Cylindrocladium crotalariae*. *Plant Disease Reporter* 57, 1035–9.

RUSH, C.M. AND LYDA, S.D. (1982) The effects of anhydrous ammonia on membrane stability of *Phymatotrichum omnivorum*. *Mycopathologia* 79, 147–52.

RUSH, C.M. AND LYDA, S.D. (1984) Evaluation of deep-chiselled anhydrous ammonia as a control for Phymatotrichum root rot of cotton. *Plant Disease* 68, 291–3.

RUSH, C.M., LYDA, S.D. AND GERIK, T.J. (1984a) The relationship between time of cortical senescence and foliar symptom development of Phymatotrichum root rot of cotton. *Phytopathology* 74, 1464–6.

RUSH, C.M., GERIK, T.J. AND LYDA, S.D. (1984b) Factors affecting symptom appearance and development of Phymatotrichum root rot of cotton. *Phytopathology* 74, 1466–9.

RUSH, C.M., GERIK, T.J. AND LYDA, S.D. (1984c) Interactions between *Phymatotrichum omnivorum* and *Sorghum bicolor*. *Plant Disease* 68, 500–1.

RUSH, C.M., GERIK, T.J. AND KENERLEY, C.M. (1985) Atypical disease symptoms associated with Phymatotrichum root rot of cotton. *Plant Disease* 69, 534–7.

SAMBANDAM, T. AND GUNASEKARAN, M. (1987) Interaction of glycogen phosphorylase, glycogen synthase and phosphorylase kinase in *Phymatotrichum omnivorum*. *Mycologia* 79, 486–8.

SHEAR, C.L. (1907) New species of fungi. *Torrey Botanical Club Bulletin* 34, 305–6.

SHEIKH, A.H. AND GHAFFAR, A. (1979) Relation of sclerotial inoculum density and soil moisture to infection of field crops by *Macrophomina phaseolina*. *Pakistan Journal of Botany* 11, 185–9.

SHORT, G.E. AND WYLLIE, T.D. (1978) Inoculum potential of *Macrophomina phaseolina*. *Phytopathology* 68, 742–6.

SHORT, G.E., WYLLIE, T.D. AND BRISTOW, P.R. (1980) Survival of *Macrophomina phaseolina* in soil and in residue of soyabean. *Phytopathology* 70, 13–17.

SMITH, R.B. AND HALLMARK, C.T. (1987) Selected chemical and physical properties of soils manifesting cotton root rot. *Agronomy Journal* 79, 155–9.

SRINIVASAN, K.V. AND KANNAN, A. (1975) Suppression of Verticillium wilt by Rhizoctonia root rot of cotton. *Current Science* 44, 354–5.

SRIVASTAVA, M.P. AND MOR, B.R. (1976) Leaf blight of cotton caused by *Macrophomina phaseolina*. *Current Science* 45, 351.

STREETS, R.B. AND BLOSS, H.E. (1973) *Phymatotrichum Root Rot*. APS Monograph No. 8, American Phytopathological Society, St Paul.

SUTTON, B.C. (1992) The genus *Glomerella* and its anamorph *Colletotrichum*. In: BAILEY, J.A. AND JEGER, M.J. (eds), Colletotrichum: *Biology, Pathology and Control*. CAB International, Wallingford, UK, pp. 1–26.

TANEJA, M. AND GROVER R.K. (1982) Efficacy of benzimidazole and related fungicides against *Rhizoctonia solani* and *R. bataticola*. *Annals of Applied Biology* 100, 425–32.

TAUBENHAUS, J.J. AND DANA, B.F. (1928) The influence of moisture and temperature on cotton root rot. *Texas Agricultural Experimental Station Bulletin* 386. 23 pp.

TAUBENHAUS, J.J. AND EZEKIEL, W.N. (1930) Studies on the overwintering of Phymatotrichum root rot. *Phytopathology* 20, 761–85.

TAUBENHAUS, J.J. AND EZEKIEL, W.N. (1932) Resistance of monocotyledons to Phymatotrichum root rot. *Phytopathology* 22, 443–52.

TAUBENHAUS, J.J., DANA, B.F. AND WOLFF, S.E. (1929) Plants susceptible or resistant to cotton root rot and their relation to control. *Texas Agricultural Experimental Station Bulletin* 393. 30 pp.

THAKAR, N.A. (1984) Influence of moisture on root rot of cotton. *Madras Agricultural Journal* 71, 629–30.

THIRUMALACHAR, M.J. (1971) Epidemiology and control of charcoal rot of some crop plants in India. *Proceedings of the Indian Natural Sciences Academy* 37, 394–8.

TIWARI, A. AND SHROFF, V.N. (1982) Differential reaction of cotton genotypes to root rot. *Indian Phytopathology* 35, 514–15.

VASUDEVA, R.S. AND ASHRAF, M. (1939) Studies on the root rot disease of cotton in the Punjab. VII. Further investigation of factors influencing incidence of the disease. *Indian Journal of Agricultural Sciences* 9, 595–608.

VILELA, V., DEL PILAR, M. AND DELGADO, J.M.A. (1987) Caracterizacion patogenica y cultural de diferentes aislemientos de *Macrophomina phaseolina* (Tassi) Goid, agente causal de la pudriccion carbonosa de laraiz del algodonero (*Gossypium barbadense* L.) en las condicioues de Piura, Peru. *Fitopatologia* 22, 1–9.

WALLA, W.J. AND BIRD, L.S. (1981) The cotton seedling disease complex and its control. *Texas Agricultural Extension Service Fact Sheet* L-2002. 4 pp.

WATANABE, T. (1972) Pycnidium formation by fifty different isolates of *Macrophomina phaseoli* originated from soil or kidney bean seed. *Annals of the Phytopathological Society of Japan* 382, 106–10.

WATKINS, G.M. (1938) Histology of Phymatotrichum root rot of field-grown cotton. *Phytopathology* 28, 195–202.

WATKINS, G.M. (ed.) (1981) *Compendium of Cotton Diseases.* American Phytopathological Society, St Paul. 87 pp.

WHEELER, J.E. and HINE, R.B. (1972) Influence of soil temperature and moisture on survival and growth of strands of *Phymatotrichum omnivorum. Phytopathology* 72, 828–32.

WHITE, T.L. (1988) The *in-situ* effects of soil water potential on *Phymatotrichum omnivorum* sclerotial formation and germination. MS Thesis, Texas A&M University. 76 pp.

WHITE, T.L., KENERLEY, C.M. AND GERIK, T.J. (1987) Quantification of hyphal strand growth and sclerotium formation of *Phymatotrichum omnivorum* using a micro-video system. *Proceedings of the Beltwide Cotton Production Research Conference* 47, 42–3.

WHITSON, R.S. AND HINE, R.B. (1986) Activity of propiconazole and other sterol-inhibiting fungicides against *Phymatotrichum omnivorum. Plant Disease* 70, 130–3.

YASSIN, A.M. AND DAFALLA, G.A. (1982) A preliminary note on the present status of cotton wilt syndrome in the Sudan. *Coton et Fibres Tropicales* 37, 379–83.

YOUSSEF, B.A., AMR, A.-M. AND HEITEFUSS, R. (1985) Interactions between herbicides and soil-borne pathogens of cotton under greenhouse conditions. *Zeitschrift für Pflanzenkrankheiten und Pflanzenschutz* 92, 55–63.

ZAKI, M.J. AND GHAFFAR, A. (1972) Effect of *Rhizobium* spp. on *Macrophomina phaseolina. Pakistan Journal of Science and Industrial Research* 30, 305–6.

6 | Fungal Diseases of the Leaf

R.J. HILLOCKS

Introduction

By the time a well-grown cotton crop reaches the stage of boll production, it has a dense canopy and large leaf surface area. The dense canopy encourages a humid microclimate within the crop, creating an environment where large numbers of phylloplane micro-organisms can flourish. Many of these organisms are non-pathogenic or they are secondary invaders of necrotic or insect-damaged tissue, but some 20 fungi have been recorded as primary leaf spot pathogens of the cotton plant. (The main leaf spot fungi are listed in Table

Table 6.1. The main cotton leaf spot pathogens.

Pathogen	Distribution	Reference
Alternaria alternata	Worldwide	Rotem *et al.*, 1988a
Alternaria gossypina	Africa	Hopkins, 1931
Alternaria macrospora	Worldwide	Cotty, 1987a
Ascochyta gossypii	USA, Africa	Thomson, 1953
Cercospora gossypina	Worldwide	Miller, 1969
Cochliobolus spp.	India	Chopra *et al.*, 1985
Colletotrichum spp.	Worldwide	Frolich and Rodewald, 1970
Leveillula taurica	Egypt, India, Peru	Nour, 1956
Myrothecium roridum	India, USA	Chopra *et al.*, 1985
Phakospora gossypii	Africa, S. America	Pineda, 1987
Phoma exigua	India	Chauhan and Yadav, 1984
Phomopsis spp.	USA	Ivey and Pinckard, 1967
Phyllosticta spp.	Africa	Hopkins, 1932
Puccinia cacabata	USA	Blank and Fisher, 1974
Ramularia areola	Worldwide	Cauquil and Sement, 1973
Rhizoctonia solani	USA	Neal, 1944

6.1.) Only some of these pathogens, in varying combinations, would be found in the crop at any one geographical location, although lesions caused by several fungi may often be present on a single leaf. Which of these fungi can be considered as major or minor pathogens often depends on environmental conditions, cultivation practice and host genotype. However, with the exception perhaps of *Alternaria macrospora* on cultivars of *G. barbadense* and *Ramularia areola* on cultivars of *G. arboreum* and *G. herbaceum*, the leaf spot pathogens of cotton generally attack senescent or weakened plants. They rarely cause significant yield loss unless the crop is predisposed to infection by pest attack, by prolonged exposure to adverse weather conditions for plant growth or by premature senescence, induced by moisture stress or nutrient deficiency.

Alternaria Leaf Spot

HISTORY AND DISTRIBUTION

Alternaria leaf spot was first reported in the USA (Faulwetter, 1918) although Atkinson (1891) had earlier reported a leaf spot complex involving *Cercospora* sp. and *Alternaria* sp. which he referred to as 'black rust'. Later, Jones (1928) described a leaf spot and boll rot in Nigeria which he attributed to *A. macrospora*. A similar leaf spot was found on Sea Island cotton in the West Indies (Hewison and Symond, 1928), on Upland cotton in the Bombay area of India (Rane and Patel, 1956), in Zimbabwe (then Rhodesia) (Hopkins, 1932), in South Africa (MacDonald *et al.*, 1945) and in China (Ling and Yang, 1941). Severe epidemics have occasionally been recorded from Senegal, Mali and Madagascar (Joly and Lagière, 1971). The disease has more recently become important in Israel (Bashi *et al.*, 1983b) and, with increasing cultivation of susceptible Pima cultivars, the disease has become important in the USA (e.g. Russell and Hine, 1978; Cotty, 1987a). In India, many of the hybrid cottons (*G. barbadense* × *G. hirsutum*) are susceptible. Alternaria leaf spot has become the most common foliar disease in cotton and is probably now found in almost every country which grows the crop.

CAUSAL ORGANISMS

Taxonomy and morphology

In his description of Alternaria leaf spot, Faulwetter (1918) could not establish the exact identity of the pathogen but noted that it was similar to *A. tenuis* Nees. More recently, in the USA, Calvert *et al.* (1964) described a leaf spot disease caused by *A. tenuis*. (*A. tenuis* is no longer recognized as a separate species but has been placed in the *A. alternata* species group, see Lucas, 1971, and will henceforth be referred to by that name.) With the apparent exception of Egypt (Kamel *et al.*, 1971a) and the former Soviet Union (Dzhamalov, 1973), where *A. alternata* (Fr.) Keissler may be the main species involved, this leaf spot is more often attributed to *A. macrospora* Zimm. However, in Zimbabwe, Hopkins (1931) concluded that, although *A. macrospora* was present in the crop, the more important leaf spot pathogen was an *Alternaria* species which was unlike *A. macrospora* in producing conidia in chains, but with spore dimensions outside the range accepted for *A. alternata* Hopkins described this species as *A. gossypina* (Thüm) Hopkins. Based on material collected in Southern and Central Africa, which is available in the herbarium collection at the International Mycological Institute, David (1988) has retained this distinction. In recent reports of investigations of the disease in Zimbabwe, the main leaf spot pathogen was described as *A. macrospora*. Also present, but much less important, was *A. alternata* (Hillocks, 1991).

Conidiophores of *A. macrospora* arise singly or in groups from infected leaf tissue. They are erect, simple straight or flexuous, almost cylindrical or tapering towards the apex and septate. They are pale brown in colour, with

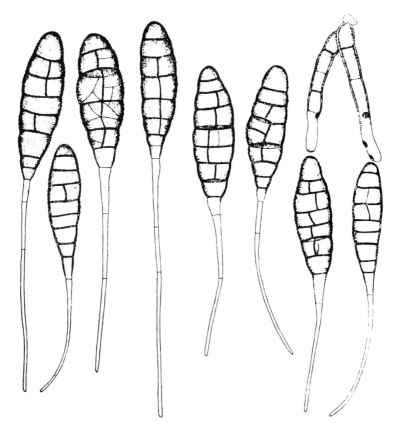

Fig. 6.1. *Alternaria macrospora*: conidia and conidiophores (× 650) (Ellis, 1971).

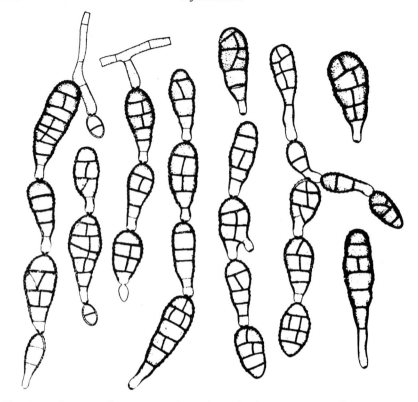

Fig. 6.2. *Alternaria alternata*: conidia and conidiophores (× 650) (Ellis, 1971).

several conidial scars, and are 4–9 μm thick and up to 80 μm in length. Conidia are solitary or occasionally in chains of two, straight or curved, obclavate or with the body of the conidium ellipsoidal tapering to a narrow beak and equal in length or up to twice as long as the body (Fig. 6.1). They are reddish brown in colour, usually minutely verruculose, with four to nine transverse septa and several longitudinal septa (Ellis, 1971). There is considerable variation in conidial length and although most descriptions of the fungus suggest a maximum length not exceeding 180 μm, conidia isolated directly from infected leaf tissue in Zimbabwe frequently reach 240 μm (Hillocks, 1991). The body of the mature conidium is 15–22 μm thick at the broadest part.

The conidia of *A. alternata* are formed in chains of up to seven spores and are pale to golden brown in colour with up to eight transverse septa and several longitudinal septa (Ellis, 1971). The conidia are smaller than those of *A. macrospora*, having an overall length of 20–63 μm and 9–18 μm in width. The beak is short, being not more than one-third of the total length (Fig. 6.2).

The conidia of *A. gossypina* are described by David (1988) as dark brown, with up to nine transverse and one or two longitudinal septa. They are 30–55 × 12–15 μm in size, excluding the beak, which is 9–52 μm long (Fig. 6.3).

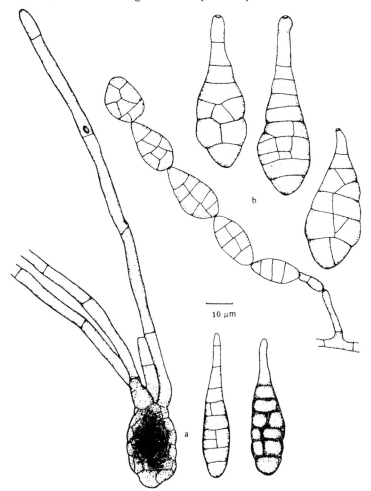

Fig. 6.3. *Alternaria gossypina*: a, conidia and conidiophores, b, conidiogenous cell and conidia (David, 1988).

Isolation and growth in culture

Alternaria spp. can be isolated from cotton leaves using standard methods and media. However, *A. macrospora* and *A. alternata* often occur together on what appears to be a single lesion, in which case it can prove difficult to isolate *A. macrospora* because *A. alternata* grows slightly faster. This might also cause the importance of the latter to be overestimated. If both species are present, it may be best to surface-sterilize the leaf, incubate for 24 hours in a moist atmosphere to stimulate sporulation, and then to select the appropriate spore type and transfer conidia directly to PDA.

    Alternaria spp. store reasonably well on PDA slopes in the refrigerator but recovery can be improved by storing on dried agar or under mineral oil.

They sporulate readily on PDA when first isolated but rapidly become non-sporulating when stored. Cultures can be encouraged to resporulate by subjecting them to mechanical damage by scraping off the mycelium and replating, followed by exposure to light and low temperature (see Shahin and Shepherd, 1979). The best medium to encourage sporulation appears to be V8 juice agar. A decoction agar made from dried mature cotton leaves and sucrose was successfully used in Zimbabwe to stimulate the production of conidia (R. J. Hillocks, unpublished).

HOST RANGE AND SUSCEPTIBILITY OF *GOSSYPIUM* SPECIES

A. macrospora is capable of causing leaf spots in all *Gossypium* spp., but there are wide variations between species in degree of susceptibility. *G. barbadense* is the most susceptible species and certain cultivars of *G. hirsutum* are probably the most resistant. Jones (1928) observed the disease on *G. peruvianum* and *G. vitifolium*. Some cultivars of *G. arboreum* grown in India are highly susceptible (Padaganur and Basavaraj, 1983), although some members of the species are resistant (Bhaskaran *et al.*, 1975). *A macrospora* may be capable of infecting other members of the Malvaceae but there does appear to be considerable host specificity exhibited by fungi in the Macrospora group. For instance, the fungus causing a leaf spot on the malvaceous weed, *Anoda cristata*, was identified as *A. macrospora*. When inoculated into *A. cristata*, this isolate caused severe disease symptoms, but it elicited a hypersensitive response in all the cotton varieties tested (Walker and Sciumbato, 1979; Walker, 1981). Conversely, strains of the fungus which were highly pathogenic to Pima cotton were virtually non-pathogenic towards *A. cristata* (Sciumbato and Walker, 1980).

Very little information is available on host range and pathogenic variation in other *Alternaria* spp. isolated from cotton. However, although isolates placed in the Alternata group have been obtained from a wide range of crops, including tobacco and apple, the evidence suggests considerable host specificity within the group. Indeed, the cotton isolates tested by Walker and Sciumbato (1979) failed to produce symptoms on several malvaceous weeds.

According to Rotem *et al.* (1988a), *G. barbadense* is more susceptible than *G. hirsutum* to *A. macrospora* but the reverse is true with respect to *A. alternata*. This difference in relative susceptibility of Pima and Upland cottons was not apparent, however, among the cultivars inoculated by Walker and Sciumbato (1979).

SYMPTOMS

The earliest symptom of the disease may be the appearance of spots on the cotyledons of the newly emerged seedling (see Plate 4A). *A. macrospora* is known to attack the seedling in Israel (Bashan and Levanony, 1987) and in India (Padaganur, 1979). In the USA, the pathogen has been implicated in

Fig. 6.4. Symptoms caused by *Alternaria macrospora* on young cotton leaf.

post-emergence damping-off (Maier, 1965). In East Africa, where the fungus attacks the cotyledons and growing tip, causing a seedling blight, this is considered by Ebbels (1976) to be the most damaging phase of the disease. In Zimbabwe under continuous cotton cultivation, the disease can also be found affecting the young cotyledons, beginning as small, brown, circular spots, bordered by a purple margin. Under favourable conditions for disease development, the spots can enlarge to 10 mm in diameter. Large numbers of spots, coalescing together, may develop on susceptible cultivars, causing the cotyledons to be shed. Cotyledons and, to a certain extent, the first-formed leaves are more susceptible than the later leaves. This is particularly true of most Upland cultivars, in which the leaves can remain free of the disease despite heavy infection of the cotyledons.

Once the canopy closes, spots similar to those found on the cotyledons can usually be found on the lower leaves. On green leaves the purple margin around the spot is very pronounced (Fig. 6.4 and Plate 4B). On older leaves the necrotic centre of the spots may be marked by a pattern of concentric zonation. Sometimes several spots coalesce to form larger areas of necrosis, particularly near the leaf margin. In humid weather conditions, the necrotic tissue turns a sooty black colour due to prolific sporulation by the fungus (Fig. 6.5 and Plate 4C). Severe infection of the upper canopy leads to premature defoliation, but this occurs only on highly susceptible cultivars, mainly *G. barbadense* or certain cultivars of *G. herbaceum* grown in India, although it can occur on normally resistant cottons if they are predisposed to infection by some form of stress. The leaf stalk and the boll can also become infected, resulting in a spherical or elliptical purple spot. Boll infection is usually superficial, being confined to the outside of the boll wall. In China, the pathogen causes a stem canker in addition to the more common leaf and boll symptoms (Ling and Yang, 1941).

LOSSES

A. macrospora causes yield losses in susceptible cultivars of *G. barbadense* which account for 20% of the crop in Israel (Rotem *et al.*, 1988b). The extent of damage caused by the leaf spot disease in these cultivars has often been underestimated in the past due to continuous regeneration of the canopy as infected leaves are shed (Bashi *et al.*, 1983b). In experiments conducted in Israel using fungicides sprayed from aircraft to control the disease, yield

Fig. 6.5. Symptoms caused by *Alternaria macrospora* on a senescent or nutrient-deficient leaf.

increases of up to 24% were obtained compared with the unsprayed crop, and this was related to reduced leaf shedding (Bashi *et al.*, 1983a). Although similar experiments have not been reported from the USA, where the same Pima cultivars are grown, the disease is recognized as an increasingly important problem (Watkins, 1981) and it is likely that similar yield losses occur.

The disease is less damaging to Upland cultivars. In Africa, the disease is usually to be found in the lower canopy of most cotton crops but is not sufficiently severe to cause losses, unless susceptibility is enhanced by some form of stress. In Zimbabwe, when the disease is associated with potassium deficiency, yield increments as high as 50% were attributed to control of the leaf spot disease and potassium deficiency with a combination of foliar fungicides and potash fertilizer (Hillocks and Chinodya, 1989).

In India, *A. macrospora* has become an important disease in *G. herbaceum*, causing 10–15% yield loss (Rane and Patel, 1956). In association with bacterial blight, it can be sufficiently damaging to justify chemical control (Padaganur and Basavaraj, 1983).

Yield losses have rarely been attributed to *A. alternata* in Upland or Sea Island cottons. However, this species is reported to be damaging in Egypt and the former Soviet Union (Kamel *et al.*, 1971b; Dzhamalov, 1973). In Gujarat State, yield increases of up to 24% were attributed to control of *A. alternata* with fungicides (Patel *et al.*, 1983). In the cotton states along the Mississippi Delta in the USA, a disease complex involving *A. alternata* and *Cercospora gossypina* was reported to cause severe premature defoliation.

Although the disease often occurs too late in the growing season to cause yield loss, it may have other important effects in reducing fibre maturity and seed quality.

DISEASE CYCLE

Maximum sporulation of *A. macrospora* occurs on leaves after they have been shed (Bashi *et al.*, 1983b) and infected trash allows the pathogen to survive between crops. The fungus is also carried within the seed. Padaganur (1979) found that 90–100% of seed from infected bolls carried the fungus and infected seed gave rise to infected seedings. The fungus is not capable of penetrating the intact boll, seed infection occurring only after damage to the boll or following normal boll-split (Bashan, 1984).

Undecomposed crop residues and infected seeds therefore provide the sources of primary inoculum, giving rise to infected cotyledons which support the early stages of an epidemic. Periods of high humidity encourage sporulation and infection spreads from the cotyledons to the lower leaves. In Upland cotton, the disease is often not much in evidence on the cotyledons or lower leaves of young plants and appears first when the canopy has closed, creating a humid microclimate, or when the lower leaves become senescent. Primary infection of lower canopy leaves can be initiated from conidia splashed up from infected crop residues or blown into the crop from other foci of infection.

Alternaria spp. also attack the bolls and grow on exposed lint if the bolls open in wet weather, giving rise to contaminated seed. The disease cycle is completed when infected leaves fall to the ground or with the planting of seed carrying the fungus.

Epidemiology

Because the cotyledons are so much more susceptible than the leaves to *A. macrospora*, they can become infected during conditions under which the leaves are resistant. They therefore provide a reservoir of inoculum to infect the lower leaves as the canopy closes, raising the humidity to a level at which the leaves become more susceptible.

Bashi *et al.* (1983a) found that, under controlled conditions, disease development was six to nine times higher on cotyledons than on leaves. The minimum temperature for the disease to occur was 10 °C, the maximum was 35 °C and the optimum was 20–25 °C. Within the optimum temperature range, cotyledons became substantially infected with a 4-hour wetting period (Spross-Blickle *et al.*, 1989), but a wetting period of 20 hours was required to obtain a similar level of leaf infection.

In susceptible cultivars of *G. barbadense*, the disease is characterized by continuous shedding of infected leaves and regeneration of new leaves, so that each new generation of leaves is exposed to increased inoculum pressure (Bashi *et al.*, 1983a).

Provided the wetting period is of sufficient duration and the temperature is in the range 20–30 °C, conidia carried on to the leaf surface by wind currents germinate and penetrate the leaf through the stomata, giving rise to mycelial growth in the mesophyll tissue. Rotem *et al.* (1989) investigated the effect of environmental variables on sporulation on Pima SJ5. Lesions appeared in five to seven days and gradually expanded. Conidia were then produced on the necrotic tissue of the lesion to initiate the secondary infection cycle. Sporulation was greatly enhanced by exposure to light before a period of moist, dark conditions. Sporulation was more prolific when dew periods were interrupted by dry periods than under a regime of continuous wetting. Spore production occurred over a wide temperature range but peak sporulation occurred at 30 °C on green leaves, 25–30 °C on chlorotic leaves and 20–30 °C on necrotic leaves, and was fives times more prolific on five-week-old leaves than on seven-week-old leaves. The conidia appear to be resistant to environmental extremes and the ability to sporulate under interrupted wetting periods and at a wide range of temperatures makes *A. macrospora* an efficient pathogen (Rotem *et al.*, 1988b). Although infection of the leaf can take place over a wide temperature range, lesion formation is progressively reduced at temperatures above 39 °C, due to reduced spore viability and increased germ-tube lysis (Cotty, 1987b). This may explain why the leaf spot is not usually severe under growing conditions such as those found in Arizona, where daily temperature maxima regularly exceed 40 °C.

   A. alternata appears to have similar requirements of temperature and humidity for infection and sporulation. Experiments conducted by Rotem *et al.* (1988a) suggest that, in addition, symptom expression is greatly enhanced by exposure of the fungus to sunlight after initial infection.

PREDISPOSITION TO INFECTION

The cotton plant becomes more susceptible to infection by *Alternaria* spp. if it suffers a period of adverse growing conditions. This is particulary true of Upland cultivars. It was noted early in the history of research into the disease that healthy, vigorous plants could not be infected by inoculation (Faulwetter, 1918). Even with cultivars of the more susceptible Sea Island cottons, disease incidence was found to be positively correlated with reduced vigour. Also, vigorous plants were more subject to attack when in a condition of low carbon : nitrogen ratio (Hewison and Symond, 1928).

   The Cercospora–Alternaria leaf blight complex in Louisiana occurs only on plants under stress from moisture deficit or mineral deficiency (Sinclair and Shatla, 1962). A similar disease complex occurs in Missouri, where Miller (1969) reported that any sort of stress at the time of heavy fruit-load increased susceptibility to the complex. Furthermore, premature defoliation of the Upland cultivar, Deltapine 16, was attributed to *A. macrospora* in Louisiana, but, in this case, only plants attacked by the reniform nematode (*Rotylenchulus reniformis*) were affected (Sciumbato and Pinckard, 1974).

   Premature defoliation associated with *A. macrospora* has also occurred at Barberton Research Station in South Africa, where the problem was found in crops growing on red clay soils but not in neighbouring crops growing on sandy soils. Extensive investigations into the problem indicated that crops growing on the clay soil were suffering from potassium deficiency, which predisposed the plants to infection (MacDonald *et al.*, 1945). A similar association between potassium deficiency and the leaf spot disease, leading to premature defoliation, has been investigated in Zimbabwe (Hillocks and Chinodya, 1989), where the leaf spot disease reached epidemic proportions only after deficiency symptoms became apparent in the upper canopy leaves. Once the upper leaves developed the discoloration indicative of potassium deficiency, an *Alternaria* epidemic followed as soon as environmental conditions became suitable for infection and sporulation of the pathogen. Moisture stress also seems to predispose the crop to infection by inducing earlier maturation, with the associated changes in carbon : nitrogen ratio of the tissues.

CONTROL

Cultural methods

A. macrospora survives between crops on infested crop residue and on ratoon and volunteer plants (Bashi *et al.*, 1983a). In Israel, the land is ploughed at the end of the growing season and the resulting burial and decomposition of crop debris which takes place greatly reduces the ability of the pathogen to survive between cropping seasons (Bashan, 1984). In Zimbabwe, seedling infection can be severe if cotton is sown in the same field two seasons in succession following a leaf spot epidemic. But a break of one season between cotton crops is sufficient to reduce seedling infection to an insignificant level (Hillocks, 1991). Ensuring that seed from an infected crop is not used to plant the next season's crop, together with the removal of ratoons and volunteers, would also help to eliminate potential sources of inoculum.

Alternaria epidemics in Upland cotton can be prevented by ensuring that the crop does not become stressed, especially during the main flowering period. If irrigation is available, moisture stress can be avoided. Care should be taken that fertilizer application is adequate to maintain soil fertility levels particularly with respect to available potassium.

Fungicides

Defoliation due to Alternaria leaf spot is severe enough in Israel and parts of India to justify chemical control. Control of the disease with foliar fungicides is usually considered economically worthwhile only on the more susceptible high-quality cottons, such as the Pima cultivars, and on the F_1 hybrids grown in India (e.g. Patel *et al.*, 1983). Chemical control has also been used on susceptible cultivars of *G. herbaceum* in India, where copper oxychloride was found to be effective (Padaganur and Basavaraj, 1983). Where the disease occurs in association with bacterial blight, chemical control is considered worthwhile for Upland cultivars, on which a combination of copper oxychloride and streptomycin sulphate has proved effective (Padaganur and Basavaraj, 1983). In addition to copper fungicides, zineb (Rasulev and Dzhamalov, 1975; Singanmathi and Ekbote, 1980), mancozeb and captafol (Patel *et al.*, 1983) have also been reported to control the disease.

The disease has been controlled by aerial applications of fentin acetate on Pima cotton in Israel, where up to ten weekly applications were required for maximum yield increases, using 500 g of formulated product in 50 l water ha^{-1} (Bashi *et al.*, 1983b). Fentin acetate applied alternately with chlorothalonil at 14-day intervals from first fruit set has proved effective against the disease on Upland cotton in Zimbabwe. However, the disease became increasingly difficult to control with foliar fungicides with increasing severity of potassium deficiency symptoms. The greatest yield increments were obtained when the fungicides were applied to cotton growing in plots which received extra potash fertilizer (Hillocks and Chinodya, 1989). In India, some control has been achieved with mancozeb and with captafol, applied at ten-day intervals, after the appearance of first symptoms (Agalarsamy *et al.*, 1989).

Where susceptible species of cotton are grown commercially, seed and in-furrow treatment with fungicides may have some potential in reducing primary inoculum. In India, Padaganur and Basavaraj (1984) reported some success in eliminating seed-borne infection by using fungicidal seed treatment. In Zimbabwe, the fungicides iprodione, prochloraz, procymidone, fentin acetate and fentin hydroxide were effective in protecting the cotyledons of emerging seedings from trash-borne inoculum (Hillocks, 1991).

Resistant varieties

Considerable variability for resistance to the leaf spot exists within the genus *Gossypium*. Among the cultivated species, *G. barbadense* is the most susceptible and some cultivars of *G. herbaceum* are also highly susceptible. *G. arboreum* and *G. hirsutum* are considered to be resistant. However, even within the Upland group, there appear to be differences in resistance which are sufficiently great to influence the degree of yield loss. In India, for example, the cultivar MCU5 is more susceptible than cv. Laxmi (Padaganur and Basavaraj, 1987). In the USA, Cotty (1987a) screened a number of Pima and Upland cultivars for resistance to *A. macrospora*. Pima SS and Pima SJ5 were the most susceptible, but quantifiable differences were also found among the Upland varieties. Deltapine-Acala 90 was more susceptible than Deltapine 61 and Acala SJ5. According to Kamel *et al.* (1971b), resistance to *Alternaria alternata* in a cross between Misella Valley and Bahatim 101, cultivars of *G. barbadense*, was inherited as a partially dominant character, governed by one or two pairs of genes.

Bashan (1986) considers that resistance to the disease might be related to higher phenol content. This view agrees with an earlier report by Bhaskaran *et al.* (1975), who found that the leaves of a resistant cultivar of *G. herbaceum* were higher in polyphenol oxidase activity and in total phenol content than were those of the more susceptible Upland cultivar MCU5.

Cercospora Leaf Spot

DISTRIBUTION AND SYMPTOMS

Cercospora leaf spot has been reported from the USA (e.g. Miller, 1969) and is common in Africa (Ebbels, 1976). It is distributed throughout the cotton-growing areas of the world and probably occurs wherever Alternaria leaf spot is found. There appears to be some difference of opinion concerning the symptoms associated with the disease, which is not surprising in view of the fact that Cercospora leaf spot is rarely to be found in the absence of other foliar diseases. Ebbels (1976) describes the symptoms as irregular brown lesions, usually surrounded by a considerable area of chlorotic tissue.

However, Watkins (1981) describes the disease as beginning with barely visible red dots, which increase in diameter to about 2 cm, retaining a narrow red margin, enclosing a white to light brown centre of dead tissue.

CAUSAL ORGANISM

Taxonomy and morphology

The causal organism is a hyphomycete fungus usually described as *Cercospora gossypina* Cooke (e.g. Ebbels, 1971). Morphologically, the conidial stage conforms to the description given by Ellis (1971) for *C. appi* (Fig. 6.6) which has a wide host range (see Johnson and Vallean, 1949). Ellis (1971) includes *C. gossypina* in a list of possible synonyms, but Little (1987) has described it as a separate species and his description follows below. In culture, colonies are amphigenous. Conidiophores are mononematous, in fascicles of 2–10, not branched and 0–5 slightly or 0–2 abruptly geniculate. They are brown in colour, 1–3 septate and 4–5 × 70– 150 μm in size, occasionally reaching 8.5 × 100 μm. Conidia are hyaline, straight or curved, 5–10 septate, bases truncate and 2.5–4.5 × 60–150 μm in size.

An ascomycete stage has also been identified and named *Mycosphaerella gossypina*. The perithecia, which are dark brown to black and nearly spherical, have been found on cotton, partially immersed in the dead tissue within older leaf spots (Watkins, 1981).

Isolation and growth in culture

The pathogen can be isolated from leaf tissue on PDA but this is often difficult because of its slow growth. The fungus is overgrown by saprophytes and other foliar pathogens, especially *Alternaria* spp., with which it is frequently associated. If conidia are present on the infected tissue, these should be transferred directly to PDA and incubated at 25–30 °C. Colonies are grey, sometimes tinged with pink, and compact, with limited aerial mycelium, and they usually fail to produce conidia on PDA under normal laboratory conditions. Sporulation can be stimulated by mycelial transfer to potato–carrot agar and incubation in the dark for three weeks at a temperature fluctuating diurnally between 21 and 29 °C (Miller, 1969).

LOSSES

Cercospora leaf spot in the USA nearly always occurs on plants in a state of physiological stress or after the onset of senescence, and is therefore not usually a cause of yield loss. In Missouri and several other cotton belt states along the Mississippi Delta, *C. gossypina* and *Alternaria* sp. act synergistically to produce severe premature defoliation of Upland cotton (Calvert *et al.*, 1964). Premature defoliation leads to decreased yields and immature fibres in seed cotton harvested from the top one-third of affected plants.

Factors affecting infection

Conditions favouring infection and sporulation and predisposing the crop to infection appear to be very similar to those described previously for Alternaria leaf spot (see Sinclair and Shatla, 1962; Miller, 1969).

Control

Calvert *et al.* (1964) were unable to prevent premature defoliation caused by the *Cercospora–Alternaria* complex using foliar fungicide treatments. Repeated application of captan and folpet throughout the season only delayed

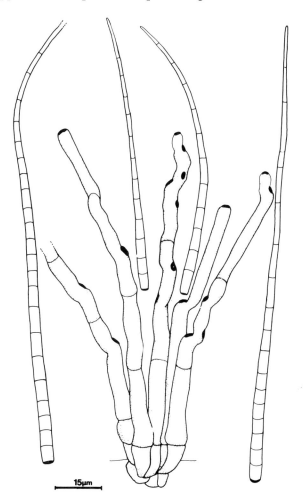

Fig. 6.6. *Cercospora gossypina*: stroma, conidiophores, conidiogenous cells and conidia (Little, 1987).

symptom expression. However, in India some success has been reported in controlling Cercospora leaf spot with foliar sprays. The most effective compounds were thiabendazole and copper oxychloride (Bhaskaran and Shanmugam, 1973).

False Mildew

DISTRIBUTION AND SYMPTOMS

False mildew was first reported in the USA (Atkinson, 1890) but has since been reported from a number of major cotton-producing countries, probably occurring in most countries where the crop is grown. The disease is known as grey mildew in India, white mould in South America and areolate mildew or false mildew in the USA. Although widespread, the disease is of little economic importance in the USA and the drier cotton-growing countries, but is sometimes troublesome in parts of India, East Africa, South America and Madagascar.

False mildew usually appears first on the lower canopy leaves after first boll-set. Lesions are 3–4 mm in width bounded by the veinlets, giving the lesions an irregular, angular outline. They are light green to yellow-green on the upper leaf surface but, on the under surface, profuse sporulation gives the lesions a white mildew-like appearance. Under very humid conditions lesions become white on the upper surface also (see Plate 4D). Once active sporulation ceases, the lesions become necrotic and dark brown in colour, at which stage they can be mistaken for the angular leaf spot phase of bacterial blight. Severe infection leads to defoliation of the crop.

Symptoms on the cotyledons, when they occur, differ from leaf symptoms, appearing as circular water-soaked patches, which become chlorotic and then reddish brown. Severe infection causes the cotyledons to wither. The fungus does not appear to sporulate on cotyledons (Rathaiah, 1976).

CAUSAL ORGANISM

Taxonomy and morphology

The conidial stage of the causal organism is known as *Ramularia areola* (Atk.) (synonyms, *Ramularia gossypii* Speg. Ciferi, *Cercosporella gossypii* Speg.) The fungus has an ascomycete sexual stage known as *Mycosphaerella areola* Ehrlich & Wolf.

Conidiophores are 25–75 × 4.5–7 μm, bearing one- or two-septate, hyaline conidia (Fig. 6.7), which are 14–30 × 4–5 μm in size. Spermogonia appear as raised black dots in lesions on the lower surface of fallen leaves. These are 28–75 μm in diameter and, when mature, rod-shaped spermatia

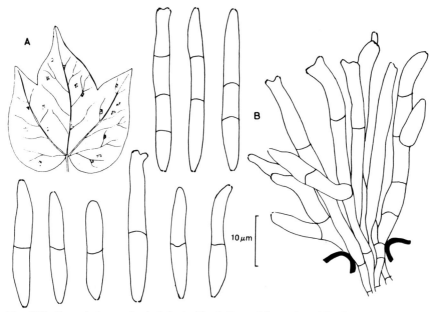

Fig. 6.7. *Ramularia areola*: A, infected leaf; B, conidia and conidiophores.

2–4 × 0.4–2.0 μm ooze out through an apical opening in a viscous mucilage. Perithecia later replace the spermogonia and are brown in colour and 70–80 μm in diameter, producing fusiform asci 35–40 × 6–8 μm in size from eight, two-celled ascospores, which are 12.4–15.6 × 3.2–3.8 μm (Ehrlich and Wolf, 1932) (Fig. 6.8).

Isolation and growth in culture

The fungus may be isolated on PDA by conventional methods from infected leaf tissue but is slow-growing in culture, producing greyish white hemispherical colonies, which reach only 4–7 mm in diameter after four weeks at 25 °C. Conidia are sparse or absent on PDA but more abundant on carrot leaf decoction agar, containing 0.2% yeast extract, V-8 juice agar or modified Richard's medium. Sporulation is best at 25 °C, at pH 5.0 and under constant light. Spermogonia, chlamydospores and sclerotia are formed on some media (Rathaiah, 1973).

Variation

Although mycelial growth is most rapid at temperatures between 20 and 28 °C, the exact optimal temperature for growth in culture and the rate of growth at optimal temperature vary among isolates. Faster-growing isolates are more pathogenic. Rathaiah (1976) found that isolates from Madagascar, Chad, Ivory Coast and India differ in virulence. Those from Madagascar were

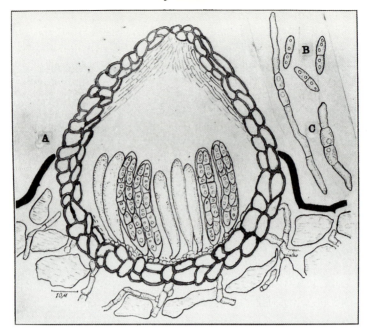

Fig. 6.8. *Ramularia areola*: A, perithecium; B, ascospores, C, germinating ascospores.

the most virulent, while isolates from Ivory Coast and India were relatively weak.

Losses

In most countries where the disease is found, it occurs late in the growing season and therefore causes little or no yield loss. It is most severe in crops having a dense canopy and in humid weather conditions, when the resulting defoliation may be beneficial in improving insecticide penetration and in reducing the risk of boll rot (Ebbels, 1976). However, in a few countries the disease is sufficiently severe and occurs early enough in the season to justify control measures. In Madagascar, severe outbreaks of false mildew occur regularly in wet weather. Encouraged by continuous cultivation, the disease has been increasing in importance since 1969 and, in 1972, large-scale protection with fungicide became necessary (Cauquil and Sement, 1973). The disease is also responsible for yield losses in India, especially on the highly susceptible *G. arboreum* cultivars (Aurangabadkar *et al.*, 1981).

Disease cycle and epidemiology

The fungus develops in three distinct stages during its life cycle. The conidial stage appears on living tissue, mainly on the underside of the leaves while they are still attached to the plant and for a short time after abscission. The spermogonial stage occurs later on fallen leaves and this is followed by the ascogenous stage, which develops on partially decayed leaves (Ehrlich and Wolf, 1932). The fungus is able to survive between crops on leaf litter.

Conidia and ascospores, produced on fallen leaves or volunteer plants, provide the primary inoculum, which is disseminated by wind and by irrigation water, where furrow irrigation is used. Conidia and ascospores germinate in free water at temperatures of 16–34 °C, with the optimum between 25 and 30 °C. The conidial germ-tube forms an appressorium along the join between closed guard cells, an infection peg is formed and penetration occurs through the stomatal aperture. Although the conidia appear to require free moisture for germination, stomatal penetration is greater under cycles of night wetting and daytime drying than with continuous wetting. Some infection occurs after two cycles of night wetting with maximum infection after four cycles. Germ-tubes can survive several 16-hour cycles at 20–60% RH (Rathaiah, 1977).

Infection hyphae arise from a substomatal vesicle formed by the swelling of the infection peg. Hyphae then ramify both between and within the spongy and palisade parenchyma (Fig. 6.9). Destruction of invaded cells accompanies the advance of the mycelium. The protoplasm is plasmolysed and the middle lamella broken down, causing cell collapse (Ehrlich and Wolf, 1932), leading eventually to leaf abscission, after the fungus has started to sporulate.

Conidiophores arise in fascicles from a substomatal stroma and protrude through the stomata. Conidia are formed in humid conditions and are liberated by air currents to establish secondary cycles of infection.

Once the leaf has fallen and died, spermogonia appear on the lower surface as slightly raised black spots within the original lesions. Spermogonia mature over a period of one to three months and numerous spermatia are extruded through an apical aperture in a mucilaginous matrix. Perithecia replace the spermogonia a few months later, arising apparently from sclerotial structures embedded in the leaf tissues. The perithecia are dark brown in colour and few in number and protrude from the leaf surface, releasing ascospores when mature (Ehrlich and Wolf, 1932).

CONTROL

Cultural practice

In view of the importance of infected leaf litter in the survival and life cycle of the pathogen, the main source of primary inoculum can be removed by the ploughing under or destruction of crop residues and by avoiding continuous cultivation of cotton.

Fungicides

Should it become necessary, the disease can be quite easily controlled by the application of fungicides to the foliage as soon as infection appears. One or

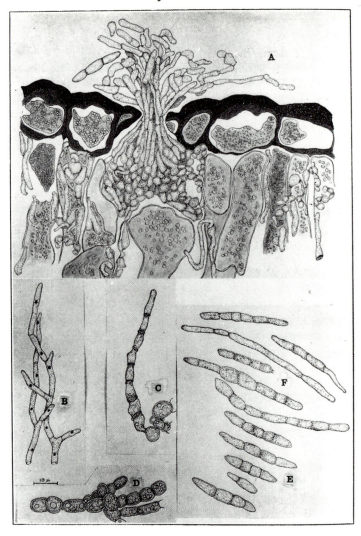

Fig. 6.9. A, Upper surface, in vertical section of cotton leaf infected by *Ramularia areola*, showing sub-stomatal vesicle; B, aerial mycelium from culture; C, hyphae with plate-like septa at surface of agar; D, hyphae from beneath agar surface; E, conidia; F, germinating conidia (Ehrlich and Wolf, 1932).

two applications of benomyl at 200–300 g active ingredient ha^{-1} is effective (Cauquil and Sement, 1973; Bell, 1981).

Resistant cultivars

There is considerable variation in the degree of susceptibility found among the cultivated cottons and, in countries where the disease is a cause of yield loss, the best approach to control is to grow resistant varieties. In India, resistance has been identified in *G. hirsutum* lines (Clauhan, 1985; Sharma *et al.*, 1986). Dake and Kannan (1982) screened a number of cotton species and Indian varieties. They found that all the *G. barbadense* cultivars tested were resistant, whereas two cultivars of *G. arboreum* and *G. herbaceum* were highly susceptible. Among the *G. hirsutum* varieties, Laxmi and MCU 5 were resistant and Gujarat 67 and Khandwat were susceptible.

Screening conducted by Rathaiah (1976) showed that many Upland cultivars were highly susceptible to a particularly virulent isolate of the pathogen from Madagascar, but two cultivars, BJA 592 and Reba BTK 12, and two cultivars of *G. barbadense*, Tadla 16 and Pima 67, exhibited a hypersensitive response to inoculation. Among the Upland cultivars which have been widely grown commercially, Acala 1517BR is less susceptible than most under the severe disease pressure to which the crop is often exposed in Madagascar (Cauquil and Sement, 1973).

Although some of the *G. arboreum* cultivars in India, such as G-47, are highly susceptible to false mildew and suffer severe yield loss as a result, 12 lines with partial resistance were identified by Aurangadbadkar *et al.* (1981). Resistant lines of arboreum cotton were also identified by Chauhan (1985) in Haryana State, where the disease has greatly increased in importance since it was first recorded in 1978.

In screening for resistance, Dake and Kannan (1982) sprayed the test plants with a spore suspension containing 10^3 conidia ml^{-1}. Kodelwar (1972) has suggested a suitable grading scale for evaluating disease resistance.

Ascochyta Blight

DISTRIBUTION AND SYMPTOMS

Ascochyta blight, also known as wet-weather blight (Smith, 1950), is widely distributed, having been recorded in most of the major cotton-producing countries (Holliday and Punithalingam, 1970; CMI Distribution Map No. 259). Since it was first described in material from Kashmir in 1908 (Sydow and Butler, 1916), the disease has most often been reported from the USA, where it became of economic importance during the 1920s. The disease is, however, also quite common in Africa (e.g. Ebbels, 1976).

The fungus causes spots on the cotyledons, leaves, stems and bolls. On cotyledons and leaves, the disease first appears as white to pale brown circular spots with a dark brown to purple border. The spots may enlarge from 2 mm in diameter to form light brown areas of necrotic tissue affecting much of the leaf surface. On the petioles and stems, the disease takes the form of elongated sunken cankers, which can be as much as 6–7 cm long. These resemble symptoms of hail damage, from which it can be distinguished by the presence of pycnidia, seen as small black dots within the canker.

CAUSAL ORGANISM

Morphology

The causal organism is *Ascochyta gossypii* Woron., of which only the conidial stage is known. The conidia are hyaline, straight or slightly curved, cylindrical or ovoid, two-celled, rounded at each end and 8 × 12 μm in size (see Fig. 6.15 below). The conidia are produced from flask-shaped phialides within pycnidia, which are globose, light brown to black in colour and with a diameter of 150–180 μm (Holliday and Punithalingam, 1970). Pycnidia may be formed on lesions on all aerial parts of the plants, being more abundant on older parts of the lesion or after infected leaves have been shed.

Isolation and growth in culture

A. gossypii can be isolated from lesions on PDA but is moderately slow-growing. Optimum temperature for growth *in vitro* is 20–25 °C and at this temperature isolates cover a 9 cm Petri dish in 11–20 days. On PDA the aerial mycelium is greyish black and pycnidial production is sparse or absent until the culture is three to four weeks old. Pycnidial production can be stimulated by growing the fungus on sterile potato cylinders at 20 °C or on modified Coon's agar and may appear after only 36 hours (Chipendale, 1929; Thompson, 1953; Crossan, 1958).

Variation

Thompson (1950) found that single spore isolates from individual pycnidia from different lesions showed considerable cultural variation. He grouped them into strains distinguished on the basis of colour and amount of aerial mycelium, growth rate and presence or absence of pycnidia on PDA. Some isolates grew best at 20 °C, others at 25 °C. Strains which were non-sporulating produced a few pycnidia on ground cotton stem agar.

Crossan (1958) isolated the pathogen from a number of different crops. These isolates were all pathogenic to cotton and exhibited a wide range of cultural variation. However, similar variation was exhibited by isolates from a single host and was not related to pathogenicity.

HOST RANGE

In the USA, *A. gossypii* appears to have a wide host range and to be synonymous with *A. phaseolorum*, a common pathogen of beans in the USA and the Netherlands. *A. phaseolorum* also causes substantial losses to green gram (*Phaseolus aureus*) and black gram (*P. munga*) in the Punjab (Sutton and Waterson, 1966).

Crossan (1958) conducted cross-inoculation experiments with isolates of *A. phaseolorum* from bean, soyabean and cowpea, *A. abelmoschi* from okra,

A. gossypii from cotton, *A. nicotiana* from tobacco, *A. lycopersici* from tomato and eggplant, and *A. capsici* from pepper. All isolates produced spores which were morphologically similar and caused similar symptoms on each of the hosts. Isolates varied somewhat in virulence towards a particular host but were generally more aggressive on okra, French bean and cotton. Isolates from cotton were more pathogenic towards beans than cotton.

Thomson (1953) was also able to produce symptoms on okra inoculated with a cotton isolate of the pathogen. Holdeman and Graham (1952) reported that isolates of *A. gossypii* from tobacco and cotton showed similar pathogenicity when inoculated into cotton cotyledons. Weimer (1951) noticed that Ascochyta canker on blue lupin (*Lupinus angustifolius*) always occurred on plots previously cropped to cotton. Inoculation experiments showed that the pathogen was more aggressive to cotton than to lupin.

Losses

Ascochyta blight is rarely now a serious disease, but severe epidemics have been recorded in the past. The disease became of economic importance in the south-east USA during the 1920s (Elliot, 1922). In the worst-affected areas of South Carolina, Armstrong (1939) reported that 50–90% of plants showed symptoms, and the tops of many of these died. Similar observations were made during a severe outbreak of the disease in Georgia in 1940 (Higgins, 1940). The disease was sporadic or locally endemic in Alabama in 1947–49 and became general in 1950, causing losses of seedlings and young plants at the six- to eight-leaf stage (Smith, 1950). Although the disease can be damaging to young seedlings, older plants quickly grow away from the disease (Armstrong, 1939).

Disease cycle and epidemiology

As the pathogen seems to have a wide host range which includes many crop species and presumably also weeds, these may serve as a reservoir of infection if in close proximity to a cotton field. The disease is also seed-borne but this is not considered to be an important source of primary inoculum, because of the long time required for the disease to develop from contaminated seed (Smith, 1950). The main source of primary inoculum is probably crop residues, as pycnidia are abundantly produced on dead leaves and stems. As the pycnidia mature, conidia are extruded from the ostiole in streams of a sticky substance and become attached to the cotyledons as they emerge through the soil. The lower leaves of young plants may also become infected by conidia splashed from the soil surface by rain. Primary infection may also be established in the crop as a result of wind-blown conidia from infested crop debris or a neighbouring infected crop. Although more damaging to young plants (Smith, 1950), the disease may attack the crop at any stage of its growth. If wet weather occurs after boll split, the fungus can grow on the lint, leading to deterioration of the fibres and contamination of the seed. Epidemics have usually been associated with prolonged periods of cloudy, wet weather and cool to moderate temperatures (20–25 °C) (Armstrong, 1939; Higgins, 1940; Smith, 1950).

CONTROL

The disease can be controlled by seed treatment (Smith 1950), rotation and deep ploughing to bury infested residues. In some areas it may be possible to alter planting dates to avoid periods of continuous cool, wet weather at the early stage of crop development when the disease is most damaging.

South-western Cotton Rust

HISTORY AND DISTRIBUTION

The first report of south-western cotton rust was based on material collected in Mexico in 1893 (Ellis and Everhart, 1897). An outbreak of the disease occurred in Texas in 1917 (Taubenhaus, 1917) and it was first reported from Arizona in 1922 (see Brown, 1939). The disease is now known in several states in the south-west of the USA, Mexico and parts of South America. The aecial stage appears to be confined to the USA and Mexico but the telial stage is more widespread, occurring in Argentina, the Bahamas, Bolivia, Brazil and the Dominican Republic (Mulder and Holliday, 1971a).

CAUSAL ORGANISM

Taxonomy and morphology

South-western cotton rust is caused by *Puccinia cacabata* Arth. & Holw. It is heteroecious and macrocyclic, producing pycnidial and aecial stages on *G. hirsutum*, *G. barbadense* and numerous wild species (Presley and King, 1943). Pycnia are amphigenous, yellow to orange in colour, becoming brown with age, are 90–120 μm in diameter and occur in raised circular groups. The aecia are mainly hypophyllous, occurring in groups surrounding the pycnia and 2–5 mm in diameter. Aeciospores are globose to oblong, 13–19 × 18–25 μm, with hyaline to yellow cells. Uredia are mainly epiphyllous, rarely amphigenous, and cinnamon brown in colour. Uredospores are globose to broadly ellipsoid, 19–24 × 23–31 μm with pale brown to cinnamon brown walls and three or four pores, which are distinct and equatorial. Telia are amphigenous, round, elliptical or linear. Teliospores are variable in shape and colour, rounded at both ends or slightly pointed at the apex, with the pedicel up to 90 μm in length (Fig. 6.10) (Mulder and Holliday, 1971a).

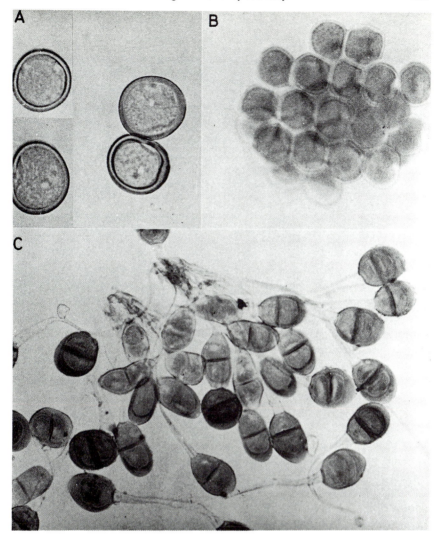

Fig. 6.10. *Puccinia cacabata*: A, uredospores × 800; B, aeciospores × 800; C, teliospores × 500 (Mulder and Holliday, 1971a).

The fungus was originally described as *Aecidium gossypii* Ell. & Ev. It was Presley and King (1943) who described the rust as a new species (*P. stakmanii* Pres.) on grama grass (*Bouteloua* spp.). The fungus was distinguished from other rusts in having uredospores with three equatorial germ pores instead of several scattered ones. As a result of comparative studies with a number of rust species, Hennen and Cummins (1956) then concluded that *P. stakmanii* was synonymous with *P. cacabata* Arth. & Holw.

Losses

The disease tends to be of sporadic occurrence, with infection being light in most years, but occasionally severe epidemics and heavy loss of yield has been reported. In Arizona in 1930, losses as high as 75% were recorded in the worst-affected fields. A severe outbreak of rust occurred in Arizona and New Mexico in 1959 with yield reductions of 50–75% (Smith, 1969). The disease appears to cause more severe losses in Mexico than in the USA. According to Duffield (1958), an epiphytotic in the Yaqui Valley of Mexico in 1956 caused losses of seed cotton estimated to be as high as 85%. Inoculation experiments in replicated trials resulted in a yield loss of 39% under conditions which gave rise to severe infection.

Symptoms and life cycle

The first symptoms usually seen in the field are small, yellow-orange spots on the upper leaf surface which contain the pycnia. Following inoculation, the first symptoms appear, after four to six days, as pale green lesions which develop into pycnial pustules. These pustules can appear on any aerial part of the plant but are more often seen on the leaves. As the pycnial pustules mature, they increase in size, ooze a thick yellow liquid which later becomes brown, and appear on the lower leaf surface. Seven to ten days after pycnium formation, yellow-orange, cup-like aecia begin to form on the underside of the leaf, erupting through the epidermis to form circular, protruding spots surrounding the pycnia (Presley and King, 1943; Percy, 1981).

The aeciospores are incapable of infecting cotton and must first encounter the alternate host, grama grass (Presley, 1942; Presley and King, 1943; Hennen and Cummins, 1956). The first symptoms on grama grass are uredial lesions which appear as elongated, brownish pustules, 12–14 days after infection. Spores from the uredial sori may then reinfect the grass. Telial lesions appear soon after the uredial lesions, mainly on the culms, but are also found on the leaves. The host epidermis becomes ruptured, exposing dark brown or black two-celled teliospores. Under favourable conditions, teliospores germinate to produce a four-celled basidium bearing a single basidiospore per cell. Basidiospores are released and carried by air currents to cotton plants, where they germinate to begin another disease cycle (Blank and Leathers, 1963). After severe attacks of the disease, the cotton leaf may be covered with several hundred lesions, causing it to curl and drop. In addition to causing defoliation, lesions on pedicels, bracts and young bolls may cause boll shedding.

Epidemiology

Disease initiation on cotton is favoured by high humidity, rainfall and a supply of teliospores which have overwintered on grama grass. In the USA, teliospores are formed in August and September. Optimal germination occurs after approximately ten months of weathering, during the later stages of which, periods of wetting and drying condition the spores for maximum germination. Teliospore germination, basidiospore production and infection of cotton occur when relative humidity levels exceed 85%, with temperatures below 28 °C for 13 hours or more (Blank and Fisher, 1974). Basidiospores produced under these conditions can infect cotton within a 5 km radius of their origin. Most basidiospores probably germinate between 18 and 24 hours after their release. They are very susceptible to desiccation, losing viability rapidly below 100% RH, and requiring free moisture for germination (Blank and Leathers, 1963).

The first attack of rust during the season is often only light and associated with the more moist, low-lying areas of the field. But, with succeeding rainfall, infection increases in severity.

The conditions required for aeciospore germination on grama grass and subsequent production of uredospores and teliospores are not as well defined as those required for infection of cotton. However, the peridium of each aecium remains intact, unless exposed to high RH or free moisture. After their release, aeciospores are highly sensitive to environmental changes and are unable to survive in the laboratory for more than 24 hours (Lewis, 1962, quoted by Blank and Leathers, 1963). Aeciospores are aerially disseminated and infection of grama grass occurs during periods of high RH, when there is free moisture on the leaves (Duffield, 1958).

The importance of uredospores in secondary infection of grama grass is not well understood but disease incidence is greatest on grass adjacent to cotton fields. This suggests that aecial lesions on cotton are the main source of inoculum (Blank and Leathers, 1963).

CONTROL

Cultural practice and fungicides

The earliest recommendations for control of south-western rust were to remove grass hosts and dispose of infested crop residues (Presley, 1942). Increasing frequency of epidemics in Mexico and the south-western USA prompted research into measures to reduce losses. Blank (1961) reported that zineb was an effective protectant fungicide and this was superseded by a zineb–maneb combination (Blank, 1971). In Mexico, bitertanol is used for rust control at 200–300 ml ha^{-1} (a.i.). To remain effective, fungicides must be applied at two-week intervals, beginning before the onset of the rainy season. It may be necessary to reduce the spray interval if favourable conditions for the disease persist for long periods. Sprays are normally required in areas prone to the disease if rust-favourable conditions persist for 13 hours or longer.

Resistant varieties

Presley and King (1943) inoculated a wide range of cultivated and wild cottons and found that nearly all the cultivated cottons were susceptible. The wild species were either susceptible or partially resistant. Duffield (1958) reported that strains of *G. arboreum*, *G. herbaceum*, *G. raimondii*, *G. gossypoides*, *G. anomalum* and *G. thurburii* were resistant. Some success was achieved in transferring resistance from *G. arboreum* and *G. anomalum* to *G. hirsutum* (Blank and Fisher, 1974), using glasshouse inoculation with teliospores from grama grass. Blank (1971) reported results of screening trials with a large number of cultivated and wild cottons, in which one strain of *G. anomalum* and several strains of *G. arboreum* were the most resistant.

In *G. hirsutum*, resistance to rust in the cotyledons is controlled by a single, dominant gene and its expression is independent of resistance in the stem and petiole, which is associated with the glandless and nectariless characters (Percy and Bird, 1985). Resistant varieties show a hypersensitive response to inoculation. The lesions appear, but quickly dry to necrotic flecks. Resistance is considered to be stable because of the absence of physiological races of the pathogen.

Tropical Rust

Distribution, host range and losses

Phakospora gossypii (Arth.) Hirat., the causal organism of tropical rust, is distributed throughout the tropics (CMI Distribution Map No. 258) but is conspicuously absent from some major cotton-growing countries, such as Egypt, Mexico, Zimbabwe, Central African Republic and the USA, except for Florida (Ebbels, 1976). The disease affects all the cultivated species of *Gossypium* and some wild species and has also been recorded on *Azanza garkeana* and *Thespesia populnea* (Punithalingam, 1968).

The disease usually appears late in the season, often after the onset of senescence, when it causes little damage to the crop, and may even be beneficial in aiding leaf fall before harvest. However, substantial losses due to the disease have occurred in Brazil, India and Jamaica, when the crop has been attacked at an earlier stage of development. In Brazil, a severe outbreak of the disease, which occurred 60 days after planting, caused losses estimated to be as high as 24% (Pineda, 1987). A similar estimate of yield loss was made in Coimbatore, India, following a severe outbreak of the disease (Johnson, 1963).

Symptoms

The uredia appear on the leaves as small (1–3 mm), pinkish brown spots which have a purple halo. On petioles and stems the uredia have an elongated shape. Severe infection causes defoliation. The disease only affects growing tissue but appears first on older tissue before spreading to the young leaves.

Causal organism

Phakospora gossypii is an autoecious rust producing primary (aecial), epiphyllous uredia and secondary (uredial) hypophyllous uredia, which are yellowish brown and 0.5–3.0 mm in diameter. The uredospores emerge in a cirrus from an apical pore. They are ellipsoidal or ovoid, hyaline to light yellow and 16–19 × 19–27 μm in size. Telia may also be produced but they are rare and inconspicuous. They are hypophyllous and light brown in colour and they bear pale brown, angular to irregularly oblong, teliospores, 10–14 × 24–32 μm in size (Punithalingam, 1968).

Factors affecting infection

Severe outbreaks of the disease have occurred during the dry season in irrigated cotton (Malaguti *et al.*, 1972), or after rain, following a long dry spell (Pineda, 1987). Disease development appears to be favoured by wide fluctuations in diurnal temperature and prolonged dew periods, under dry weather conditions. The disease does not become severe during wet weather (Sterne, 1981).

Control

In areas where severe outbreaks of tropical rust occur sufficiently early in crop development to cause yield loss, seed from an infected crop should not be used for planting. Contaminated residues should be burned or ploughed in at the end of the season.

There is some evidence that cultivars differ in degree of susceptibility to this rust. In Venezuela, some cultivars of *G. barbadense* had the lowest disease ratings in commercial plantings (Malaguti *et al.*, 1972).

Cotton Rust

A third rust fungus, *Puccinia shedonnardi* Kellerm. & Sw., affects cotton, but, although in the USA it is more widely distributed than *P. cacabata*, it is less important. The last report of severe defoliation attributed to this fungus was from Arizona in 1938.

The rust occurs on cotton mainly in Lower California, New Mexico, Oklahoma, Texas and Mexico. The aecial stage affects cotton and some other malvaceous plants. Uredial and telial stages occur on species of *Limnodea*, *Muhlenbergia*, *Shedonnardus* and others. The pathogen is morphologically similar to *P. cacabata* but is distinguished by producing amphigenous uredia and uredospores which have six to eight scattered pores and a thinner spore wall (Mulder and Holliday, 1971b).

Powdery Mildews

DISTRIBUTION

Powdery mildew on cotton can be caused by two different fungal species, *Leveillula taurica* (Lev.) Arn. and *Salmonia malachrae* (Seaver) Blumer & Muller. *L. taurica* has a wide host range and is of worldwide distribution (see CMI Distribution Map No. 217), occurring on cotton in the more arid cotton-growing areas, such as the Sudan and parts of India, Peru, the West Indies and the former Soviet Union (Bell, 1981). It has also been reported from China (Jia *et al.*, 1986).

S. malachrae has been recorded on cotton only in South America and the Antilles. In Peru, both mildews are found, sometimes affecting the same leaf, with the conidial stage of *L. taurica* mainly on the lower surface and *S. malachrae* mainly on the upper surface (Bell, 1981).

LEVEILLULA TAURICA

Symptoms

The first symptom of infection is the appearance of powdery white patches, mainly on the underside of the leaf. These patches are at first only about 2 mm in size and angular or irregularly rounded, depending on the cultivar. The affected leaf tissue turns yellow and, as the spots increase in size and number, the whole leaf becomes chlorotic and falls. Sometimes the fungus does not appear on the leaf surface, remaining within the mesophyll tissue, when leaf yellowing is the only visible symptom of infection.

Losses

In severe attacks, the host may be completely covered with the fungus, causing defoliation. However, infection usually occurs late in the season, causing little yield loss (Abbot, 1932). The disease has been known to occur sufficiently early in India, on a cultivar of *G. barbadense*, to reduce significantly the number of bolls set (Raychaudhuri, 1949).

Causal organism, morphology and variation

In some areas where *L. taurica* occurs on cotton, the sexual fruiting bodies (cleistothecia) are not produced and consequently the causal organism is sometimes referred to as *Oidiopsis gossypii*. Cleistothecia, when formed, are scattered and embedded in a dense superficial mycelium. They are initially globose, becoming concave at maturity with a diameter of 135–250 µm. Each ascocarp contains 20–35 two-spored asci, which are ovate, distinctly stipitate and 70–110 × 25–40 µm in size. Ascospores are cylindrical to pyriform, sometimes curved and 20–40 × 12–22 µm in size.

Conidia are borne singly on short hyphal branches. They are of two distinct shapes, cylindrical and navicular, and vary in size on different hosts,

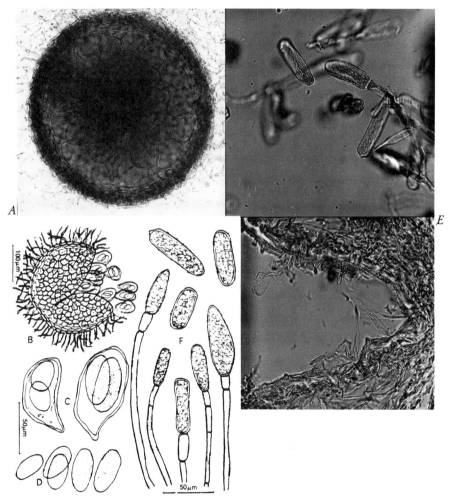

Fig. 6.11. *Leveillula taurica*: A, cleistothecium; B, cleistothecium and asci; C, asci; D ascospores; E, conidiophores from leaf surface and endophytic mycelium; F, conidiophores and conidia (Mukerji, 1968).

from 25–29 μm in length and from 14–20 μm in width (Fig. 6.11) (Mukerji, 1968).

Raychaudhuri (1949) suggested that the powdery mildew occurring on cotton in Dacca should be considered as a new form, because the conidia differed in size from those measured by Wakefield (1920) on cotton from the West Indies infected with *L. taurica*. However, the fungus produces conidia of different sizes on different hosts and there appears to be no connection between conidial size and pathogenic specialization. For instance, Nour (1958) found that *G. barbadense* and *G. hirsutum* both became infected with inoculum derived from the common weed species, *Euphorbia heterophylla*, or from *Vicia faba*. He considered *L. taurica* to be less specialized than other powdery mildews. Although Kamat and Patel (1949) had previously reported that there was considerable pathogenic specialization within this fungus, the view of Nour was supported by the work of Correll *et al.* (1988). They found that the tomato isolate of the pathogen had a wide host range, which included cotton, onion and chilli pepper, among the crop species tested, as well as a number of common and widely distributed weeds such as *Sonchus oleraceus*, *Senecio vulgaris* and *Xanthium strumarium*.

Isolates from different hosts do, however, vary in their pathogenicity to cotton. Although, in Sudan, isolates from *Gossypium* spp. and from *E. heterophylla* readily cross-infect, isolates from *Abutilon figarianum* failed to infect cotton (Nour, 1958). Furthermore, an isolate from Texas which infects the wild cottons, *G. tomentosum* and *G. logicalyx*, is avirulent towards the cultivated species (Bell, 1981).

Disease cycle and epidemiology

Under cool climatic conditions the fungus produces cleistothecia, but these are rarely seen in tropical and sub-tropical areas. Nour (1958) noted the absence of cleistothecia in Sudan, which presented him with difficulties in identifying the source of primary inoculum, as the conidia are short-lived. However, it is likely that, under hot, arid conditions, survival of the fungus is aided by its wide host range amongst common weed species. In the Sudan, for instance, *E. heterophylla* is widespread and a common alternative host for the pathogen. It is probably the main source of primary inoculum for the infection of cotton and broad beans (Nour, 1956, 1958). In one case of the disease in California, the source of infection of a cotton crop was traced to a nearby tomato crop (Correll, 1986).

Conidia of *L. taurica* are able to germinate under a wide range of humidities, from very dry to 100%. Maximum germination, with vigorous germ-tube growth, occurs at 85–100% RH. Germination rates are low and germ-tube growth poor at RH levels below 30%. Conidia can germinate in free moisture, although germination of submerged conidia is poor, compared with those floating on a film of water on the leaf surface (Nour, 1958).

Penetration occurs through the stomata. Nour (1958) was able to find no evidence of direct penetration through the leaf cuticle. However, the germ-tubes branch extensively, forming appressoria on the leaf surface which give rise to haustoria or allow penetration of the stomata. Stomatal penetration may be completed 24–48 hours after the start of germination when the RH exceeds 80% and temperature is in the range 20–30 °C (Bell, 1981).

The pathogen is an endoparasite and much of the mycelial development after initial penetration is confined to the mesophyll tissues, with only the conidiophores protruding through the stomata, to produce the white powdery symptom seen on the leaf surface.

Salmonia malachrae

Symptoms

Scattered circular patches of white mycelium usually appear on the upper leaf surface. Under conditions favourable to the disease, the spots enlarge and coalesce to affect the whole leaf surface and the leaf then becomes chlorotic and may be shed. Although the symptoms are similar to those caused by *L. taurica*, the white spots produced by *S. malachrae* are due to mycelial growth, with conidia rarely being seen. Cleistothecia are numerous and seen as small white to pale brown specks scattered through the mycelium (Abbot, 1932; Bell, 1981).

Causal organism

Conidia, which are 20–22 × 10–13 µm, are rarely found on cotton leaves. The cleistothecia are characteristic, being much smaller than those of *L. taurica*. They are usually in the range of 52–67 µm in diameter, hyaline or light brown in colour and without appendages. Each cleistothecium contains three asci, 30–40 µm in size, and each ascus contains five, or rarely six, ellipsoid hyaline ascospores, 14–20 × 10–14 µm in size (Bell, 1981).

Disease cycle and epidemiology

Because of its minor pathogen status, there is little information concerning the epidemiology of *S. malachrae*, but it appears to be favoured by weather conditions similar to those which favour *L. taurica*. In Peru, both mildews can be found on the same leaf. The fungus survives in the absence of cotton by infecting alternative hosts among common weed species and was first recorded on a malvaceous weed in Puerto Rico (Bell, 1981).

Control of powdery mildews

Neither of the powdery mildew diseases infecting cotton affect yield sufficiently to warrant control measures. Infection by *L. taurica* of other crops, such as beans and tomato, may require chemical control. Effective control has been achieved with lime–sulphur, thiram, cuprous oxide, benomyl, thiophanate, dinocap, fenaminosulph and other fungicides (e.g. Nour, 1957).

Myrothecium Leaf Spot

DISTRIBUTION AND HOST RANGE

Myrothecium roridum Tode ex Fr. is an important leaf spot pathogen of cotton in India (e.g. Munjal, 1960; Srinivasan and Kannan, 1974). It is also reported as a minor disease of the crop in the USA (Cognee and Bird, 1964).

The fungus has a wide host range, affecting solanaceous vegetable crops and cucurbits. Although it is widely distributed in temperate and tropical regions of the world (CMI Distribution Map No. 458), it is rarely reported as a pathogen of *Gossypium* spp. outside the Indian subcontinent.

SYMPTOMS AND LOSSES

First symptoms of the disease usually appear on the leaves of young plants, four to six weeks old (Chauhan and Suryanarayana, 1970), but it is also capable of causing pre-emergence and post-emergence damping-off of seedlings (Cognee and Bird, 1964). The leaf spots are at first circular and tan-coloured with violet-brown margins. They may enlarge to 3 cm in diameter and are surrounded by translucent areas which are concentrically zoned and which eventually bear sporodochia, appearing as black, pinhead-sized dots. Under optimal conditions for disease development, the spots increase in size and number, coalescing to affect large areas of the leaf, resulting in defoliation. The fungus can also attack both young and more woody stem tissue, causing stem lesions and dieback in a number of hosts, including cotton (Cognee and Bird, 1964; Fitton and Holliday, 1970).

Myrothecium leaf spot is considered to be one of the most important cotton diseases in the Indian states of Delhi and Rajasthan (Arya, 1960; Munjal, 1960). During the late 1960s, the disease became prominent in Punjab and Haryana States also (Chauhan and Suryanarayana, 1970). No estimates of yield losses are available but the disease may sometimes be sufficiently damaging in India to require chemical control (e.g. Sharma and Chauhan, 1985).

CAUSAL ORGANISM

Morphology

The sporodochia on host leaf or stem tissue are sessile, rarely slightly stipitate and very variable in size, ranging in diameter from 16 to 750 μm, with a depth of 50 to 200 μm. The spore mass is wet at first, then drying to form a hard, shining, black mass, usually surrounded by a fringe of entangled white hyphae. Conidiophores arise directly from epidermal cells of the host and are hyaline and cylindrical, dividing into two or three branches which bear the phialides. The phialides are in whorls of two to five, most cylindrical, hyaline but sometimes dilute olivaceous, occasionally darker at the apex and 9–27 × 1–1.8 μm in size. They are closely compacted into parallel rows forming a dense 'hymenial' layer. Spores are rod-shaped with rounded ends, hyaline or olivaceous, appearing black in mass and 4.4–10.8 × 1.3–2.7 μm in size (Fig. 6.12) (Fitton and Holliday, 1970).

Physiology

The fungus is able to utilize a wide range of carbon and nitrogen sources *in vitro*. D-glucose and raffinose were found to be the best carbon sources and glutamic acid the best source of nitrogen. Tyrosine and alanine were the best amino acids for stimulating mycelial growth and sporulation (Chauhan and Suryanarayana, 1970).

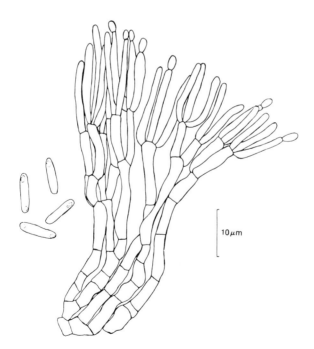

10 μm

Fig. 6.12. *Myrothecium roridum*: conidiophores and conidia (Fitton and Holliday, 1970).

Isolation and growth in culture

The fungus can be isolated from infected leaf tissue on standard culture media. Cultures on PDA reach 40–60 mm in diameter after 14 days at 25 °C. The colonies are white, floccose, wrinkled, often raised in the centre and pink to pinkish cinnamon in colour when viewed from the reverse. Sporulation spreads through the colony in concentric zones, with small groups of conidia forming rudimentary sporodochia. The sporodochial surface is at first white, becoming olivaceous-black, shiny and wet. Growth and sporulation are better in light than in dark (Fitton and Holliday, 1970).

EPIDEMIOLOGY

Epidemiological aspects of Myrothecium leaf spot are not well understood for cotton, but the pathogen is probably a common soil saprophyte, with the capacity to become pathogenic under certain conditions, not yet clearly defined (Fitton and Holiday, 1970). The disease cycle is probably similar to that of the anthracnose pathogen, the fungus being apparently capable of producing symptoms at all stages in the development of the cotton plant, including infection of the boll (Cognee and Bird, 1964). Primary sources of inoculum are therefore likely to be infested soil and infected weed species, as the pathogen has a wide host range. Optimum temperature for spore germination is reported to be 29 °C but tropical isolates may have higher temperature optima than do the temperate isolates.

Control

Only in India is the disease sufficiently severe to require control measures. In fields where the disease has occurred in the past, infested plant residues should be destroyed and susceptible weeds removed to prevent inoculum build-up. Copper oxychloride has been used successfully to control the disease (Sharma and Chauhan, 1985).

Cochliobolus Leaf Spot

DISTRIBUTION AND SYMPTOMS

Cochliobolus spicifer Nelson is widely distributed, especially in the tropics. It mainly causes seedling diseases but also leaf spots and dieback symptoms on a large number of hosts. An isolate from *G. hirsutum* caused symptoms when inoculated into 17 common crop species (Bedi *et al.*, 1969a).

Reports of serious disease in cotton caused by this pathogen are confined to India, although a leaf spot disease found on cotton plants in Malawi was attributed to *Helminthosporium* sp. (Peregrine and Sidiqui, 1972) and may be the same fungus, which is sometimes referred to as *H. spiciferum*.

This is a common leaf spot disease in India and has been serious in the Punjab, where it causes seed rot and pre-emergence damping-off, as well as premature defoliation in the adult plants. The leaf spot appears first on the seedling leaves, beginning as light yellow areas which increase in size, becoming dark brown and spreading to the leaf stalk in wet weather (Bedi *et al.*, 1967).

In addition to *C. spicifer*, a second species, *C. lunatus* Nelson & Haasis, has also been known to cause a leaf spot of cotton in India (Sharma and Chauhan, 1985). This fungus is found throughout the tropics and has a wide host range, but is a less important cotton pathogen than *C. spicifer*.

CAUSAL ORGANISMS

Cochliobolus spicifer (conidial state: *Curvularia spicifer* (Bainier) Boedijn) produces asci in a globose pseudoperithecium, which is ellipsoidal, black in colour, up to 710 μm high and 650 μm broad, often developing from a columnar stroma. It has a well-defined ostiolar beak, up to 670 μm in length. Asci are cylindrical to clavate, straight or slightly curved, bitunicate, short-stalked, one- to eight-spored and 130–260 × 12–20 μm. Ascospores are filiform, hyaline, tapered at the ends, six- to 16-septate, 20–40 × 9–14 μm in size and coiled in a helix within the ascus. Pseudoparaphyses are hyaline, filiform and branched.

Conidiophores arise singly or in small groups. They are flexuous, repeatedly geniculate, mid- to dark brown in colour and up to 300 μm in length and 4–9 μm in width. Conidia are straight, oblong or cylindrical, rounded at the ends, golden brown in colour and 20–40 × 9–14 μm in size (Fig. 6.13) (Sivanesan and Holliday, 1981).

C. lunatus can be readily distinguished from the other species by the shape of its conidia (see Fig. 6.14). Conidia are three-septate, almost always curved at the third cell from the base and 20–32 × 9–15 μm in size. Cells at each end are subhyaline or pale brown, while intermediate cells are darker brown (Ellis and Gibson, 1975).

CONTROL

Bedi *et al.* (1969b) evaluated several fungicides applied as seed-dressings and foliar sprays for control of *C. spicifer*. PCNB as seed treatments controlled pre-emergence losses, while zineb, ziram and captan, applied as foliar sprays, were all effective in limiting secondary spread of the fungus.

During a severe outbreak of the disease in the Punjab, a number of cultivars and lines were evaluated for disease resistance. Many appeared to be immune under conditions of natural infection and artificial inoculation. Immune cultivars included Abel 51–5, Acala SJ1 and Bobdel (Dutt *et al.*, 1973).

Fig. 6.13. *Cochliobolus spicifer.* conidiophores and conidia (× 409) (Sivanesan and Holliday, 1981).

Minor Leaf Spots

ANTHRACNOSE

The cotton anthracnose fungus, *Colletotrichum gossypii*, is known primarily as a pathogen of the seedling and of the boll, but it may also produce spots on the leaf when inoculum is abundant and conditions warm and humid within a dense canopy. The pinkish-brown spots appear mainly on the underside of the leaves, as a result of infection from contact with infested crop residues on the soil surface or by conidia carried by rain-splash from infected seedlings lying on the soil surface. If the infection becomes severe, large areas of leaf tissue around the main veins become necrotic. Acervuli containing fusiform conidia may be present as black dots on infected tissues. (A more detailed description of the pathogen is given in Chapter 1.)

RAMULOSE (WITCHES BROOM)

A fungus described as *Colletotrichum gossypii* South. var. *cephalosporioides* causes necrotic spots on the leaves and growth abnormalities (witches broom) of the stem (see Chapter 5). It is sufficiently important in the northwest of Brazil to warrant selection for resistance (Carvalho *et al.*, 1988). There is a correlation between pilosity and resistance. Smooth leaved lines were found to be more resistant than pubescent lines. Resistance is controlled by two genes which are partially dominant in their expression (Carvalho *et al.*, 1985).

RHIZOCTONIA LEAF SPOT

Although principally a seedling pathogen of cotton, *Rhizoctonia solani* (*Thanatephorus cucumeris* Donc.) is occasionally the cause of damage to the foliage, particularly in rank cotton in areas of high rainfall. The pathogen produces light brown irregular spots which may be 1–2 cm in diameter with a pronounced purple margin. Dead areas within the spot crack and fall out giving the leaf a ragged appearance (Neal, 1944). In very humid weather, the fungus is capable of growing up the external surface of the stem and under these conditions the sexually formed basidiospores have been found on cotton (Ullstrup, 1939). Basidiospores discharged into the air may be a source of inoculum for leaf infection. Kotila (1945) also noted the fact that *R. solani* could produce leaf spotting of cotton and that this cotton isolate was pathogenic to sugarbeet seedlings, on which it readily produced basidiospores.

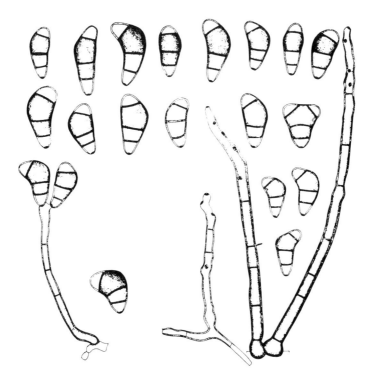

Fig. 6.14. *Cochliobolus lunatus*: conidiophores and conidia (× 516) (Ellis and Gibson, 1975).

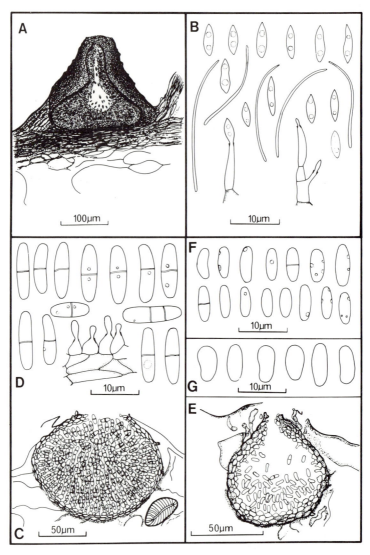

Fig. 6.15. *Phomopsis malvacearum* (Westend.) Died., IMI 22605. A, VS of partially erumpent conidioma on cotton boll; B, conidiophores, conidiogenous cells, alpha and beta conidia. C–D *Ascochyta gossypii* Woronichin on cotton leaf, IMI 111903. C, VS of immersed/partially erumpent pycnidial conidiomata within cotton leaf; D, conidiogenous cells and conidia. E–G *Phoma* spp. on cotton leaves referred to in the past as '*Phyllosticta*' species (e.g. *Phyllosticta malkoffii* Bubak). E, VS of immersed/partially erumpent pycnidial conidiomata, IMI 84774d. F, Conidia from specimen IMI 98920. G, Conidia from IMI 50074. (Courtesy Dr E. Punithalingam.)

PHOMOPSIS LEAF SPOT

A leaf spot caused by *Phomopsis* sp. appeared in Louisiana for the first time in 1963 and has not been reported elsewhere in the USA. The fungus attacks all above-ground parts of the plant. The first symptom is the appearance of slightly sunken, greyish, water-soaked areas on the leaf and stem. Later the lesions become covered with pycnidia, which produce conidia of two distinct shapes (see Fig. 6.15). The main damage is caused to the flower-bud, which fails to open if infected at an early stage of development (Ivey and Pinckard, 1967).

A fungus identified as *Phomopsis malvacearum* Western has been isolated from bolls in Zimbabwe (Rothwell, 1983) and in Tanzania (Ebbels and Allen, 1979).

PHYLLOSTICTA LEAF SPOT

Two species of *Phyllosticta* have been reported as causing leaf spots on cotton in Africa. Rothwell (1983) recorded *P. gossypina* Ell. & Mart. and *P. malkoffii* bub. in Zimbabwe. Riley (1960) also recorded *P. malkoffii* in Tanzania. Hopkins (1932) reported that *P. gossypina* was a common cause of leaf spotting in Zimbabwe, often occurring together with *Alternaria macrospora*. Recent examination of specimens in the herbarium at the International Mycological Institute suggests that different species of *Phoma* may have been described in the past as *Phyllosticta* spp. There appear to be no true *Phyllosticta* spp. occurring on cotton (E. Punithalingam, personal communication).

PHOMA LEAF SPOT

Phoma spp. occasionally cause leaf spots on cotton in Africa (R.J. Hillocks, unpublished). *P. exigua* has been recorded as a leaf spot pathogen of cotton in Haryana and Maharashtra States in India (Chauhan and Yadav, 1984). Symptoms are very similar to those of leaf spots caused by *Phomopsis* and *Ascochyta*, which produce circular necrotic spots (See Plate 4E) similar to those caused by *A. macrospora*, with the addition of pycnidia embedded in the necrotic tissue. These fungi are therefore difficult to distinguish on the basis of symptoms alone but can be separated on the basis of spore size and shape (see Fig. 6.15).

References

ABBOT, E.V. (1932) A powdery mildew on cotton from Peru. *Mycologia* 24, 4–6.
ALAGARSAMY, G., KRISHNAMOHAN, G. AND JEYARAJAN, R. (1989) Field evaluation of fungicides against leaf spot of cotton caused by *Alternaria macrospora*. *Indian Journal of Crop Protection* 17, 105–6.
ARAYA, H.C. (1960) On the occurrence of physiological strains of *Myrothecium roridum*. *Indian Phytopathology* 12, 164–7.

ARMSTRONG, G.M. (1939) Outbreak of Ascochyta blight on cotton in South Carolina. *Plant Disease Reporter* 22, 324–5.

ATKINSON, G.F. (1890) A new *Ramularia* on cotton. *Botanical Gazette* 15, 166–8.

ATKINSON, G.F. (1891) Black rust of cotton: a preliminary note. *Botanical Gazette* 16, 61–5.

AURANGABADKAR, J.H., SHUKLA, V.N. AND WANGIKAR, P.D. (1981) Reaction of some cotton varieties against grey mildew caused by *Ramularia areola*. *Indian Phytopathology* 34, 244.

BASHAN, Y. (1984) Transmission of *Alternaria macrospora* in cotton seeds. *Phytopathologische Zeitschrift* 110, 110–18.

BASHAN, Y. (1986) Phenols in cotton seedlings resistant and susceptible to *Alternaria macrospora*. *Journal of Phytopathology* 116, 1–10.

BASHAN, Y. AND LEVANONY, H. (1987) Transfer of *Alternaria macrospora* from cotton seed to seedling: light and scanning electron microscopy of colonization. *Journal of Phytopathology* 120, 60–8.

BASHI, E., SACHS, Y. AND ROTEM, J. (1983a) Relationship between disease and yield in cotton fields affected by *Alternaria macrospora*. *Phytoparasitica* 11, 89–98.

BASHI, E., ROTEM, J., PINNSCHMIDT, H. AND KRANZ, J. (1983b) Influence of controlled environment and age on development of *Alternaria macrospora* and on shedding of leaves in cotton. *Phytopathology* 73, 1145–7.

BEDI, P.S., TRIPATHI, M.N. AND SURYANARYANA, D. (1967) Outbreaks and new records–India: a new disease of cotton in Punjab. *FAO Plant Protection Bulletin* 15, 77.

BEDI, P.S., TRIPATHI, M.N. AND SURYANARYANA, D. (1969a) Host range of *Cochliobolus spicifer* from *Gossypium hirsutum*. *Indian Phytopathology* 22, 270–1.

BEDI, P.S., TRIPATHI, M.N. AND SURYANARYANA, D. (1969b) Studies on the chemical control of *Helminthosporium* blight of cotton. *Review of Plant Pathology* (1971) 50, 116.

BELL, A.A. (1981) Aerolate mildew. In: Watkins, G.M. (ed.), *Compendium of Cotton Diseases*. American Phytopathological Society, St Paul, Minnesota, pp. 32–5.

BHASKARAN, R. AND SHANMUGAM, N. (1973) Preliminary studies on the efficacy of aureofungin-sol against foliar diseases of cotton. *Review of Plant Pathology* (1974) 53, 112.

BHASKARAN, R., NATARAJAN, C. AND MOHARAJ, D. (1975) Biochemistry of resistance and susceptibility in cotton to *Alternaria macrospora*. *Acta Phytopathologica Academiae Scientiarum Hungaricae* 10, 33–40.

BLANK, L.M. (1961) Preliminary studies on control of southwestern cotton rust. *Plant Disease Reporter* 45, 241–3.

BLANK, L.M. (1971) Southwestern cotton rust. *Proceedings of the Beltwide Cotton Production Research Conference*, Memphis, Tennessee, pp. 76–7.

BLANK, L.M. AND FISHER, W.D. (1974) Southwestern cotton rust. *Proceedings of the Beltwide Cotton Production Research Conference*, Memphis, Tennessee, p. 20.

BLANK, L.M. AND LEATHERS, C.R. (1963) Environmental and other factors influencing development of southwestern cotton rust (*Puccinia stakmannii*). *Phytopathology* 53, 921–8.

BROWN, J.G. (1939) Cotton rust in Arizona. *Plant Disease Reporter* 22, 380–2.

CALVERT, O.H., SAPPENFIELD, W.P., HICKS, R.D. AND WYLLIE, T.D. (1964) The *Cercospora–Alternaria* leaf blight complex of cotton in Missouri. *Plant Disease Reporter* 48, 466–7.

CARVALHO, L.P. de, LIMA, E.F., RAMALHO, F. de S., LUKEFAHR, M.J. AND CARVALHO, J.M.F.C. (1985) Influencia da pliosidade do algodoeiro na expressao de sintomas de ramulose. *Fitopatologia Brasileira* 10, 649–54.

CARVALHO, L.P. de, LIMA, E.F., CARVALHO, J.M.F.C. AND MOREIRA, J.A.N. (1988) Heranca da resistencia a ramulose do algodoeiro (*Colletotrichum gossypii* var *cephalosporioides*). *Fitopatologia Brasileira* 13, 10–15.

CAUQUIL, J. AND SEMENT, G. (1973) Le faux mildiou du cotonnier (*Ramularia areola* Atk.) dans le sud-ouest de Madagascar. *Coton et Fibres Tropicales* 28, 279–86.

CHAUHAN, M.S. (1985) Grey mildew disease of arboreum cotton in Haryana. *Indian Journal of Mycology and Plant Pathology* 13, 214–15.

CHAUHAN, M.S. AND SURYANARAYANA, D. (1970) Physiological studies of the fungus causing Myrothecium leaf spot in Haryana, India. *Cotton Growing Review* 47, 29–35.

CHAUHAN, M.S. AND YADAV, J.P.S. (1984) leaf spot of cotton caused by *Phoma exigua*. *Haryana Agricultural University Journal of Research* 14, 92–3.

CHIPENDALE, H.G. (1929) The development in culture of *Ascochyta gossypii* Syd. *Transactions of the British Mycological Society* 3/4, 201–15.

CHOPRA, B.L., BEDI, P.S. AND SOMDU, T.T. (1985) Studies on Helminthosporium leaf blight of cotton in the Punjab. *Indian Journal of Mycology and Plant Pathology* 15, 155–8.

COGNEE, M. AND BIRD, L.S. (1964) *Myrothecium roridum* Tode as a cotton pathogen. *Annual Meeting of the American Phytopathological Society*, Atlanta.

CORRELL, J.C. (1986) Powdery mildew of cotton caused by *Oidiopsis taurica* in California. *Plant Disease* 70, 259 (Abstr.).

CORRELL, J.C., GORDON, T.R. AND ELLIOT, V.J. (1988) The epidemiology of powdery mildew on tomatoes. *California Agriculture* 42, 8–10.

COTTY, P.J. (1987a) Evaluation of cotton cultivar susceptibility to Alternaria leaf spot. *Plant Disease* 71, 1082–4.

COTTY, P.J. (1987b) Temperature induced suppression of Alternaria leaf spot of cotton in Arizona. *Plant Disease* 71, 1138–40.

CROSSAN, D.F. (1958) The relationship of seven species of *Ascochyta* occurring in North Carolina. *Phytopathology* 48, 248–55.

DAKE, G.N. AND KANNAN, A. (1982) Reaction of cotton species and varieties to *Ramularia areola*. *Indian Phytopathology* 35, 156–8.

DAVID, J.C. (1988) *Alternaria gossypina*. CMI Descriptions of Pathogenic Fungi and Bacteria No. 953. CAB International Mycological Institute, Kew, UK.

DUFFIELD, P.C. (1958) Biology of *Puccinia stakmannii*. *Dissertation Abstracts* 19, 419.

DUTT, S., BEDI, P.S. AND SINGH, T.H. (1973) Evaluation of genetic stock of cotton vis-à-*vis* attack of blight in Punjab state. *Journal of Research of Punjab Agricultural University* 10, 401–3.

DZHAMALOV, A. (1973) Irrigation and Alternaria leaf spot of cotton. *Zaschita Rastenii* 12, 48.

EBBELS, D.L. (1976) Diseases of Upland cotton in Africa. *Review of Plant Pathology* 55, 747–63.

EBBELS, D. AND ALLEN, D.J. (1979) A supplementary and annotated list of plant diseases, pathogens and associated fungi in Tanzania. *Phytopathological Paper* No. 22. Commonwealth Agricultural Bureaux, Farnham Royal, Bucks, UK, 89 pp.

EHRLICH, J. AND WOLF, F.A. (1932) Areolate mildew of cotton. *Phytopathology* 22, 229–40.

ELLIOT, J.A. (1922) A new *Ascochyta* disease of cotton. *Arkansas Agricultural Experimental Station Bulletin* 178. 12 pp.

ELLIS, J.B. AND EVERHART, B.M. (1897) New West American fungi. III. *Erythea* 5, 5–7.

ELLIS, M.B. (1971) *Dematiaceous Hyphomycetes*. CAB International, Wallingford, UK, 608 pp.

ELLIS, M.B. AND GIBSON, I.A.S. (1975) *Cochliobolus lunatus*. *Descriptions of Plant Pathogenic Fungi and Bacteria* No. 302. Commonwealth Agricultural Bureaux, Farnham Royal, Bucks, UK.

FAULWETTER, R.C. (1918) The Alternaria leaf spot of cotton. *Phytopathology* 8, 98–105.

FITTON, M. AND HOLLIDAY, P. (1970) *Myrothecium roridum*. *CMI Descriptions of Pathogenic Fungi and Bacteria* No. 253. Commonwealth Agricultural Bureaux, Farnham Royal, Bucks, UK.

FROLICH, G. AND RODEWALD, W. (1970) Diseases and pests of fibre crops and their control. In: *Pests and Diseases of Tropical Crops and Their Control*. Pergamon Press, Oxford p. 371.

HENNEN, J.F. AND CUMMINS, G.B. (1956) Uredinales parasitizing grasses of the tribe Chloridae. *Mycologia* 48, 126–62.

HEWISON, H.K. AND SYMOND, J.E. (1928) Observations on a fungus disease and an insect pest of cotton. *Empire Cotton Growing Review* 5, 48–51.

HIGGINS, B.B. (1940) Outbreak of Ascochyta blight of cotton in Georgia. *Plant Disease Reporter* 24, 327.

HILLOCKS, R.J. (1991) Alternaria leaf spot of cotton with special reference to Zimbabwe. *Tropical Pest Management* 37, 124–8.

HILLOCKS, R.J. AND CHINODYA, R. (1989) The relationship between Alternaria leaf spot and potassium deficiency causing premature defoliation of cotton. *Plant Pathology* 38, 502–8.

HOLDEMAN, Q.L. AND GRAHAM. T.W. (1952) Ascochyta leaf spot in tobacco plant beds in South Carolina. *Plant Disease Reporter* 36, 8.

HOLLIDAY, P. AND PUNITHALINGHAM, E. (1970) *Aschochyta gossypii*. *CMI Descriptions of Pathogenic Fungi and Bacteria* No. 27, Commonwealth Mycological Institute, Kew, Surrey, UK.

HOPKINS, J.C.E. (1931) *Alternaria gossypina* (Thum) comb. nov. causing a leaf spot and boll rot of cotton. *Transactions of the British Mycological Society* 16, 136–45.

HOPKINS, J.C.E. (1932) Some diseases of cotton in Southern Rhodesia. *Empire Cotton Growing Review* 9, 109–18.

IVEY, J.L. AND PINCKARD, J.A. (1967) A new disease of cotton caused by *Phomopsis*. *Phytopathology* 57, 462 (Abstr.).

JIA, Z.H., QIU, R.F., TANG, B. AND JIA, J. (1986) The powdery mildew of cotton in Xinjang, a new record in China. *Acta Phytopathologica Sinica* 16, 159–60.

JOHNSON, A. (1963) *Quarterly Report for July–September 1963 of the Plant Protection Committee for South East Asia and Pacific Region*. FAO, Bangkok. 21 pp.

JOHNSON, E.M. AND VALEAU, W.D. (1949) Synonymy in some common species of *Cercospora*. *Phytopathology* 39, 763–70.

JOLY, P. AND LAGIRE, R. (1971) A propos d'*Alternaria macrospora* Zim. parasite des feuilles de cotonier (*G. hirsutum* L.). *Coton et Fibres Tropicales* 26, 259–62.

JONES, G.H. (1928) An *Alternaria* disease of the cotton plant. *Annals of Botany* 42, 935–47.

KAMAT, M.N. AND PATEL, M.K. (1949) Some new hosts of *Oidiopsis taurica* (Lev.) Salmon in Bombay. *Indian Phytopathology* 1, 153–8.

KAMEL, M., IBRAHIM, A.N., KAMEL, S.A. AND EL-FAHL, A.M. (1971a) Spore germination of *Alternaria tenuis* in juice of susceptible and resistant cotton cultivars to Alternaria leaf spot disease. *United Arab Republic Journal of Botany* 14, 245–54.

KAMEL, M., IBRAHIM, A.N., KAMEL, S.A. AND EL-FAHL, A.M. (1971b) Inheritance of resistance to *Alternaria tenuis*, causing Alternaria leaf spot disease in a cross between Missella Valley and Bahatim 101 cotton cultivars. *United Arab Republic Journal of Botany* 14, 255–63.

KODELWAR, R. (1972) Grades for evaluating grey mildew caused by *Ramularia areola* Atk. in *Gossypium arboreum*. *Indian Journal of Agricultural Science* 42, 913–15.

KOTILA J.E. (1945) Cotton leaf spot *Rhizoctonia* and its perfect stage on sugar beet. *Phytopathology* 35, 741–3.

LIANG, P.Y. (1964) Identification of *Phytophthora* species causing cotton boll blight and castor bean blight in North China. *Acta Phytopathologica Sinica* 7, 11–20.

LING, L. AND YANG, J.Y. (1941) Stem blight of cotton caused by *Alternaria macrospora*. *Phytopathology* 31, 664–71.

LITTLE, S. (1987) *Cercospora gossypina*. *CMI Descriptions of Pathogenic Fungi and Bacteria* No. 914. CAB International, Wallingford, UK.

LUCAS, G.B. (1971) *Alternaria alternata* (Fries) Keissler, the correct name for *Alternaria tenuis* and *Alternaria longipes*. *Tobacco Science* 15, 37–42.

MacDONALD, D., RUSTON, D.F. AND KING, H.E. (1945) Barberton reports. *Reports from Experimental Stations 1943–1944*. Empire Cotton Growing Corporation.

MAIER, C.R. (1965) The importance of *Alternaria* spp. in the cotton seedling disease complex in New Mexico. *Plant Disease Reporter* 49, 904–9.

MALAGUTI, G., LOPEZ PINTO, O. AND ALFONZO, M. (1972) Rust caused by *Phakospora gossypii* in commercial cotton fields. *Review of Plant Pathology* (1974) 53, 300.

MILLER, J.W. (1969) The effect of soil moisture and plant nutrition on the Cercospora–Alternaria leaf blight complex of cotton in Missouri. *Phytopathology* 59, 767–9.

MUKERJI, K.G. (1968) *Leveillula taurica*. *CMI Descriptions of Pathogenic Fungi and Bacteria* No. 182. Commonwealth Agricultural Bureaux, Farnham Royal, Bucks, UK.

MULDER, J.L. AND HOLLIDAY, P. (1971a) *Puccinia cacabata*. *CMI Descriptions of Pathogenic Fungi and Bacteria* No. 294. Commonwealth Agricultural Bureaux, Farnham Royal, Bucks, UK.

MULDER, J.L. AND HOLLIDAY, P. (1971b) *Puccinia shedonnardi*. *CMI Descriptions of Pathogenic Fungi and Bacteria* No. 294. Commonwealth Agricultural Bureaux, Farnham Royal, Bucks, UK.

MUNJAL, R.L. (1960) A commonly occurring leaf spot disease caused by *Myrothecium roridum* Tode & Fr. *Indian Phytopathology* 3, 150–5.

NEAL, D.C. (1944) Rhizoctonia leaf spot of cotton. *Phytopathology* 34, 599–602.

NOUR, M.A. (1956) *Leveillula taurica* on cotton in the Sudan. *Commonwealth Phytopathological News* 2, 44.

NOUR, M.A. (1957) Control of powdery mildew diseases in the Sudan with special reference to broad bean. *Empire Journal of Experimental Agriculture* 25, 119–31.

NOUR, M.A. (1958) Studies on *Leveillula taurica* (Lev) Am. and other powdery mildews. *Transactions of the British Mycological Society* 41, 17–38.

PADAGANUR, G.M. (1979) The seed borne nature of *Alternaria macrospora* Zimm. in cotton. *Madras Agricultural Journal* 66, 325–6.

PADAGANUR G.M. AND BASAVARAJ, M.K. (1983) Spray schedule for the control of important foliar diseases and bollworms of cotton in the transition belt of Karnataka. *Indian Journal of Agricultural Science* 53, 725–9.

PADAGANUR, G.M. AND BASAVARAJ, M.K. (1984) Elimination of seed borne inoculum of *Alternaria macrospora* Zimm. causing blight of cotton by fungicidal seed treatment. *Plant Pathology Newsletter* 2, 20–1.

PADAGANUR, G.M. AND BASAVARAJ, M.K. (1987) Fungicidal control of Alternaria blight of cotton in transition belt of Karnataka. *Indian Journal of Agricultural Science* 57, 445–7.

PATEL, J.G., PATEL, J.K. AND PATEL, A.J. (1983) Efficacy of certain systemic and non-systemic fungicides in control of Alternaria leaf spot of cotton. *Indian Journal of Mycology and Plant Pathology* 13, 227–8.

PERCY, R.G. (1981) Southwestern cotton rust. In: Watkins, G.M. (ed.) *Compendium of Cotton Diseases*. American Phytopathological Society, St Paul, Minnesota, USA, pp. 37–9.

PERCY, R.G. AND BIRD, L.S. (1985) Rust resistance expression in cotyledons, petioles and stems of *Gossypium hirsutum* L. *Journal of Heredity* 76, 202–4.

PEREGRINE, W.T.H. AND SIDIQUI, M.A. (1972) *A Revised and Annotated List of Plant Diseases in Malawi*. Commonwealth Mycological Institute, Kew, Surrey, UK. 51pp.

PINEDA, L.B. (1987) Tropical rust of cotton *ASCOLFI Informa* 13, 40–2.

PRESLEY, J.T. (1942) Cotton rust in Arizona. *Plant Disease Reporter* 24, 144–5.

PRESLEY, J.T. AND KING, C.J. (1943) A description of the fungus causing cotton rust, and a preliminary survey of its hosts. *Phytopathology* 33, 382–9.

PUNITHALINGAM, E. (1968) *Phakospora gossypii*. *CMI Descriptions of Pathogenic Fungi and Bacteria* No. 172. Commonwealth Agricultural Bureaux, Farnham Royal, Bucks, UK.

RANE, M.S. AND PATEL, M.K. (1956) Diseases of cotton in Bombay. 1. *Alternaria* leaf spot. *Indian Phytopathology* 9, 106–13.

RASULEV, U.U. AND DZHAMALOV, A. (1975) Macrosporiosis and alternariosis of fine fibre cotton and their control. *Review of Plant Pathology* (1975) 54.

RATHAIAH, Y. (1973) Study of false mildew of cotton caused by *Ramularia areola* Atk. 1. Growth and sporulation of the fungus in culture. *Coton et Fibres Tropicales* 28, 287–92.

RATHAIAH, Y. (1976) Reaction of cotton species and cultivars to four isolates of *Ramularia areola*. *Phytopathology* 66, 1007–9.

RATHAIAH, Y. (1977) Spore germination and mode of cotton infection by *Ramularia areola*. *Phytopathology* 67, 351–7.

RAYCHAUDHURI, S.P. (1949) *Oidiopsis gossypii* (Wakef) Raychauhudri f. indica f. nov. on cotton. *Transactions of the British Mycological Society* 32, 288–90.

RILEY, E.A. (1960) A revised list of plant diseases in Tanganyika Territory. *Mycological Paper* No. 75. Commonwealth Agricultural Bureaux, Farnham Royal, Bucks, UK 42 pp.

ROTEM, J., WENDT, U. AND KRANZ, J. (1988a) The effect of sunlight on symptom expression of *Alternaria alternata* on cotton. *Plant Pathology* 37, 12–15.

ROTEM, J., EIDT, J., WENDT, U. AND KRANZ, J. (1988b) Relative effects of *Alternaria alternata* and *Alternaria macrospora* on cotton in Israel. *Plant Pathology* 37, 16–19.

ROTEM, J., BLICKLE, W. AND KRANZ, J. (1989) Effect of environment and host on sporulation of *Alternaria macrospora* in cotton. *Phytopathology* 79, 263–6.

ROTHWELL, A. (1983) A revised list of plant diseases occurring in Zimbabwe. *Kikikia* 12, 233–351.

RUSSELL, T.E. AND HINE, R.B. (1978) *Alternaria* leaf spot of Arizona-grown Pima ELS cotton. In: *Proceedings of the Beltwide Cotton Production Research Conference*, Memphis, Tennessee, p. 32.

SCIUMBATO, G.L. AND PINCKARD, J.A. (1974) *Alternaria macrospora* leaf spot of cotton in Louisiana in 1972. *Plant Disease Reporter* 58, 201–8.

SCIUMBATO, G.L. AND WALKER, H.L. (1980) A comparison of four *Alternaria* isolates associated with cotton and spurred anoda. *Proceedings of the Beltiwde Cotton Production Research Conference* (Abstr.), Memphis, Tennessee.

SHAHIN, E.A. AND SHEPHERD, J.F. (1979) An efficient technique for inducing profuse sporulation of *Alternaria* spp. *Phytopathology* 69, 618–20.

SHARMA, B.K. AND CHAUHAN, M.S. (1985) Studies on the chemical control of foliar diseases of cotton in Haryana State. *Agricultural Science Digest* 5, 153–6.

SHARMA, Y.R., SANDHU, B.S. AND GILL, M.S. (1986) Screening of cotton germplasm for resistance to grey mildew. *Plant Disease Research* 1, 82.

SIMMONS, E.G. (1967) Typification of *Alternaria*, *Stemphylium* and *Ulocladium*. *Mycologia* 59, 67–92.

SINCLAIR, J.B. AND SHATLA, M.N. (1962) Stemphylium leaf spot of cotton reported from Louisiana. *Plant Disease Reporter* 46, 744.

SINGANMATHI, B.V. AND EKBOTE, M.V. (1980) The chemical control of Alternaria leaf blight of cotton in Maharashtra State. *Journal of the Maharashtra Agricultural Universities* 5, 87–8.

SIVANESAN, A. AND HOLLIDAY, P. (1981) *Cochliobolus spicifer. CMI Descriptions of Pathogenic Fungi and Bacteria* No. 702. Commonwealth Agricultural Bureaux, Farnham Royal Bucks, UK.

SMITH, A.L. (1950) Ascochyta blight of cotton in Alabama in 1950. *Plant Disease Reporter* 34, 233–5.

SMITH, T.E. (1969) Observations on cotton rust (*Puccinia stakmanii*) under severe disease conditions. *Plant Disease Reporter* 42, 77–9.

SPROSS-BLICKLE, B., ROTEM, J., PERL, M. AND KRANZ, J. (1989) The relationship between infections of the cotyledons of *Gossypium barbadense* and *Gossypium hirsutum* with *Alternaria macrospora* and cotyledon abscission. *Physiological and Molecular Plant Pathology* 35, 293–9.

SRINIVASAN, K.V. AND KANNAN, A. (1974) Myrothecium and Alternaria leaf spots of cotton in South India. *Current Science* 43, 489–90.

STERNE, R. (1981) Tropical Rust In: Watkins, G.M. (ed.), *Compendium of Cotton Diseases*. American Phytopathological Society, St Paul, Minnesota, p. 39.

SUTTON, B.C. AND WATERSON, J.M. (1966) *Ascochyta phaseolorum. CMI Descriptions of Pathogenic Fungi and Bacteria* No. 81. Commonwealth Agricultural Bureaux, Farnham Royal, Bucks, UK.

SYDOW, H. AND BUTLER, E.J. (1916) Fungi Indiae Orientalis, 5. *Annals of Mycology* 14, 177–85.

TAUBENHAUS, J.J. (1917) On a sudden outbreak of cotton rust in Texas. *Science* 46, 267–9.

THOMPSON, G.E. (1950) Variability of cultures of *Ascochyta gossypii* and its pathogenicity on cotton seedlings. *Phytopathology* 40, 791 (Abstr.).

THOMPSON, G.E. (1953) A comparison of *Ascochyta abelmoschi* Hunter and *Ascochyta gossypii* Sydow in culture and inoculation experiments. *Phytopathology* 43, 293–4 (Abstr.).

ULLSTRUP, A.J. (1939) The occurrence of the perfect stage of *Rhizoctonia solani* in plantings of diseased cotton seedlings. *Phytopathology* 29, 373–4.

WAKEFIELD, E.M. (1920) On two species of *Ovulariopsis* from the West Indies. *Kew Bulletin*, 235–8.

WALKER, H.L. (1981) Host range studies on four *Alternaria* isolates pathogenic to cotton (*Gossypium* spp.) or spurred anoda (*Anoda cristata*). *Plant Science Letters* 22, 71– 5.

WALKER, H.L. AND SCIUMBATO, G.L. (1979) Evaluation of *Alternaria macrospora* as a potential biocontrol agent for spurred anoda (*Anoda cristata*): host range studies. *Weed Science* 27, 612–14.

WATKINS, G.M. (1981) Leaf spots. In: Watkins, G.M. (ed.), *Compendium of Cotton Diseases*. American Phytopathological Society, St Paul, Minnesota, USA, pp. 28–30.

WEIMER, J.L. (1951) *Ascochyta* canker of blue lupin. *Plant Disease Reporter* 35, 81–2.

7

Fungal Diseases of the Boll

R.J. Hillocks

Introduction

Microbial decay of the boll before normal dehiscence occurs to some extent in all areas of the world where cotton is grown. However, primary boll rot makes a significant contribution to yield loss only in areas of persistently high humidity or where the crop produces particularly dense, vegetative growth.

In the USA, boll rot is the most important cotton disease, in terms of total national crop loss, causing an estimated 5% loss of lint in 1988 (Blasingame, 1989). Losses in individual states may be considerably higher than this when the end of the season is wet, reaching as much as 20% in Georgia and North Carolina.

Over 100 micro-organisms have been isolated from rotted bolls, at one time or another, in different parts of the world. Most of these are wound pathogens causing boll rot after insect damage or premature rupture of the boll suture, while others are secondary invaders of already infected tissue. Bagga (1970a) inoculated green bolls with a number of micro-organisms isolated from infected bolls. All organisms tested caused boll decay following wound inoculation by puncturing the boll wall with a needle. Only *Diplodia gossypina*, *Glomerella gossypii*, *Myrothecium roridum*, *Xanthomonas campestris* and *Bacillus subtilis* caused decay after contact inoculation. Additional fungi have also been described as primary boll rot pathogens by other authors and results of pathogenicity tests doubtless vary considerably according to the conditions of the test, the virulence of the fungal isolate, the age of the boll and the cotton cultivar used. A list of fungi reported as primary boll rot pathogens is given in Table 7.1. For a more complete list of micro-organisms isolated from rotted bolls, see Roncadori (1969), Lagière (1972) and Krishnamurty and Verma (1974).

Colletotrichum gossypii was at one time the most important boll rot

Table 7.1. Fungi recorded as primary boll rot pathogens.

Pathogen	Reference
Alternaria alternata	Cauquil and Ranney, 1969
Alternaria spp.	Roncadori, 1969
Ascochyta gossypii	Pinckard *et al.*, 1981
Colletotrichum capsici	Chopra *et al.*, 1975
Colletotrichum gloeosporioides	Leakey and Perry, 1966
Diplodia gossypina	Wang and Pinckard, 1973
Fusarium moniliforme	Belliard, 1972
Phytophthora spp.	Allen and West, 1986
Rhizoctonia solani	Pinckard and Luke, 1967
Myrothecium roridum	Bagga, 1970a

fungus in many cotton-growing states of the USA (e.g. Miller, 1943) but it has become less important with the widespread use of seed-dressing fungicides. The main fungi associated with boll rot in the USA today are *Fusarium* spp., *Alternaria* spp. *Diplodia gossypina* and *Aspergillus* spp. (e.g. Bagga and Ranney, 1969a, b; Cotty, 1988). In Africa and other parts of the developing world, *Colletotrichum* spp. probably remain the most important boll rot pathogens, together with *Fusarium moniliforme* and *Diplodia gossypina* (Lagière, 1972; Krishnamurty and Verma, 1974).

Mode of Infection

The means by which boll rot pathogens gain entry to the boll varies to some extent with the particular micro-organism involved. Ashworth and Hine (1971) have divided them into three groups based on mode of infection: (i) those which are capable of direct penetration of the intact boll, such as *D. gossypina*; (ii) those which are introduced by insects, e.g. *Nematospora* spp.; and (iii) those which gain entry only after the boll wall has been damaged by insects, or after the suture has ruptured, such as *Rhizopus* spp. and *Penicillium* spp.

Damage to the boll or premature rupture of the suture will enhance infection by all potential boll rot pathogens, but the wound pathogens and secondary invaders may be the more vigorous colonizers of exposed, immature lint than are the primary pathogens. Among the primary pathogens, not all are capable of penetrating the wax cuticle and epidermis of the boll wall. They are none the less able to invade the undamaged boll by infection of the epidermal hairs (Kay *et al.*, 1975) or the bracts (Pinckard and Wang, 1972). Internal infection can also take place by way of the nectaries (Bagga and Laster, 1969) and stomata. By 30 days after anthesis, as many as 70% of boll stomata may be senescent and unable to close, providing entry sites for micro-organisms (e.g. Pinckard and Montz, 1971). In addition, *Aspergillus flavus* and certain fusaria can enter the boll through the vascular system.

Conditions Favouring Infection

Prolonged periods of high atmospheric humidity are the main requirement for a boll rot epidemic, but Ranney *et al.* (1971) listed four conditions which favour boll infection: (i) long periods with free moisture on the plants; (ii) long periods when the relative humidity exceeds 75%; (iii) low light intensity; and (iv) high temperature. In Georgia in 1968, which was a dry year, yield loss due to boll rot was estimated at 1.5%, compared with 14% in the following season, which was warm and wet. If the crop is particularly vegetative, relative humidity within the canopy may be sufficiently high to favour boll rot, even when general atmospheric humidity is below the optimum for infection by boll rot micro-organisms (Ranney, 1971). This effect was seen in the USA during the 1970s when cotton-growers experienced increased losses due to boll rot, which was attributed to greater use of nitrate fertilizers.

To some extent, the composition of the boll rot microflora varies in different climatic conditions. Arndt (1950) reported that cool wet weather followed by high temperatures was conducive to boll rot in general but the main pathogens of the boll rot complex differed according to the amount of rainfall in a given area. *Colletotrichum gossypii*, *Diplodia gossypina* and *Fusarium moniliforme* predominated in very wet weather, while under drier conditions *Ascochyta gossypii*, *C. gossypii* and other *Fusarium* spp. predominated. *Aspergillus flavus* tends to infect the boll under drier conditions than are suitable for other boll rot organisms (Marsh *et al.*, 1973).

The distribution of boll rot fungi in the USA appears to be influenced by temperature as well as by rainfall. In the cooler central valley of California, where air temperature in the summer averages about 26 °C, the most common boll rot fungi are *Nigrospora oryzae* and *F. moniliforme*. However, in the southern valley, with average temperatures of 32 °C, the predominant fungi associated with boll rot are *Aspergillus* spp. and *Rhizopus nigricans* (Halisky *et al.*, 1961).

Control

Methods for reducing boll rot incidence usually involve agronomic practices aimed at maintaining low humidity within the crop canopy or measures to reduce the level of primary inoculum.

Fallen crop debris, particularly flowers and bracts from the current crop, are the main source of inoculum for boll infection (Sanders and Snow, 1977). Some success has been achieved, therefore, in reducing the percentage of rotted bolls either by burning trash under the canopy with tractor-mounted burners or by spraying an antifungal bacterium on to the ground (Sanders and Snow, 1978).

Seed harvested from infected bolls often carries boll rot fungi and bacteria as surface contaminants and some of these organisms may also be carried internally. Most of the surface contaminants can be eliminated by the practice of acid delinting (e.g. Bain, 1939). One of the main internal invaders of the seed is *C. gossypii*, but the incidence of boll rot attributed to this fungus has declined dramatically in the USA since the widespread adoption of the practice of treating seed with a fungicide to control seedling disease. However, *Fusarium* species are also common internal contaminants of cotton seed and seed-dressings currently in use have little effect on the fusaria. As a result, *Fusarium* spp. have become the most common fungi isolated from diseased bolls in the USA (e.g. McCarter *et al.*, 1970).

Numerous trials have been conducted in the USA to test the effect of foliar-applied fungicides on boll rot incidence. Results have generally proved unsatisfactory, partly because of the difficulty of obtaining spray penetration to the lower canopy, where most boll rot occurs and also because the chemical is washed off the leaves and bolls by heavy rainfall. Furthermore, numerous sprays would be required to provide adequate control because latent infection of the boll often occurs soon after flowering, developing into boll rot only when conditions become sufficiently humid to favour the pathogen and/or the boll matures to a stage when it is physiologically predisposed to infection. Significant reduction in boll rot has sometimes been achieved under experimental conditions with compounds such as benomyl and thiabendazole, but their use on a commercial scale has not proved economically viable (e.g. Ranney, 1971).

A chemical approach to control may be successful where boll rot occurs mainly as a result of insect damage, by using insecticide to reduce the population of late-season pests. Tu and Cheng (1970) advocated the use of insecticide combined with a fungicide and obtained effective control of Diplodia boll rot with a combination of captafol and EPN. Unfortunately, one of the main insect pests associated with boll rot is the pink bollworm (*Pectinophora gossypiella*), which is not easy to control with insecticides at the larval stage.

Various planting systems have been devised in an attempt to increase air flow and reduce humidity within the crop canopy. The most common of these practices is the use of skip rows (e.g. Roncadori *et al.*, 1975). Although skip rows can reduce humidity within the canopy, there is no evidence that yield gains from reduced boll rot can compensate for losses due to reduced plant population and the costs incurred through reduced weed suppression.

Certain heritable characteristics of the cotton plant have shown promise for boll rot control. Significant reductions in boll rot have been achieved using cultivars with the okra leaf shape, which allows greater air flow through the canopy (e.g. Andries *et al.*, 1969). It has been shown that the bract plays an important role in boll infection (Luke and Pinckard, 1970) and the frego bract character appears to confer some resistance to boll rot (Jones and Andries, 1969). Roncadori (1977) found that frego bract reduced boll rot due to *Fusarium* spp. but was ineffective against *Diplodia*. The nectaries are also important in boll rot epidemiology, providing entry sites for micro-organisms and attracting insects which damage the boll. Roncadori *et al.* (1975)

investigated the effect on boll rot incidence of various combinations of skip rows, reduced nitrogen application and using a susceptible cultivar compared with a line which combined the characters of frego bract and okra leaf with the absence of nectaries. Boll rot incidence was lower in the nectariless line grown in the conventional manner than in the susceptible cultivar grown with skip rows and reduced nitrogen and to which fungicide was applied to the developing bolls.

Boll Rot in Relation to Insect Damage

Although usually still requiring conditions of high relative humidity to completely rot the boll, the most common mode of entry by micro-organisms into the boll is through insect punctures. Bagga (1970b) examined green bolls for sites of infection in a number of different cotton varieties. When levels of insect damage were low there were some differences between varieties with respect to the main entry site for micro-organisms. However, in the absence of insect control measures, microbial infection associated with insect damage predominated over all other modes of entry in all varieties.

Once the boll wall is breached, the immature lint is susceptible to microbial decay at much lower relative humidity than is required for the invasion of the intact boll. For instance, Leakey and Perry (1966) noted that *Colletotrichum* was capable of causing extensive rot to the boll wall and lint when introduced into the boll through insect feeding punctures, irrespective of humidity level.

High incidence of aflatoxin in cotton seed resulting from invasion of the boll by *A. flavus* is normally associated with pink bollworm attack (Ashworth *et al.*, 1971; Lee *et al.*, 1987). Internal boll rot caused by *Nematospora* spp. only occurs following penetration of the boll by the stainer bug (*Dysdercus* spp.).

Some of the heritable characteristics that are reported to reduce boll rot incidence, such as high gossypol content, frego bract and absence of nectaries, are effective, at least in part, because plants carrying these characters are less attractive to insect pests.

Causal Organisms

DIPLODIA GOSSYPINA

Distribution

D. gossypina is of worldwide distribution but mainly confined to an area 40° north and south of the equator (Punithalingam, 1976). It is particularly damaging in the more humid cotton-growing areas, notably parts of the USA

(Pinckard and Wang, 1972), South America (Pizzinato *et al.*, 1983), West Africa (Follin and Goebel, 1973) and India (Krishnamurty and Verma, 1974).

Taxonomy

D. gossypina is an ubiquitous tropical to sub-tropical plant pathogen, popularly known as *Botryodiplodia theobromae*, although Sutton (1980) considers that the correct generic name for the fungus should be *Lasiodiplodia*. Punithalingam (1976) has listed *Diplodia gossypina* as a synonym. The thick-walled, striate, slowly maturing conidia are diagnostic. The perfect state has been described as *Physalospora rhodina* Berk & Curt, but ascospores play a minor role in infection.

Morphology

Colonies on oat agar are grey to black, with abundant aerial mycelium. Pycnidia are simple or compound, stromatic, ostiolate, frequently setose and up to 5 mm in diameter. Conidia are initially unicellular, hyaline, subovoid to ellipsoid–oblong, thick-walled and truncate at the base. Mature conidia are uniseptate, cinnamon to fawn in colour, often longitudinally striate and 18–30 × 10–15 μm in size. Pycnidia are immersed in host tissue, becoming erumpent, simple or grouped, 2–4 mm wide, ostiolate and frequently pilose, with conidia extruding in a black mass (Punithalingam, 1976) (Fig. 7.1).

Isolation and growth in culture

The fungus can be isolated from infected tissue using standard procedures and media. Punithalingam (1976) describes colony morphology on oat agar as grey to black and fluffy, with abundant aerial mycelium. The reverse of the colony is black. Pycnidia are readily produced on this medium.

Symptoms

Diplodia boll rot is known in the USA as black boll rot because the surface of the decayed boll becomes covered in a sooty black layer of compacted fungal filaments, containing pycnidia and extruded conidia. Initial infections are evident on capsules and bracts as small brown spots, which expand in moist conditions, becoming black in colour (Pinckard *et al.*, 1981).

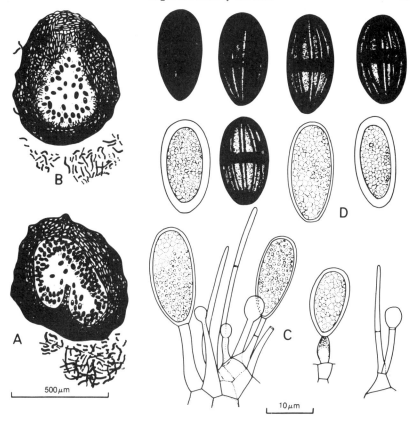

Fig. 7.1. *Diplodia gossypina (Botryodiplodia theobromae)*: A, B, V.S. pycnidia; C, conidiogenous cells and conidia; D, conidia (Punithalingam, 1976).

Host range

Host range is very wide, with more than 500 known hosts. Some pathogenic specialization occurs, with nine races differentiated on 25 jute varieties in Bangladesh (Haque *et al.*, 1973).

Infection cycle

The main source of primary inoculum is partially decomposed crop residues. Infection derived from this source causes shedding of squares, flowers and young bolls up to the age of about seven days after anthesis (Pinckard and Wang, 1972). Crop debris therefore provides an increasingly abundant source of inoculum as the season progresses (Sanders and Snow, 1978). The fungus is reported to be seed-borne in cotton (Noble and Richardson, 1969), but it was not one of the fungi identified in cotton seed tested for internal infection in the USA by Simson *et al.* (1973).

Inoculation experiments with green bolls indicate that they are most susceptible to infection by *D. gossypina* between five and 15 days after anthesis (Wang and Pinckard, 1973). Under moist conditions, spores are extruded from the pycnidia in large numbers and germinate in free water. The germ-tube becomes attached to the boll surface and, provided conditions remain humid, the fungus produces cell wall-degrading enzymes to penetrate the subepidermal layers. In the young boll, the guard cells are the first to become infected, as they have a thinner cuticle. Although the pathogen is capable of direct penetration through the cuticle of young bolls, it is the older bolls which are more subject to invasion, despite having a thicker cuticle. This is because, as the boll ripens, increasing numbers of stomata become dys-functional and remain permanently open, allowing the entry of micro-organ-isms (Pinckard and Montz, 1971; Pinckard and Baehr, 1973; Snow and Sachdev, 1977). The epidermal hairs (Kay *et al.*, 1975) and bracts (Luke and Pinckard, 1970) also provide a route into the capsule which bypasses the cuticle. After initial penetration, the fungus grows into the boll along the sutural parenchyma. Hyphal growth may be as much as 10–15 μm hour^{-1} at 30 °C and at that temperature boll decay is rapid (Wang and Pinckard, 1972). The fibrous endocarp surrounding the lock is almost impervious to fungal penetration unless damaged by an insect, and therefore infection may be confined to one or two locks (Pinckard and Wang, 1972).

COLLETOTRICHUM SPP.

Distribution

C. gossypii is found mainly in the tropics and sub-tropics. At one time the main boll rot pathogen in the south-eastern USA (e.g. Weindling *et al.*, 1941), it has become much less important in recent years but is still a major com-ponent of the boll rot complex in Africa (e.g. Belliard, 1972), India (Chopra *et al.*, 1975) and Australia (Evenson and Doepel, 1969).

Taxonomy and morphology

The taxonomy of *Colletotrichum* remains rather confused because of the extreme variability of the genus. Many reports of Colletotrichum boll rot refer to the causal organism as *Colletotrichum gossypii*, without reference to spore shape. However, it appears that two distinct species have been associ-ated with cotton boll rot, one having straight or cylindrical conidia, the other falcate conidia (Follin, 1969; Follin and Goebel, 1973). In West Africa, Belliard (1972) identified the species as *C. gossypii* and *C. indicum*. In El Salvador, Raynal (1970) also referred to the species with straight conidia as *C. gossypii* and the species with falcate conidia as *C. indicum*. In Senegal, *C. gossypii* was isolated more frequently from bolls than was *C. indicum*, but both were considered to be primary pathogens (Lagière, 1972).

Southworth (1891) first described the fungus causing anthracnose of cotton in America, naming it *C. gossypii*. Since then, isolates from cotton having straight conidia have been known by this name. However, isolates from cotton with straight spores are not host-specific and Arx (1957) reduced *C. gossypii* to synonomy with *C. gloeosporioides*, the conidial form of *Glomerella cingulata* (Stonem.) Spauld. & von Shrenck. The perfect stage of a *Colletotrichum* species isolated from diseased bolls in Uganda was referred to as *G. cingulata* (Leakey and Perry, 1966).

C. gloeosporioides is a very variable fungus, including saprophytes and parasites of a wide range of hosts. As a consequence, Sutton (1980) considers that no standardized description can be given other than that the conidia are straight, obtuse at the apex and 9–24 × 3–4.5 µm in size, with appressoria which are 6–20 × 4–12 µm in size.

Isolates from cotton having curved conidia have been referred to either as *C. indicum* (e.g. Raynal, 1970) or as *C. capsici* (e.g. Chopra *et al.*, 1975). Assuming that these are not separate species, the preferred name would be *C. capsici*, according to Sutton (1980). The two *Colletotrichum* species associated with boll rot therefore appear to be the same as those causing seedling disease and are described in more detail in Chapter 1. In addition to the two most common species, a third species was identified by Snow (1978) as *C. falcatum* and reported to be responsible for a previously unreported boll rot in Louisiana. This species is normally associated with graminaceous species (see Sutton, 1980).

Symptoms

Colletotrichum boll rot can be recognized by the thick grey mycelium produced on the surface of the rotted boll. The grey mycelium often has a pink tinge, due to the presence of conidial masses. When the boll dries and the mycelium disappears, the black, setate, acervuli (see Fig. 7.2) can usually be seen, with the aid of a microscope, adhering to the boll epidermis.

The first sign of infection is the appearance on the capsule of small reddish brown spots, which enlarge and blacken in moist conditions. Larger spots are sometimes concentrically zoned, the outer zone being reddish brown, the middle zone black and the inner zone pink, due to conidial masses oozing from the acervuli (Pinckard *et al.*, 1981).

Infection cycle

C. gossypii (e.g. Arndt, 1953) and *C. capsici* (R.J. Hillocks, unpublished) are internally seed-borne in cotton. Primary infection derives from infested seed and from partially decomposed crop residues in the soil. The lower bolls become infected by conidia carried by rain-splash from the soil or, less

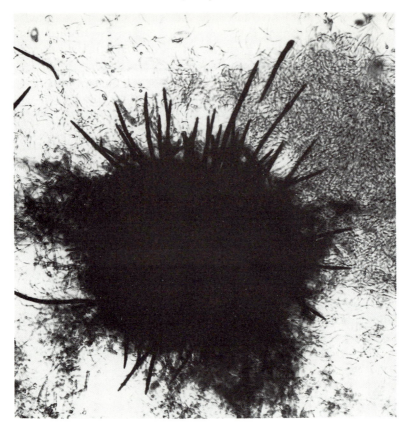

Fig. 7.2. *Colletotrichum capsici*: acervulus and conidia from surface of cotton boll (× 200).

frequently, by secondary conidia produced on stem or leaf lesions, which may develop on the plant as part of the anthracnose syndrome. *C. gloeosporioides* produces appressoria on bolls under relatively dry conditions, but these develop into latent subcuticular infections. Under humid conditions, the fungus grows from the latent infection sites through the boll wall into the locule, causing the boll to rot. The fungus is apparently capable of causing boll rot in dry conditions if conidia enter the boll through insect feeding punctures (Leakey and Perry, 1966).

Control

Experience in the USA has shown that the disease can be more or less eliminated by the use of fungicides applied to the seed or furrow. Other methods of reducing primary inoculum, such as rotation, might also be expected to reduce disease incidence.

FUSARIUM SPP.

Distribution and causal organisms

Fusarium spp. are among the most commonly isolated fungi from rotted bolls in every country where the crop is grown. In the USA, fusaria have replaced *Colletotrichum* spp. as the most common boll rot fungi (e.g. Bagga and Rannay, 1969a; McCarter *et al.*, 1970). In Africa, *F. moniliforme* is the main species involved and equals *Colletotrichum* in importance as a boll rot pathogen (e.g. Belliard, 1972). *F. roseum* and *F. solani* are also associated with boll decay in Africa (Follin and Goebel, 1973). In the USA, *F. oxysporum* is probably the main species isolated from rotted bolls but it is usually found together with other fusaria such as *F. roseum*, *F. solani*, *F. moniliforme* (McCarter *et al.*, 1970), *F. lateritium* and *F. semitectum* (Mertley and Snow, 1978). *F. equisiti* has been reported to cause a boll rot of *G. arboreum* in India (Sharma and Sandhu, 1985).

Symptoms

Infection of the boll usually begins with necrotic lesions on the margins of the bracts. These lesions enlarge in wet conditions to cover most of the bract, allowing the fungus to penetrate the capsule through the receptacle, causing a brown rot.

Under humid conditions, bolls decayed by *F. moniliforme* become covered initially with a white or grey mycelium (Fig. 7.3a,b and Plate 5A), which disappears when the fungus sporulates. Conidia are produced in large numbers on the surface of the boll and have a pink colour in mass. See Booth (1971) and Tousson and Nelson (1968) for comparative morphology of the fusaria.

Infection cycle

Fusarium species can be carried internally on cotton seed (e.g. Hillocks, 1983) and are involved in the seedling disease complex (see Chapter 1). Partially decayed seedlings on the soil surface therefore provide a reservoir of inoculum to infect the lower bolls. Also, it appears that *F. moniliforme* and other fusaria may be present in the vascular tissue of healthy plants, becoming damaging only under certain conditions.

Belliard (1972) found that isolates of *F. moniliforme* varied considerably in their pathogenicity towards bolls inoculated by immersion in a spore suspension. The most virulent isolate was capable of invading the boll through the nectaries and apex, causing complete rot after 35 days. Boll rot occurred much faster than this if penetration was aided by puncturing the carpel. Decay progressed more rapidly at 25 °C than at 28 °C. Studies conducted in the USA by Sparnicht and Roncardi (1972) showed that *Fusarium* spp. do not penetrate

(a)   (b)

Fig. 7.3. (a) Cotton boll showing early symptoms of infection by *Fusarium moniliforme*. (b) Cotton boll completely rotted after infection by *Fusarium moniliforme*.

directly through the pericarp. *F. oxysportum* and *F. roseum* initially colonize the bracts and then penetrate the capsule base through the receptacle.

ASPERGILLUS FLAVUS

Distribution and damage

Aspergillus flavus is a common saprophyte in soil and decaying organic matter. A number of *Aspergillus* species are among the many secondary invaders associated with rotted bolls in most cotton-growing countries (e.g. Belliard, 1972; Krishnamurty and Verma, 1974). *A. flavus* is an important cause of fibre deterioration in parts of the USA, notably California, Texas and Arizona (Marsh and Simpson, 1968). The fungus may sometimes induce carpel rot but the main damage is caused by the presence of a fluorescent stain on the fibres, which is produced by *A. flavus*. The fungus also produces aflatoxin in the seed, rendering it unsuitable for animal feed. This is a chronic problem in Arizona, where it significantly reduces the market value of the crop (Cotty, 1988).

Cultural characteristics and morphology

A. flavus grows and sporulates readily on common culture media. On Czapek and malt agar the colony is at first yellow-green, the reverse being colourless or dark red-brown. Sclerotia are produced by most isolates on Czapek agar. They are brown in colour, becoming purple, and they are 400–700 µm in diameter. Conidial heads are light yellow in colour, becoming greenish yellow with age, radiate, splitting into several poorly defined columns and 300–600 µm in diameter. The conidiophore is colourless, usually less than 1 mm in length and 10–20 µm in width. Both metulae and phialides are present. Metulae are 6–16 × 4–9 µm and phialides are 6–10 × 3–5 µm in size. Conidia are globose, conspicuously echinulate and 3–6 µm in diameter (Onions, 1966).

Conditions favouring infection

Boll decay and colonization of the lint and seed occur over a wide temperature range but are maximal at high temperature and humidity (Pinckard *et al.*, 1981). However, Aspergillus boll rot occurs under drier conditions than would be suitable for other boll rot pathogens (Marsh *et al.*, 1973). Aflatoxin contamination of the seed is greatest at 25–35 °C and usually occurs in the field rather than in storage because seed moisture at harvest is usually below the 15% required for growth of the fungus (Pinckard *et al.*, 1981).

Mode of infection

A. flavus is normally a wound pathogen, invading the boll through insect holes or through the suture after the boll begins to split. Infection by other pathogens such as *Rhizopus* spp. also enhances infection by *A. flavus* by inducing premature separation of the carpels to expose the lint (Ashworth *et al.*, 1971). However, the fungus is capable of entering the cotyledonary node after abscission and can then enter the seed through the xylem vessels. Around the time of anthesis, invasion of the boll can also occur through the nectaries (Klich *et al.*, 1984, 1986).

Most aflatoxin-contaminated seed is derived from bolls damaged by pink bollworm (Cotty and Lee, 1989). Pink bollworm larvae enter the young bolls, where they develop for 10–14 days. The third instar then emerges, leaving a large open exit hole through which insect-borne and air-borne propagules of *A. flavus* can enter (Ashworth *et al.*, 1971).

NEMATOSPORA SPP. (STIGMATOMYCOSIS)

Distribution and causal organisms

Internal infection of the cotton boll by yeast-like fungi was first described by Nowell (1917) in the West Indies. The most important of these fungi on cotton and other Malvales is *Nematospora gossypii* (Pearson, 1958), which is commonly found in the bolls of Sea Island cotton throughout the West Indies (Ashby and Nowell, 1926). The same fungus also causes lint discoloration in Uganda (Hansford, 1929), Malawi, Nigeria, Burma, India, the USA and Central America (CMI Distribution Map No. 153).

A second species, *N. coryli*, is found on the seed of legume crops, on coffee beans and sometimes on cotton lint, mainly in Africa and Asia but it has also been recorded in California (Pinckard *et al.*, 1981) and Europe (CMI Distribution Map No. 163).

N. phaseoli is more common on legume seeds but has been recorded on cotton in Africa (Pearson, 1948). *N. nagpuri* appears to be confined to India (Dastur and Singh, 1930).

Two other similar fungi have been associated with internal boll rot and lint stain. Ashby and Nowell (1926) described *Spermophthora gossypii*, which predominated in bolls collected from a number of West Indian islands where the crop was heavily infested with the stainer bug (*Dysdercus* spp.). Nowell (1939) noted that *N. gossypii* was the main fungus isolated from bolls collected on Montserrat but that *S. gossypii* predominated in bolls from St Vincent.

The other fungus similar to *Nematospora* spp. is *Erymothecium cymbalariae*, which was responsible for 90% of boll infection in the West Indies in 1918. Nowell (1939) mentions *E. ashbyi*, which occurs in cotton bolls in the Sudan.

Symptoms

Bolls infected one to two weeks after anthesis become completely rotted (see Plate 5B) and are shed. Bolls infected when three to four weeks old may be reduced in size and the lint is heavily stained to a dirty brown or yellow colour and does not fluff out at maturity. Infection by *Nematospora* causes no growth reduction once the bolls reach five weeks old but, at maturity, the lint is found to be stained yellow. The intensity of the stain decreases with increasing age of the boll when infected. Bolls infected more than eight weeks after anthesis show virtually no effect (Pearson, 1948, 1958). The yellow stain is caused by the production of riboflavin, a characteristic of some strains of *Nematospora* spp. (Arragon *et al.*, 1946).

Morphology of N. gossypii

Hyphae are hyaline, often vacuolated or containing granular material, at first non-septate but becoming septate at maturity, and they are dichotomously branched. Vegetative reproduction is by lateral budding or transverse fission. Abundant asci develop from hyphal segments. They are cylindrical or sigmoid, single or in groups or chains and $100–200 \times 10–20\,\mu m$ in size. The ascospores are grouped in parallel into two or more fascicles of two to six, arranged lengthwise in the sacs, with 4–32 asci sac^{-1}. They are acicular to fusiform, often with a thin septum in the centre, and $25–37 \times 2–5\,\mu m$ in size.

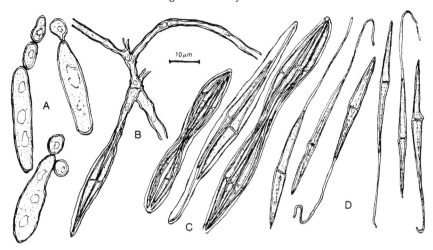

Fig. 7.4. *Nematospora coryli*: A, vegetative budding cells; B, hyphae with an ascus; C, asci; D, ascospores (Mukerji, 1968b).

N. coryli (Fig. 7.4) is distinguished from *N. gossypii* (Fig. 7.5) in possessing a yeast-like vegative phase and septate ascospores which are broader at the septum (Mukerji, 1968b). *Erymothecium ashbyi* (Fig. 7.6) is distinguished from *Nematospora* by having ascospores which are shorter, broader and

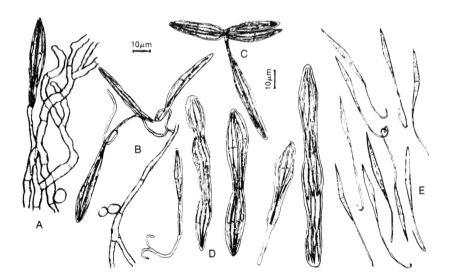

Fig. 7.5. *Nematospora gossypii*: A, hyphae with a single ascus; B, hyphae with three asci; C, four asci grouped together; D, asci containing one, two and three groups of ascospores; E, ascospores with appendage (Mukerji, 1968c).

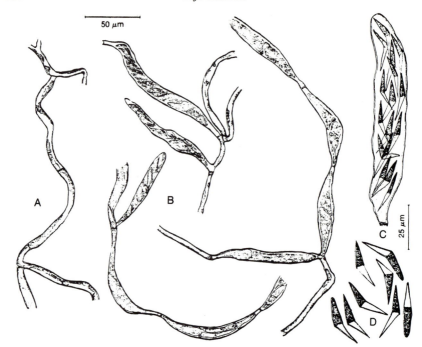

Fig. 7.6. *Erymothecium ashbyi*: A, vegetative hyphae; B, developing and mature asci; C, mature ascus; D, ascospores (Mukerji, 1968a).

curved at the needle-like extremity and they lack any elongate appendage (Mukerji, 1968a).

Association with Dysdercus *spp.*

Stigmatomycosis or internal boll rot occurs without any visible evidence of infection on the outside of the boll. Transmission of the disease cannot be effected by applying inoculum to the outer surface of the boll and then pricking the carpel wall with a needle. The disease must apparently be transmitted to the boll during feeding by hemipterous insects, mainly the cotton stainer bug. It is likely that the insect acquires the fungus and introduces it initially into the cotton crop after feeding on wild Malvales near by, which are alternative hosts of the fungus. The green bug (*Nezara* spp.) and the irridescent bug (*Callidea* spp.) are among the other insects known to transmit *Nematospora* spp. (Pearson, 1985).

The spindle-shaped ascospores are sucked up by stainer bugs while feeding on the seeds of an infected plant. Spores of the fungus have been found in the gut and salivary glands of *Dysdercus*, but the fungus does not pass through any developmental stages within the insect. Spores are then introduced through the feeding stylet into a new boll with the salivary fluid (Frazer, 1944).

PHYTOPHTHORA SPP.

Distribution, symptoms and causal organisms

Boll rot caused by *Phytophthora* spp. has been reported from the USA (Pinckard and Guidoz, 1966), Australia (Allen and West, 1986) and China (Liang, 1964). Pinckard and Guidoz (1966) reported that in the USA the fungus produces a soft, watery, black rot under wet conditions. The mode of entry was considered to be by direct penetration of the bract, followed by invasion of the base of the boll. No attempt was made to determine the species involved.

In China, a cotton boll rot and castor bean blight were attributed to a similar *Phytophthora* species (Liang, 1964). Following a nationwide survey of boll rot diseases in China in 1984, it was concluded that the most common causal organism was *P. bohemeri* (Joint Survey Team, 1986). Boll rot incidence in the year of the survey ranged from 5 to 25% with *Phytophthora* spp., accounting for between 28 and 100% of cases. Other fungi responsible were *F. moniliforme*, *C. gossypii*, *D. gossypina* and *Cephalothecium roseum*. Pinckard *et al.* (1981) refer to *P. capsici*, which is considered to be a primary boll rot pathogen in the Louisianna–Mississippi Delta of the USA.

Phytophthora boll rot was recorded for the first time in Australia in 1985 at a site in New South Wales and one in Queensland (Allen and West, 1986). Infected bolls became blackened and opened prematurely and the lint was discoloured. The causal organism was identified as *P. nicotiana* Breda de Haan var. *parasitica* (Dastur) Waterh.

The pathogen is a common soil-inhabiting fungus with a wide host range, plants normally becoming infected when spores produced in wet soil are splashed by heavy rain on to the lower leaves and fruits.

Morphology of P. nicotiana

Hyphae are up to 9 μm wide, but irregular in width without hyphal swellings. Sporangiospores are more slender than the mycelial hyphae and irregularly or sympodially branched. The sporangia are broadly ovoid to spherical, sometimes intercalary, with a short pedicel, having an average size of 38 × 30 μm. Chlamydospores form after one to two weeks in culture. They are up to 60 μm in diameter and become yellow-brown with age. Oogonia, which are 24–26 μm in diameter on average, are usually produced in single culture, although sparsely and not until after some weeks. However, they are readily produced when the fungus is grown with an opposite strain. The antheridia are spherical or oval and 10 × 12 μm while the oospore is 18–20 μm with a thick wall (Waterhouse and Waterston, 1964) (Fig. 7.7).

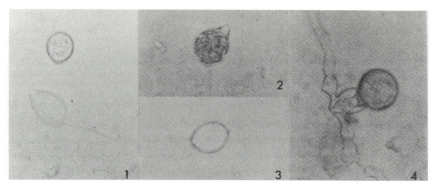

Fig. 7.7. *Phytophthora nicotiana* var. *parasitica*; 1–3, sporangia; 4, oogonium with amphigenous antheridium (× 450) (Waterhouse and Waterson, 1964).

Other Boll Rot Pathogens

RHIZOPUS SPP.

Rhizopus stolonifer (Ehrenb.) and *R. arrhizus* were originally considered to be wound pathogens, but lint decay associated with these fungi has been observed in California in the absence of insect damage (Halisky and Satour, 1964). In Arizona, *R. arrhizus* was the most pathogenic of all the fungi isolated from bolls, following injection of a spore suspension into the boll (Simbwa-Bunnya and Boyle, 1969).

ALTERNARIA SPP.

A. macrospora and *A. alternata* both cause small purple spots (see Plate 5C), which usually remain superficial on older bolls. *A. macrospora* is capable of penetrating the carpel wall of young bolls soon after pollination, causing them to be shed (Jones, 1928; Hopkins, 1931). However, in Africa at least, *Alternaria* spp. are responsible more for lint rots after boll dehiscence than for boll rot prior to dehiscence (see Chapter 8). In the USA, *Alternaria* spp. are among the most common fungi isolated from rotted bolls (e.g. Simbwa-Bunnya and Boyle, 1969; McCarter *et al.*, 1970), occurring mainly as a result of secondary invasion.

RHIZOCTONIA SOLANI

The perfect stage of *R. solani* (*Thanatephorus cucumeris*) occurs on cotton plants near soil level and basidiospores are capable of direct penetration of the bract under conditions of high humidity. Mycelium grows inside the bract, penetrating the base of the boll and producing a brown, rather dry rot (Pinckard and Luke, 1967).

Myrothecium Roridum

This fungus is a common leaf spot pathogen of cotton in India and has been isolated from rotted bolls in the USA. Bagga (1970a) isolated numerous micro-organisms from diseased bolls but *M. roridum* was one of only three fungi which caused bolls to rot following contact inoculation, without the requirement of wounding. Despite its apparent aggressiveness as a boll invader following artificial inoculation, there is little information available on the role of this fungus in boll disease in the field.

Xanthomonas Campestris pv. Malvacearum

The bacterial blight pathogen (see Chapter 2) is capable of causing damage to the boll. However, the bacterium penetrates the carpel wall only in the most susceptible cultivars and usually the lesion remains superficial. Internal boll rot occurs when the bacterium is introduced through the carpel wall on the stylet of insect pests such as *Dysdercus* spp.

References

ALLEN, S.J. AND WEST, K.L. (1986) Phytophthora boll rot of cotton. *Australian Plant Pathology* 15, 34.

ANDRIES, J.A., JONES, J.E., SLOANE, L.W. AND MARSHALL, J.G. (1969) Effects of okra leaf shape on boll rot, yield and other important characters of Upland cotton (*Gossypium hirsutum* L.). *Crop Science* 9, 705–10.

ARNDT, C.H. (1950) Boll rots of cotton in South Carolina in 1949. *Plant Disease Reporter* 34, 176.

ARNDT, C.H. (1953) Survival of *Colletotrichum gossypii* on cotton seeds in storage. *Phytopathology* 43, 220.

ARRAGON, G., MAINLI, J., REFRAIT, R. AND VELU, H. (1946) La flavogenesis par *Eremothecium ashbyii* Guilliermond: étude du facteur surface/volume. *Annales d'Institut Pasteur* 72, 300–5.

ARX, J.A. VON (1957) Die Arten der Gattung *Colletotrichum* Cord. *Phytopathologische Zeitschrift* 29, 413–68.

ASHBY, S. F. AND NOWELL, N. (1926) The fungi of stigmatomycosis. *Annals of Botany* 40, 69–83.

ASHWORTH, L.J. AND HINE, R.B. (1971) Structural integrity of the cotton fruit and infection by microorganisms. *Phytopathology* 61, 1245–8.

ASHWORTH, L.J., RICE, R.E., McMEANS, J.L. AND BROWN, C. M. (1971) The relationship of insects to infection of cotton bolls by *Aspergillus flavus*. *Phytopathology* 61, 488–93.

BAGGA, H.S. (1970a) Pathogenicity studies with organisms involved in the cotton boll rot complex. *Phytopathology* 60, 158.

BAGGA, H.S. (1970b) Mode of entry of boll rot pathogens in selected cotton varieties and strains. *Plant Disease Reporter* 54, 719–21.

BAGGA, H.S. AND LASTER, M.L. (1969) Evaluation of several cotton varieties with specific genetic characteristics against boll rot. *Proceedings of the Beltwide Cotton Production Research Conference*, Memphis, Tennessee, p. 26.

BAGGA, H.S. AND RANNEY, C.D. (1969a) Boll rot potential and actual boll rot in seven cotton varieties. *Phytopathology* 59, 255–6.

BAGGA, H.S. AND RANNEY, C.D. (1969b) Boll rot in six cotton varieties. *Proceedings of the Beltwide Cotton Production Research Conference*, Memphis, Tennessee, pp. 23–4.

BAIN, D.C. (1939) Effects of sulphuric acid on fungi and bacteria present on cotton seed from diseased bolls. *Phytopathology* 29, 878.

BELLIARD, J. (1972) Sur les pourritures des capsules du cotonnier en Afrique: quelques champignons responsable: inoculations avec *Fusarium moniliforme* Sheld. *Coton et Fibres Tropicales* 27, 243–50.

BLASINGAME, D. (1989) Disease loss estimate committee report. *Proceedings of the Beltwide Cotton Production Research Conference*, Memphis, Tennessee, p. 4.

BOOTH, C. (1971) *The Genus* Fusarium. Commonwealth Mycological Institute, Kew, Surrey, UK, 237 pp.

BOULANGER, J. (1966) Duration of boll growth cycle of the cotton plant. 1. Contribution to the study of Upland cotton resistance to stigmatomycosis. *Coton et Fibres Tropicales* 21, 173–82.

CAUQUIL, J. AND RANNEY, C.D. (1969) Etude sur l'infection interne des capsules vertes de cotonnier et sur les possibilités d'une selection génétique pour réduire l'incidence de pouritture capsulaire. *Coton et Fibres Tropicales* 24, 193–204.

CHOPRA, B.L., SINGH, T.H. AND PRAKESH, R. (1975) Effect of the boll rot phase of anthracnose on the germination and technological properties of cotton fibre. *Journal of Research of Punjab University* 12, 1–5.

COTTY, P.J. (1988) Greenhouse evaluation of cotton cultivar susceptibility to aflatoxin contamination via colonization of wounds by *Aspergillus flavus*. *Proceedings of the Beltwide Cotton Production Research Conference*, Memphis, Tennessee, pp. 31–2.

COTTY, P.J. AND LEE, L.S. (1989) Aflatoxin contamination of cottonseed: comparison of pink bollworm damage and undamaged bolls. *Tropical Science* 29, 273–7.

DASTUR, J.F. AND SINGH, J.A. (1930) A new *Nematospora* on cotton bolls in Central Provinces (India). *Annals of Mycology* 28, 191–3.

EVENSON, J. P. AND DOEPEL, R. (1969) Fungal root rots, leaf spots and boll rots of cotton in the Ord Valley, north-west Australia. *Cotton Growing Review* 46, 49–52.

FOLLIN, J.C. (1969) Sur les différentes formes de *Glomerella* Spaul. et Schr. et de *Colletotrichum* Cda. isoles du cotonnier. 1. Localisations et étude morphologique. 2. Etude du pouvoir pathogène, premières conclusions. *Coton et Fibres Tropicales* 24, 337–50.

FOLLIN, J.C. AND GOEBEL, S. (1973) Les pourritures de capsules du cotonnier en culture irrigué en Côte d'Ivoire. *Coton et Fibres Tropicales* 28, 401–7.

FRAZER, H.L. (1944) Observations on the method of transmission of internal boll disease of cotton by the cotton stainer bug. *Annals of Applied Biology* 31, 271–90.

HALISKY, P.M. AND SATOUR, M.M. (1964) Evaluating fungicides for the control of Rhizopus boll rot of cotton. *Plant Disease Reporter* 48, 359–63.

HALISKY, P.M., SCHNATHORST, W.C. AND SHAGRUN, M.A. (1961) Severity and distribution of cotton boll rots as related to temperature. *Phytopathology* 51, 501–5.

HANSFORD, C.G. (1929) Cotton diseases in Uganda 1926–1928. *Empire Cotton Growing Review* 6, 10–26.

HAQUE, M.A., CHOWDHURY, S.N.A. AND AHMED, Q.A. (1973) Physiological specialization in *Botrydiodiplodia theobromae*, the causal organism of black hand of jute, *Corchorus* species. *Bangladesh Journal of Botany* 2, 83–92.

HILLOCKS, R. J. (1983) Infection of cotton seed by *Fusarium oxysporum* f.sp. *vasinfectum* in cotton varieties resistant and susceptible to Fusarium wilt. *Tropical Agriculture* 60, 141–3.

HOPKINS, J.C.F. (1931) *Alternaria gossypina* (Thum) comb. nov. causing a leaf spot and boll rot of cotton. *Transactions of the British Mycological Society* 16, 136–44.

JOINT SURVEY TEAM (1986) A nationwide survey on recent incidence of cotton boll disease. *China Cottons* 1, 42–4.

JONES, G.H. (1928) An *Alternaria* disease of the cotton plant. *Annals of Botany* 32, 935–7.

JONES, J.E. AND ANDRIES, J.A. (1969) Effect of frego bract on the incidence of cotton boll rot. *Crop Science* 9, 426–8.

KAY, J.P., SACHDEV, M. AND SNOW, J.P. (1975) Possible role of epidermal hairs in cotton boll rots.

KLICH, M.A. AND CHMIELEWSKI, M.A. (1985) Nectaries as entry sites for *Aspergillus flavus* in developing cotton bolls. *Applied and Environmental Entomology* 50, 602–4.

KLICH, M.A., THOMAS, S.H. AND MELLON, J.E. (1984) Field studies on mode of entry of *Aspergillus flavus* into cotton seeds. *Mycologia* 76, 665–9.

KLICH, M.A., LEE, L.S. AND HUIZAR, H.E. (1986) The occurrence of *Aspergillus flavus* in vegetative tissue of cotton plants and its relation to seed infection. *Mycopathologia* 95, 111–74.

KRISHNAMURTY, V. AND VERMA, J.P. (1974) Preliminary studies on boll rot of cotton in India. *Cotton Growing Review* 51, 226–9.

LAGIÈRE, R. (1972) Cotton boll rots in Senegal. 2. Fungi isolated from bolls in the process of rot. *Coton et Fibres Tropicales* 28, 493–508. (English supplement.)

LEAKEY, C.L. AND PERRY, D.A. (1966) The relation between damage caused by insect pests and boll rot associated with *Glomerella cingulata* (*Colletotrichum gossypii*) on Upland cotton in Uganda. *Annals of Applied Biology* 57, 337–44.

LEE, L.S., LACEY, P.E. AND GOYES, W.R. (1987) Aflatoxin in Arizona cottonseed: a model study of insect vectored entry of cotton bolls by *Aspergillus flavus*. *Plant Disease* 71, 297–301.

LIANG, P.Y. (1964) Identification of *Phytophthora* species causing cotton boll blight and castor bean blight in North China. *Acta Phytopathologia Sinica* 7, 11–20.

LUKE, W.J. AND PINCKARD, J.A. (1970) The role of the bract in boll rots of cotton. *Cotton Growing Review* 47, 20–8.

McCARTER, S.M., RONCARDI, R.W. AND CRAWFORD, J.L. (1970) Microorganisms associated with cotton boll rots in Georgia. *Plant Disease Reporter* 54, 586–90.

MARSH, P.B. AND SIMPSON, M.E. (1968) Occurrence of fibre containing fluorescent spots associated with *Aspergillus flavus* boll rot in the US cotton crop of 1967. *Plant Disease Reporter* 52, 671.

MARSH, P.B., SIMPSON, M.E. AND FILSINGER, E.C. (1973) *Aspergillus flavus* boll rot in the US cotton belt in relation to high aqueous extract pH of fibre: an indication of exposure to humid conditions. *Plant Disease Reporter* 57, 664–7.

MERTLEY, J.C. AND SNOW, J.P. (1978) The boll rotting *Fusarium* species in Louisiana. *Proceedings of the Beltwide Cotton Production Research Conference*, Memphis, Tennessee, pp. 29–30.

MILLER, P.R. (1943) A summary of four years cotton seedling and boll rot disease surveys. *Plant Disease Reporter* Suppl. 141, 54–8.

MUKERJI, K.G. (1968a) *Erymothecium ashbyi. CMI Descriptions of Pathogenic Fungi and Bacteria* No. 181. Commonwealth Agricultural Bureaux, Farnham Royal, Bucks, UK.

MUKERJI, K.G. (1968b) *Nematospora coryli. CMI Descriptions of Pathogenic Fungi and Bacteria* No. 184. Commonwealth Agricultural Bureaux, Farnham Royal, Bucks, UK.

MUKERJI, K.G. (1968c) *Nematospora gossypii. CMI Descriptions of Pathogenic Fungi and Bacteria* No. 185. Commonwealth Agricultural Bureaux, Farnham Royal, Bucks, UK.

NOBLE, M. AND RICHARDSON, M.J. (1969) An annotated list of seed borne diseases. *Phytopathological Paper* No. 8. Commonwealth Agricultural Bureaux, Farnham Royal, Bucks, UK.

NOWELL, W. (1917) Internal boll disease of cotton in the West Indies. *West Indian Bulletin* 16, 152–9.

NOWELL, W. (1939) Internal boll disease. *Cotton Growing Review* 16, 18–24.

ONIONS, A.H.S. (1966) *Aspergillus niger. CMI Descriptions of Pathogenic Fungi and Bacteria* No. 94. Commonwealth Agricultural Bureaux, Farnham Royal, Bucks, UK.

PEARSON, E.O. (1948) The development of internal boll disease of cotton in relation to time of infection. *Annals of Applied Biology* 34, 527–45.

PEARSON, E.O. (1958) *The Insect Pests of Cotton in Tropical Africa.* Empire Cotton Growing Corp. and Commonwealth Institute of Entomology, pp. 355.

PINCKARD, J.A. AND BAEHR, L.F. (1973) Histological studies on the development of cotton bolls in relation to microbial infection and decay. *Cotton Growing Review* 50, 115–29.

PINCKARD, J.A. AND GUIDOZ, G. (1966) A parasitic *Phytophthora* boll rot of cotton found in Louisiana. *Plant Disease Reporter* 52, 780–1.

PINCKARD, J.A. AND LUKE, W.J. (1967) *Pellicularia filamentosa* (pat.) Rogers. A primary cause of cotton boll rot in Louisiana. *Plant Disease Reporter* 51, 67–70.

PINCKARD, J.A. AND MONTZ, G.N. (1971) Behaviour of boll stomata in *Gossypium hirsutum* in relation to age, dehiscence and decay. *Cotton Growing Review* 49, 153–9.

PINCKARD, J.A. AND WANG, S.C. (1972) The nature of cotton boll rots in Louisiana. *Proceedings of the Beltwide Cotton Production Research Conference*, Memphis, Tennessee, p. 22.

PINCKARD, J.A., ASHWORTH, L.J., SNOW, J.P., RUSSELL, T.E., RONCADORI, R.W. AND SCIUMBATO, G.L. (1981) Boll rots. In: Watkins, G.M. (ed.), *Compendium of Cotton Diseases.* American Phytopathological Society, St Paul. Minnesota, pp. 20–4.

PIZZINATO, M.A., SOAVE, J. AND CIA, E. (1983) Pathogenicity of *Botrydiodiplodia theobromae* Pat. on cotton (*Gossypium hirsutum* L.) plants of different ages and on cotton bolls. *Fitopatologia Brasileira* 8, 223–8.

PUNITHALINGAM, E. (1976) *Botryodiplodia theobromae. CMI Descriptions of Pathogenic Fungi and Bacteria* No. 519. Commonwealth Agricultural Bureaux, Farnham Royal, Bucks, UK.

RANNEY, C.D. (1971) Studies with benomyl and thiabendazole on control of cotton disease. *Phytopathology* 61, 783–6.

RANNEY, C.D., HURSHE, J.S. AND NEWTON, O.H. (1971) Effect of bottom defoliation on microclimate and the reduction of boll rot of cotton. *Agronomy Journal* 63, 259–63.

RAYNAL, G.(1970) Contribution à l'étude des pourritures des capsules du cotonnier en El Salvador. 2. Comparison des caractères culturaux et de la morphologie de soncles de *Colletotrichum indicum* Dast. provenant d'El Salvador et de Thailand. *Coton et Fibres Tropicales* 25, 433–8.

RONCADORI, R.W. (1969) Fungal invasion of developing cotton bolls. *Phytopathology* 59, 1356–9.

RONCADORI, R.W. (1977) Comparative susceptibility of cotton bolls with standard and frego bract to rot fungi. *Plant Disease Reporter* 61, 132–4.

RONCADORI, R.W., McCARTER, S.M. AND CRAWFORD, J.L. (1975) Evaluation of various control measures for cotton boll rot. *Phytopathology* 65, 567–70.

SANDERS, D.E. AND SNOW, J.P. (1977) A possible source of inoculum of boll rotting fungi. *Proceedings of the Beltwide Cotton Production Research Conference*, Memphis, Tennessee, p. 27.

SANDERS, D.E. AND SNOW, J.P. (1978) Control of cotton boll rot by inoculum reduction. *Proceedings of the Beltwide Cotton Production Research Conference*, Memphis, Tennessee, pp. 21–2.

SHARMA, Y.R. AND SANDHU, B.S. (1985) A new fungus associated with boll rot of *G. arboreum* cotton. *Current Science* 54, 936.

SIMBWA-BUNNYA, M. AND BOYLE, A.M. (1969) Cotton boll rots in Arizona. *Phytopathology* 59, 667–8.

SIMSON, M.E., MARSH, P.B., MEROLA, G.V., FERRETTI, R.J. AND FILSINGER, C.E. (1973) Fungi that infect cottonseed before harvest. *Applied Microbiology* 26, 608–13.

SNOW, J.P. (1978) Previously unreported boll rot in Louisiana. *Proceedings of the Beltwide Cotton Production Research Conference*, Memphis, Tennessee, pp. 10–11.

SNOW, J.P. AND SACHDEV, M.G. (1977) Scanning electron microscopy of cotton boll invasion by *Diplodia gossypina*. *Phytopathology* 76, 589.

SOUTHWORTH, E.A. (1891) Anthracnose of cotton. *Journal of Mycology* 6, 100–5.

SPARNICHT, R.H. AND RONCARDI, M.W. (1972) Fusarium boll rot of cotton: pathogenicity and histopathology. *Phytopathology* 62, 1381–6.

SUTTON, B.C. (1980) *The Coelomycetes*. Commonwealth Mycological Institute, Kew, Surrey, England, 696 pp.

TOUSSON, T.A. AND NELSON, P.E. (1968) *A Pictorial Guide to the Identification of* Fusarium *Species* Pennsylvania State University Press, University Park and London.

TU, C. AND CHENG, Y.H. (1970) Experiments for the control of cotton boll black rot. *Plant Disease Reporter* 54, 29–31.

WANG, S.C. AND PINCKARD, J.A. (1972) Some biochemical factors associated with infection of cotton fruit by *Diplodia gossypina*. *Phytopathology* 62, 460–5.

WANG, S.C. AND PINCKARD, J.A. (1973) Cotton boll cuticle, a potential factor in boll rot resistance. *Phytopathology* 63, 315–19.

WATERHOUSE, G.M. AND WATERSTON, J.M. (1964) *Phytophthora nicotianae* var. *parasitica. CMI Descriptions of Pathogenic Fungi and Bacteria* No. 35. Commonwealth Agricultural Bureaux, Farnham Royal, Bucks, UK.

WEINDLING, R., MILLER, P.R. AND ULLSTRUP, A.J. (1941) Fungi associated with diseases of cotton seedlings and bolls with special consideration to *Glomerella gossypii. Phytopathology* 31, 158–67.

8

Microbial Contamination of the Lint

R.J. Hillocks

Introduction

Many of the boll rot micro-organisms described in Chapter 7 are found as contaminants of the lint if the boll is only partially destroyed but still capable of dehiscence. Alternatively, their air-borne conidia may become trapped among the fibres of the seed cotton after boll dehiscence. If the seed cotton is not promptly harvested and is exposed to rainfall or prolonged periods of high relative humidity, fungal growth may then be sufficient to cause discoloration of the lint. Many of the fungi occurring as contaminants of the lint produce cellulolytic enzymes. Given a sufficiently long period of conditions suitable for fungal growth prior to harvesting, they are capable of degrading the cellulose fibres, making them too weak to be spun without breaking. However, even in the absence of cellulose degradation, discoloration lowers the market value of seed cotton.

Causal Organisms

Component members of the lint-rotting complex (see Table 8.1) vary according to geographical location, origin of the contamination, maturity of the lint when it became contaminated, temperature, humidity, light intensity and whether the lint has been wetted by rainfall or has remained dry.

Lagière (1974) examined samples of raw cotton from 13 countries in Africa, Latin America and Asia, identifying 39 species of fungus. The most cosmopolitan of these were *Aspergillus niger*, *Aspergillus flavus* and *Rhizopus stolonifer*, followed by *Fusarium roseum* and *Fusarium moniliforme*. Reviewing the available literature, Lagière (1974) listed 95 species isolated from cotton fibres.

Table 8.1. Some fungi associated with lint discoloration and their ability to degrade cellulose[a].

Fungus	Worldwide importance[b] as lint contaminant	Ability to decompose cellulose[c]
Alternaria spp.	+++	+++
Aspergillus flavus	++	+
Aspergillus niger	++	+
Cladosporium herbarum	+++	++
Curvularia spp.	+	+++
Diplodia gossypina	+++	++
Fusarium moniliforme	++	+++
Fusarium oxysporum	+	++
Fusarium roseum	+	++
Glomerella gossypii	+	+++
Nematospora spp.	++	+
Nigrospora oryzae	++	++
Penicillium spp.	+	+
Rhizopus spp.	+++	+
Trichoderma spp.	+	+++
Verticillium spp.	+	++

[a] *Source:* Marsh *et al.* (1949), except *Nigrospora* (Waked *et al.*, 1981).
[b] Worldwide importance: +++ Common and of widespread distribution;
                                      ++ Common in certain areas;
                                      + Generally of lesser importance.
[c] Ability to decompose cellulose: +++ Excellent; ++ Good; + Poor.

In the USA, the fungi most commonly found on lint exposed to weathering after dehiscence are *Alternaria* spp., *Cladosporium herbarum* and *Fusarium moniliforme* (Marsh and Bollenbacher, 1949). These three were found growing on weathered seed cotton across the cotton belt, although, in samples collected from Texas, Oklahoma and the West, *Aspergillus* spp. and *Rhizopus stolonifer* were the most common lint contaminants.

Conditions Favouring Lint Contamination

Investigations carried out in Zimbabwe (Hillocks and Brettell, 1993) indicated that discoloration of seed cotton caused by the growth of microbial contaminants occurred in the field under three distinct circumstances.

1. As a result of incomplete boll rot, where the boll is able to open normally but one or more locules fail to fluff out due to fungal or bacterial infection (see Plate 5D). This condition is known as 'tight loc' in the USA, where it can occur following invasion of the boll by several organisms, particularly species of *Diplodia*, *Alternaria* or *Fusarium* (Marsh and Bollenbacher, 1949; Marsh *et al.*, 1950).

2. When the boll opens and fluffs out normally but is then exposed to prolonged periods of moist conditions, conducive to fungal growth. Seed cotton which becomes discoloured in this way is referred to as 'weathered'.
3. When the crop becomes heavily infested with aphids or whiteflies towards the end of the growing season, resulting in the deposition on the exposed lint of honeydew, a sugar-rich secretion from the insects. Under humid conditions, these sugars stimulate the growth of fungal contaminants.

Furthermore, a crop which appears clean at harvest may become discoloured by microbial growth if it is stored damp or in a humid atmosphere.

Different Symptoms of Lint Rot

Many of the organisms involved in the boll rot complex can be responsible for discoloured fibres in the harvest, if the boll opens before it has been completely rotted. This may occur if infestation occurred near to boll maturity or if the infection is confined to a single locule. Lint in the affected locules is discoloured black or grey by microbial growth (Fig. 8.1).

The bacterial blight organism, *Xanthomonas campestris* pv. *malvacearum*, and the internal boll rot fungi, *Nematospora* spp., cause a yellow or

Fig. 8.1. Discoloured lint following field exposure after boll dehiscence.

Table 8.2. Conditions leading to lint discoloration, visual appearance of affected lint and some causal organisms.

Conditions	Colour	Possible causal organisms
Infection of boll before normal dehiscence leading to 'tight loc'	black/grey	*Diplodia gossypina* *Alternaria* spp. *Nigrospora oryzae* *Curvularia* spp.
	pink yellow/brown	*Fusarium moniliforme* *Nematospora* spp. *Xanthomonas campestris*
Contamination of lint due to weathering after normal dehiscence	grey	*Alternaria* spp. *Cladosporium herbarum* *Rhizopus stolonifer* *Botrytis* spp.
Deposition of honeydew by insect pests after normal dehiscence	black grey	*Cladosporium herbarum* *Alternaria* spp. *Rhizopus stolonifer* *Aspergillus niger*
Storage of seed cotton or lint under damp conditions	green/blue pink	*Trichoderma* spp. *Penicillium* spp. *Fusarium* spp.

yellow-brown stain under such circumstances. *Aspergillus flavus* causes a fluorescent yellow stain. These yellow stains may be found on seed cotton which has fluffed out normally, if conditions after infection are not suitable for disease development or if infection occurs sufficiently close to the time of boll dehiscence. To some extent, the colour of the stain is a guide to the identity of the main contaminating organism (see Table 8.2).

TIGHT LOC

The tight loc symptom is caused most frequently in the USA by *Diplodia gossypina*, but similar symptoms may be caused by *Alternaria*, *Cladosporium*, *Colletotrichum*, *Curvularia*, *Nigrospora* and *Rhizoctonia* (Marsh *et al.*, 1950). Both *Diplodia* and *Colletotrichum* can cause tight loc after primary invasion through the boll wall. However, most of the other fungi usually do so only after the boll wall has been breached, either by insect damage or by premature rupture of the suture.

Of special interest among the fungi responsible for tight loc is *Nigrospora oryzae* (Berck & Br.) Petch. A common cause of lint rot in the western USA, this fungus cannot directly penetrate the carpel wall, usually invading the lint at an early stage in boll dehiscence (Houston and Garber, 1959). It has a wide host range and is commonly associated with graminaceous species. It is transmitted by the mite *Siteroptes reniformis* (Laemmlen, 1969). *N. oryzae* is also responsible for lint rot in the former Soviet Union, where it is carried on to the boll by *Siteroptes graminis* (Bondarovich, 1961). There appears to be a mutualistic symbiosis between the mite and the fungus. Laemmlen and Hall (1973) demonstrated that under experimental conditions only 15% of bolls could be successfully inoculated with the fungus on its own, but a 99% success rate was achieved in the presence of the mite. Soon after mating, females attempt to place fungal spores into specialized sacks on their backs, to deposit them on the boll surface. The optimum temperature for development of the fungus is 21–27 °C and for the mite, 27 °C.

WEATHERING

Seed cotton which remains unpicked after normal dehiscence has a dull grey appearance and may become dark grey or black if exposed to rainfall and prolonged periods of high humidity. Prindle (1934) identified the fungi involved in the greying of weathered seed cotton in the USA as *Alternaria* sp., *Cladosporium* and *Fusarium*. Similar results were reported by Jaczewski (1929) in the former Soviet Union, where *Cladosporium herbarum* was the main fungus isolated but *Fusarium* spp. and *Alternaria* spp. were also present. Marsh and Bollenbacher (1949) added yeast species and a bacterium to this list.

In Zimbabwe, the main fungi responsible for greying of weathered seed cotton are *Alternaria alternata*, *A. macrospora*, *R. stolonifer* and *C. herbarum* (Hillocks and Brettell, 1993).

Marsh and Bollenbacher (1949) considered the fungi growing on cotton fibres to be a small and distinct group. Most of them have dark-pigmented spores which enable them to withstand the destructive effect of ultraviolet light on the exposed lint (e.g. Fulton and Coblentz, 1929). They are common inhabitants of the upper layers of field soils and are cellulose degraders. By contrast, the fungi commonly associated with cotton stored under damp or humid conditions are predominantly those with thin-walled spores having little or no pigmentation and which are usually poor producers of cellulolytic enzymes (see Table 8.1 above). The most common members of this group are species in the genera *Aspergillus*, *Penicillium* and *Trichoderma*.

STICKY COTTON

A number of homopterous insects produce an excretion rich in sugars, which is referred to as entomological sugar or honeydew. The main insect pests of cotton which produce honeydew are the whitefly, *Bemisia tabaci*, and the aphid, *Aphis gossypii* (Couilloud, 1986). If the crop becomes infested with large populations of either of these insects late in the growing season, honeydew drips from the infested leaves to contaminate any exposed seed cotton.

Since the mid-1970s, sticky cotton, contaminated with honeydew, has

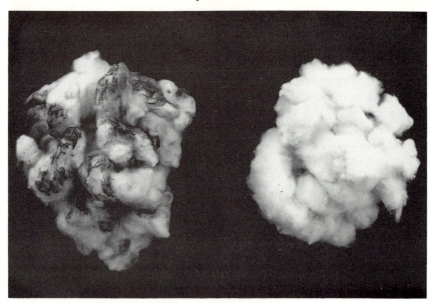

Fig. 8.2. Discoloured (compared with clean) seed cotton caused by growth of *Cladosporium herbarum* and other fungi on honeydew deposited by whitefly.

become an increasingly serious problem for the spinning industry. Whitefly populations have gradually built up where organophosphate and, later, synthetic pyrethroid insecticides have been used on a wide scale, such as in the Sudan (e.g. Abdeldaffie *et al.*, 1987) and, to a lesser extent, in Israel (Elsner, 1984; Gutknecht, 1988). Although aphids were adequately controlled with organophosphate insecticides, where the change has been made to pyrethroids, sucking pests in general have been less effectively controlled. In addition to changes in crop protection chemicals, Gutknecht (1988) listed several other reasons for increased whitefly populations. These included increased use of nitrate fertilizers, leading to more vegetative growth, and later harvesting, due to shortage of labour. Furthermore, modern high-speed spinning machinery is more sensitive to the presence of sticky fibres.

The presence of honeydew on seed cotton encourages the growth of certain micro-organisms, producing a black growth, sometimes referred to as 'sooty mould' (Fig. 8.2). In Israel, sooty mould on leaves and lint is regarded as the most serious problem associated with whitefly infestation of the cotton crop (Horowitz *et al.*, 1984). High populations of whitefly in the crop towards the end of the season always result in honeydew deposition on the exposed lint, which then supports the growth of several mould fungi under conditions of at least 70% relative humidity. These fungi often consume the honeydew so that it may not be detected on seed cotton samples despite the presence of sooty mould. There is some evidence from Israel that certain defoliants and insecticides inhibit the growth of sooty mould fungi (Elsner, 1984).

The main fungus associated with honeydew contamination in Zimbabwe is *Cladosporium herbarum* (Hillocks and Brettell, 1993). In Israel it is *Aspergillus niger* (Elsner, 1984). The presence of sugar in the form of honeydew stimulates the growth of a number of fungi with air-borne conidia which become trapped among the lint fibres after boll dehiscence. Experiments conducted in Zimbabwe indicated that *Alternaria* spp. and *R. stolonifer* tended to predominate on damp seed cotton, in the absence of sugar, but that *Cladosporium* became dominant when sugar was present, even on dry seed cotton, provided relative humidity was high. However, if the seed cotton was soaked with water, *Alternaria* spp. again predominated over *C. herbarum*.

Cladosporium mould growth associated with honeydew contamination is distinct in appearance from microbial growth resulting from partial boll rot or the grey appearance of weathered cotton. It is black, encrusted and confined to patches, corresponding to areas where honeydew drops have fallen. The presence of this type of mould is therefore indicative of honeydew contamination. However, under dry conditions seed cotton may become contaminated with honeydew but remains free of mould growth (Hillocks and Brettell, 1993).

Cellulose Degradation

Many of the fungi associated with lint contamination are capable of producing cellulolytic enzymes in sufficient quantity to degrade cotton fibres if they are able to grow on the seed cotton for long enough. The most efficient cellulose degraders are *Alternaria* spp., *Curvularia*, *Fusarium moniliforme* and a *Glomerella* species isolated from cotton bolls. Most species of *Aspergillus*, including *A. niger*, are poor cellulose degraders (White *et al.*, 1948; Marsh *et al.*, 1949). It should be noted however, that there is considerable variation with respect to production of cellulolytic enzymes between species of the same genus and, indeed, among isolates of the same species. Marsh *et al.* (1949) found that *C. herbarum* was a moderately good cellulose degrader at 20–25 °C but much less good at 30 °C, and suggested that this explained the negative results obtained in tests of cellulolytic activity conducted by other investigators. The same authors tested three species of *Rhizopus*, finding them to be poorly cellulolytic. But in a later paper, Marsh and Simson (1965) reported that, when fibres taken from unopened bolls were incubated with *R. stolonifer* or *R. arrhizus*, the fibre lost strength. Also, in apparent contradiction to previous findings, Mousa *et al.* (1948) reported that *A. niger* and *R. stolonifer* caused more degradation of cotton fibres than did *Alternaria alternata*.

With such evidence of varying degrees of cellulolytic activity between isolates of the same fungus and wide differences in experimental conditions, it is difficult to draw conclusions about the likely extent of fibre damage simply on the basis of identifying the contaminating organism. In any event, the proportion of damaged fibre in the harvest needs to be high in order to have a

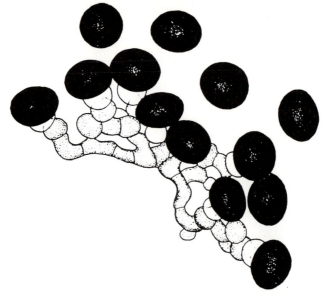

Fig. 8.3. *Nigrospora oryzae*: conidial state of *Khuskia oryzae* (× 650) (Ellis, 1971).

deleterious effect on spinning properties of the lint. In the case of cotton contaminated with *Diplodia gossypina*, Sands *et al.* (1962) found that greater than 10% of damaged fibres was required to have a significant effect on spinning efficiency, and yet, in the USA, the level of *Diplodia*-contaminated fibres rarely exceeded 2.5%.

Description of Some Causal Organisms

Many of the lint-rotting fungi are common members of the soil-borne and air-borne microflora of cotton fields causing leaf spot and boll rot diseases. Many of these fungi have been described in previous chapters. Some of the more common members of the lint rot complex which have not been described in Chapters 6 or 7 are described briefly below.

NIGROSPORA ORYZAE

Greying of cotton lint by *N. oryzae* is caused by the presence of conidial heads (Fig. 8.3) among the fibres. The conidia are subglobose and 12–17 × 13–18 μm in size (Ou, 1984). Although only the conidial state is found on cotton lint, the teleomorph has been described by Hudson (1963) and named *Khuskia oryzae*.

CLADOSPORIUM HERBARUM

Colonies on natural substrata and in culture are olive-green or olivaceous brown and velvety and the reverse on malt agar appears black. The

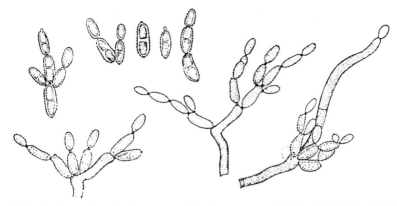

Fig. 8.4. *Cladosporium herbarum*: conidiophores and conidia (× 650) (Ellis, 1971).

conidiophores are mostly macronematous, straight or flexuous, pale brown, smooth and up to 250 μm long and 3–6 μm wide. Conidia are often in branched chains. They are ellipsoidal or oblong with rounded ends (Fig. 8.4), pale or olivaceous brown in colour with thick walls, distinctly verruculose, aseptate or one-septate and 5–23 × 3–8 μm in size, with scars at one or both ends (Ellis, 1971).

RHIZOPUS STOLONIFER

Colonies on PDA at 25 °C are at first white, becoming speckled by the presence of sporangia. The dark sporangia are also produced on moist lint, giving it a grey appearance. Sporangiophores are up to 34 μm in diameter and 1000–3500 μm in length, arising in groups of three to five from stolons

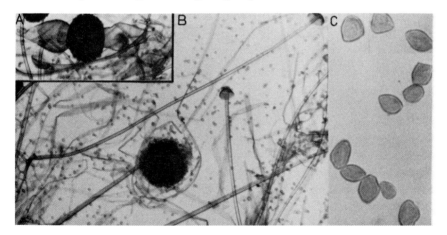

Fig. 8.5. *Rhizopus stolonifer*: A, zygospore formed between two compatible isolates (× 93); B, sporangiophore, rhizoids, columella and sporangium (× 54); C, sporangiospores (× 930) (Lunn, 1977).

opposite rhizoids. Sporangia are 100–350 µm in diameter and globose in shape, with a flattened base. They are at first white in colour, becoming black, and contain many spores. The spores are usually 8–29 µm in diameter, irregularly round or ovate in shape and brownish-black in colour. Zygospores may be produced when compatible isolates are grown together (Fig. 8.5) (Lunn, 1977).

References

ABDELDAFFIE, E.Y.A., ELHAG, E.A. AND BASHIR, N.N.H. (1987) Resistance in the cotton whitefly, *Bemisia tabaci* to insecticides recently introduced into Sudan Gezira. *Tropical Pest Management* 33, 283–7.

BONDAROVICH, M.Y. (1961) *Nigrospora* of cotton and its control. *Review of Applied Mycology* 41, 306.

COUILLOUD, R. (1986) Bibliographical data on honeydew producing insects. *Coton et Fibres Tropicales* 49, 226–8.

ELLIS, M.B. (1971) *Dematiaceous Hyphomycetes* Commonwealth Agricultural Bureaux, Farnham Royal, Bucks, UK. 608pp.

ELSNER, I.O. (1984) The effect of *Bemisia tabaci* on cotton quality. *International Cotton Testing Conference Proceedings*, Bremen, 1983.

FULTON, H.R. AND COBLENTZ, W.W. (1929) Fungicidal action of ultra-violet radiation. *Journal of Agricultural Research* 38, 159–68.

GUTKNECHT, J. (1988) Assessing cotton stickyness in the mill. *Textile Month*, May, 53–7.

HILLOCKS, R.J. AND BRETTELL, J.H. (1993) The association between honeydew and growth of *Cladosporium herbarum* and other fungi on cotton lint. *Tropical Science* (in press).

HOROWITZ, A.R., PODLER, H. AND GERLING, D. (1984) Life table analysis of the tobacco whitefly, *Bemisia tabaci* in cotton fields in Israel. *Acta Oecologia Applicata* 5, 222–33.

HOUSTON, B.R. AND GARBER, R.H. (1959) A lint rot of cotton in California caused by *Nigrospora oryzae*. *Plant Disease Reporter* Supplement 259, 233–5.

HUDSON, H.J. (1963) The perfect stage of *Nigrospora oryzae*. *Transactions of the British Mycological Society* 46, 355–60.

JACZEWSKI, A.A. (1929) Some diseases of cotton fibres. *Review of Applied Mycology* 9, 307.

LAEMMLEN, F.F. (1969) The association of the mite *Siteroptes reniformis* and *Nigrospora oryzae* in *Nigrospora* lint rot of cotton bolls. *Phytopathology* 59, 1036–7. (Abstr.).

LAEMMLEN, F.F. AND HALL, D.H. (1973) Interdependence of a mite *Siteroptes reniformis* and a fungus *Nigrospora oryzae* in the *Nigrospora* lint rot of cotton. *Phytopathology* 63, 308–15.

LAGIÈRE, R. (1974) Contributions to the study of the mycological flora of dry cotton fibres. *Coton et Fibres Tropicales* 29, 437–45.

LUNN, J.A. (1977) *Rhizopus stolonifer*. *CMI Descriptions of Pathogenic Fungi and Bacteria* No. 524. Commonwealth Agricultural Bureaux, Farnham Royal, Bucks, UK.

MARSH, P.B. AND BOLLENBACHER, K. (1949) Fungi concerned in fibre deterioration. 1. Their occurrence. *Textile Research Journal* 19, 313–24.

MARSH, P.B. AND SIMSON, M.E. (1965) Effects produced in cotton fibre by boll rotting species of *Rhizopus. Phytopathology* 55, 52–6.

MARSH, P.B., BOLLENBACHER, K., BUTLER, M.L. AND RAPER, K.B. (1949) The fungi concerned in fibre deterioration. 2. Their ability to decompose cellulose. *Textile Research Journal* 19, 462–84.

MARSH, P.B., GUTHRIE, L.R., BOLLENBACHER, K. AND HARRELL, D.C. (1950) Observations on microbial deterioration of cotton fibre during the period of boll opening in 1949. *Plant Disease Reporter* 34, 165–75.

MOUSA, O.M., NOMEIR, A.A. AND ALI, A.Y. (1984) Fibre properties of some cotton cultivars inoculated with some phytopathogenic fungi under different levels of relative humidity. *Annals of Agricultural Science Moshtohor* 20, 29–38.

OU, O.H. (1984) *Rice Diseases.* CAB International, Wallingford, UK. 380pp.

PRINDLE, B. (1934) Microbiology of textile fibres. *Textile Research Journal* 5, 11–31.

SANDS, J.E., FIORI, L.E., GROVES, N.H. AND MARSH, P. B. (1962) The utilization of *Diplodia*-damaged boll rot cotton. Part 1. Effects on yarn and spinning efficiency. *Textile Research Journal* 32, 1013–22.

WAKED, M.Y., EL-SAMRA, I.A. AND NOUMAN, K.A. (1981) Microscopic studies on Egyptian cotton fibres inoculated with some rotting fungi. *Phytopathologia Mediterranea* 20, 169–73.

WHITE, W.L., DARBY, R.T., STWCHERT, G.M. AND SANDERSON, K. (1948) Assay of cellulolytic activity of moulds isolated from fabrics and related items exposed in the tropics. *Mycologia* 40, 34–84.

Plate 1

Symptoms of bacterial blight *(Xanthomonas campestris* pv. *malvacearum)*

A, Seedling blight on cotyledons; **B,** angular leaf spot; **C,** foliar symptoms of systemic infection; **D,** stem infection 'blackarm'; **E,** infection of bracts; **F,** bacterial boll rot.

Plate 2

Symptoms of Verticillium wilt *(Verticillium dahliae)*

A, Infection of young plant; **B,** leaf symptoms showing 'tiger striping'; **C,** leaf symptoms showing chlorosis and reddening; **D,** longitudinal section of stem showing vascular discoloration in an infected plant (V), compared to a healthy plant (H).

Plate 3

Symptoms of Fusarium wilt caused by *Fusarium oxysporum f.sp. vasinfectum* and of root rot caused by *Phymatotrichum omnivorum*

A, Advanced symptoms of wilt on a young plant (J. Bridge); **B,** early wilt symptoms showing stunting and chlorosis of leaf margins, compared to surrounding symptomless plants; **C,** longitudinal section of infected stem stained with safranin and fast green, showing *Fusarium* conidia and mycelium in the xylem vessels; **D,** field of cotton infected by Phymatotrichum root rot (C.M. Rush, Texas Agricultural Experimental Station, Bushland); **E,** root rot symptoms at flowering (C.M. Rush); **F,** *P. omnivorum* on the surface of the tap root of an infected plant (C.M. Rush).

Plate 4

Symptoms caused by some common leaf spot pathogens

A, *Alternaria macrospora* on cotyledons; **B,** *A. macrospora* on lower-canopy leaf, showing necrotic spots with purple halo; **C,** *A. macrospora* on upper-canopy leaf of a potassium-deficient plant, showing black, sooty spots; **D,** false mildew caused by *Ramularia areola*; **E,** necrotic spots caused by *Phoma* sp.

Plate 5

Symptoms caused by some common boll rot pathogens

A, *Fusarium moniliforme*; **B,** internal boll rot caused by *Nematospora* spp. after boll damage by *Dysdercus* spp.; **C,** *Alternaria macrospora*; **D,** 'tight loc' can be caused by several fungi responsible for boll rot and lint contamination.

Plate 6

Symptoms of some virus diseases

A, Mosaic, Brazil (A.S. Costa, Campines, Brazil); **B,** vermelhão, showing foliar chlorosis (A.S. Costa); **C,** vermelhão, showing foliar reddening (late symptoms) (A.S. Costa); **D,** leaf crumple symptoms (CLCV) on *G. hirsutum* cv Deltapine 70 (J.K. Brown); **E,** CLCV symptoms on cotton flower (J.K. Brown); **F,** CLCV symptoms caused by severe (R) and mild (L) isolates (D.C. Erwin, University of California, Riverside).

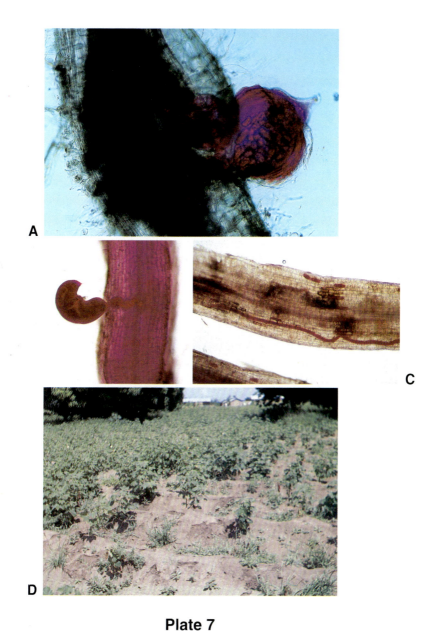

Plate 7

Nematodes and nematode damage to cotton (J. Bridge)

A, Root gall with mature female and egg sack of *Meloidogyne incognita*; **B,** mature female of *Rotylenchulus reniformis* on cotton root; **C,** *Hoplolaimus seinhorsti* in cotton root; **D,** patchy stand of cotton in Tanzania, caused by the Fusarium wilt/root-knot nematode disease complex.

Plate 8

Symptoms of nutrient deficiency (S.C. Hodges)

A, Nitrogen deficiency symptoms, showing stunted plants with foliar chlorosis which is more pronounced in the lower leaves; **B,** early symptoms of potassium deficiency on lower leaves; **C,** late symptoms of potassium deficiency on upper leaves; **D,** late potassium deficiency, grading scale (1–6) for symptom severity used in Zimbabwe (R.J. Hillocks); **E,** magnesium deficiency showing foliar reddening with main veins remaining green; **F,** sulphur deficiency showing pale green leaves in upper canopy.

9

Virus Diseases

J.K. BROWN[*]

Introduction

Our knowledge of and ability to investigate plant virus diseases of food and fibre crops has increased substantially over the past several decades. However, on a worldwide basis, relatively few advances have been made concerning the development of methods for identification, aetiology or molecular characterization of the viruses and virus-like disorders affecting cotton (*Gossypium* spp.).

Several factors have probably contributed to the situation. First, there is a general shortage of laboratory facilities equipped to work with plant viruses, as well as an understaffing of virological expertise in tropical regions where cotton is cultivated. Hence, most virus-like disorders of cotton have received inadequate attention and little concerted effort with respect to characterization of the aetiological agent. Second, even with the most sophisticated equipment, the cotton plant, a woody perennial, is difficult to manipulate under laboratory conditions. The leaves of the infected cotton plant, which potentially serve as a source of virus for investigative purposes, contain interfering polysaccharides and primary and secondary metabolites. Upon disruption of the leaf cells, these compounds are subject to oxidation and subsequent formation of insoluble complexes, composed of virions, nucleic acids and other cellular constituents, which impede extraction of plant viruses using traditional methodologies. A third factor involved is that, among the suspect virus-like agents which affect cotton, most are either associated with or known to be transmissible in a semi-persistent or persistent manner by homopteran insect vectors (aphids, leafhoppers or whiteflies)

* Department of Plant Sciences, College of Agriculture, University of Arizona, Forbes Building, Tucson, Arizona 85721, USA.

Table 9.1. Summary of partially characterized virus and virus-like diseases of cotton.

Disease	Vector/transmission	Geographical distribution	References
African cotton mosaic	*Bemisia tabaci*	Africa (central, western)	Bink, 1973
Cotton anthocyanosis[a]	*Aphis gossypii*	Brazil	Costa, 1957
Cotton blue disease	*A. gossypii*	Africa, South America	Cauquil & Follin, 1983
Cotton leaf crumple[a]	*B. tabaci*	USA (south-western)	Dickson *et al.*, 1954
Cotton leaf curl	*B. tabaci*	Africa (Chad, central,	Farquharson 1912
(cotton leaf crinkle)		Nigeria, Sudan, Togo, western)	Golding, 1930
			Kirkpatrick, 1931
Cotton yellow vein[a]	*Scaphytopius albifrons*	USA (Texas)	Rosberg, 1957
(Texas vein-clearing virus)			

[a] Diseases for which virus-like particles have been observed.

(Costa, 1976). This suggests that experimental transmission by mechanical means is difficult, if not impossible in many cases, and that the viruses may be limited or restricted to the phloem cells. Both of these latter attributes also contribute to difficulties in experimental manipulation of virus cultures, in the isolation of virus particles from infected plants and demonstration of Koch's postulates, and in the subsequent characterization of the viruses on a molecular level.

An alternative strategy which could be devised to circumvent these problems would be the identification of annual, herbaceous alternate hosts of the viruses from which infectious virions could be purified and further characterized. Either this kind of information is lacking for many of the virus-like disease agents described to date, or workers have been unable to identify alternate, non-woody hosts with which to work. Hence it is not surprising that virus-like particles (VLPs) have been associated with only two diseases of cotton to date. Spherical or isometric VLPs (23 nm) were observed by electron microscopy in thin sections of cotton leaves infected by cotton yellow vein virus (CYVV) (Halliwell *et al.*, 1980), but aetiology has yet to be formally demonstrated. For cotton leaf crumple virus (CLCV), virus-like inclusion bodies (2–10 μm) containing VLPs were observed by light microscopy in phloem parenchyma of infected cotton in 1963 (Tsao, 1963). Not until the mid-1980s, when bean *(Phaseolus vulgaris* L.) was found to be an alternate, nearly asymptomatic host of CLCV, were virus particles purified and visualized by electron microscopy (Brown and Nelson, 1984). Following the visualization of characteristic paired or geminate particles (20 × 30 nm), the virus was tentatively designated as a member of the geminivirus group of plant viruses (established in 1978). Recent studies indicate the presence of infectious viral DNA in both cotton and bean infected with virus (J.K. Brown, unpublished); however, the aetiology of the disease remains only a matter of conjecture until Koch's postulates are completed. For the remainder of the mosaic, leaf curl and other symptomatic disorders of cotton, plant viruses are suspected, but aetiologies remain to be determined.

Despite these drawbacks, however, a substantial amount of new information is available regarding biological (host range and transmission) characteristics of several previously described virus-like disease agents, and also on management of the diseases through the use of tolerant/resistant cotton varieties and control of the insect vectors.

With the increasing availability of diagnostic tools involving serological and molecular hybridization methodologies, there is also great potential for rapid progress to be achieved in establishing methods for virus detection and identification to support efforts directed toward defining aetiologies and disease epidemiologies, and subsequently to develop resistant cotton varieties using classical and/or genetically engineered methods. A brief summary of suspect or putative viral and mycoplasma-like (MLO) diseases of cotton, the vectors and geographical distribution is provided in Tables 9.1–9.4.

Table 9.2. Summary of suspect virus-like disorders of cotton.

Disease	Vector/transmission	Geographical distribution	Original references
Acromania (crazytop)	Unknown	USA (Arizona), China, Haiti, Sudan	Bedford, 1938
			Cook, 1920, 1923
Cotton (common) yellow mosaic	*Bemisia tabaci*	Brazil	Costa and Forster, 1938; Costa, 1955
Cotton mosaic	*B. tabaci*	Central America	Quant, 1977
	Unknown	USA (Louisiana)	Neal, 1946
Cotton leaf curl	*B. tabaci*	Philippines	Reyes *et al.*, 1959; Nour, 1960
Cotton leaf mottle	Grafting	Sudan	Nour, 1959
Cotton leaf roll	*Aphis gossypii*	Thailand	Delalande, 1970
	A. gossypii	Paraguay	
Cotton leaf roll/leaf curliness	*A. gossypii*	USSR (former)	Moskovetz, 1941
Infectious variegation	Grafting, *B. tabaci*	India	Mali, 1978b
Indian leaf crumple	Grafting, *B. tabaci*	India	Mali, 1977
Murcha vermelhao/purple wilt	*A. gossypii*	South America	Costa and Sauer, 1954
Terminal stunt	Unknown	USA (Texas)	Sleeth *et al.*, 1963
Tobacco streak	Possibly thrips	Brazil	Costa, 1961
Viral wilt	Grafting	India	Mali, 1979

Table 9.3. Summary of virus-like disease agents which infect cotton as an alternate host.

Disease	Vector/transmission	Geographical distribution	References
Infectious variegation (AbMV)	*Bemisia tabaci*	Brazil	Costa and Carvalho, 1968
Mosaic of jute	*B. tabaci*	Pakistan	Ahmad, 1978
Mosaic of kenaf	*B. tabaci*	El Salvador	Granillo et al., 1974
Mosaic of *Wissadula amplissima* (cotton mottle)	*B. tabaci*	USA (Texas)	Schuster, 1964

Table 9.4. Summary of diseases of cotton associated with MLOs or MLO-like symptoms.

Disease	Vector/transmission	Geographical distribution	References
Flavescence	Grafting; scale insect (*Margarodes* sp.)	Africa	Cauquil and Follin, 1983
Flower virescence (phyllody)	*Orosius cellulosus*	Africa	Dellatre, 1965; Bink, 1975
Psyllosis (psyllose)	*Paurocephala gossypii*	Africa (Malawi, Congo)	Soyer, 1947 Dineur, 1957
Stenosis (small leaf)	Grafting	Pakistan, India	Kottur and Patel, 1920; Uppal et al., 1944

Fig. 9.1. Adult whitefly, *Bemisia tabaci* Genn., the most common whitefly vector of plant viruses (J.K. Brown).

A survey of the literature indicates that at least 20 apparently distinct viral and four MLO diseases have been reported since the early 1900s. These diseases occur in sub-tropical and tropical regions, including Africa, Asia, the Caribbean Basin and Central and South America, and in sub-tropical and fringe temperate regions of North America (Mexico and the USA). In many of these regions cotton is ratooned, and potential alternate hosts (in the Malvaceae, Fabaceae and Euphorbiaceae) of the virus disease agents occur as indigenous or introduced weed species. In addition, climatic conditions are mild enough to potentially support nearly year-round populations of the most important virus vectors (aphids, leafhoppers and whiteflies), resulting in chronic virus-like disease problems in all cotton-growing areas.

Historically, and out of necessity, when virus diseases are described, the names are based upon the host plant species affected and on symptoms caused in that particular host. Although there is little doubt that this system is necessary, especially when an unknown pathogen is involved, it is somewhat problematic in that symptoms incited by the same pathogen may vary from region to region, due to genetic differences in cultivars grown and the variability in environmental conditions. This may result in several different designations for a disease caused by the same plant virus. Virus-like symptoms described in cotton include mosaics, mottles, leaf curls, discoloration and deformation, stunting and reddening of leaves and other plant parts. Without further characterization of these agents in biological and biochemical terms, however, it remains impossible to definitely state whether diseases with like and/or dissimilar symptoms have the same aetiology.

In most cases reported, virus-like diseases of cotton are primarily transmitted by insects, with homopterans like aphids, leafhoppers and whiteflies predominating in importance. Among these, the cotton (or melon) aphid, *Aphis gossypii* Glover, and the cotton (tobacco or sweet potato) whitefly, *Bemisia tabaci* Genn. (Fig. 9.1), are the most prevalent vectors, and the viruses they vector are transmitted in a semi-persistent or persistent manner.

In general, the *Bemisia tabaci*-transmitted viruses have characteristically narrow host ranges (Table 9.5), despite the broad range of plant species which the insect is capable of infesting (Mound, 1969, 1983; Butler *et al.*, 1986b). In contrast, the cotton aphid has a relatively narrow host range (Isely, 1946; Kring, 1959; Khalid and Al-Zarari, 1982), thus limiting the species of plants which could serve as alternate hosts for aphid-transmitted viruses of cotton, such as cotton blue disease (Cauquil *et al.*, 1982). A single leafhopper vector has been identified, *Scaphytopius albifrons* Hepner, which transmits cotton yellow vein virus (CYVV) (Halliwell *et al.*, 1980). Psyllids, and possibly scale insects, are important in transmission of several MLOs and MLO-like diseases (Tarr, 1964; Watkins, 1981; Cauquil and Follin, 1983). Although cotton has been experimentally infected with tobacco mosaic virus, tobacco ringspot and some viruses which infect *Prunus* spp., natural occurrence of these viruses in cotton has not been documented (Watkins, 1981). Tobacco streak virus, now known to be transmitted by thrips, was reported on cotton in Brazil (Costa, 1961). Field-related transmission of viruses to or from cotton by mechanical means has not been reported, nor is transmission through seed generally a problem.

In general, and in pathological terms, much information derived from past and present studies is available and useful in devising disease control measures. Although current disease diagnosis continues to rely on observations of characteristic disease symptoms which can be associated with the presence of a recognized insect vector population, there is substantial information available which can be applied towards defining and implementing approaches for disease control. These currently include the cultivation of tolerant or resistant cotton varieties and rigorous control of insect vector populations and alternate hosts which serve as known or suspected virus reservoirs.

Changing agricultural practices, the introduction of higher-yielding, high-quality cotton varieties, and the increased activity relative to transport of plants and germplasm materials throughout the world will undoubtedly create new challenges in understanding the pathology, epidemiology and management of cotton diseases incited by viruses, MLOs and other pathogenic agents on a worldwide basis. In this chapter, therefore, an effort is made

Table 9.5. Host range[a] comparison of viruses and virus-like agents of cotton.

Host test plant	Geminivirus/ Abutilon mosaic virus*	Unknown/ African cotton mosaic	Possible luteovirus/ cotton anthocyanosis	Unknown cotton blue disease	Geminivirus/ cotton leaf crumple	Unknown/ cotton leaf curl	Isometric, virus unknown/ cotton yellow vein	Unknown/ Indian infectious variegation	Unknown/ mosaic of kenaf	Unknown/ mosaic of Wissadula
Geographical location	Brazil	Africa	Brazil	Africa	USA (AZ, CA)	Africa (Sudan, Nigeria)	USA (TX)	India	El Salvador	USA (TX)
Vector transmission	Bemisia tabaci[b]	Bemisia tabaci[c]	Aphis gossypii[d]	Aphis gossypii[e]	Bemisia tabaci[f]	Bemisia tabaci[g]	Scaphytopius albifrons[h]	Bemisia tabaci[i]	Bemisia tabaci[k]	Bemisia tabaci[l]
Abutilon crispum	+									
A. indicum		+ (unknown species)				+ (spp. ?)		+		
A. striatum	+									
A. theophrasti	+				+					–
Althaea rosea	+				+	+				+
Cucumis spp.					–					–
Corchorus fascicularis						+				
Datura stramonium	+				–	+				
Glycine max	+				–	–			+	
Gossypium arboreum	– (most resistant)			–	+/– (highly resistant)					
G. barbadense	+		+	+	+ (mild)	+				
G. herbareum					+ (except var. 'Wightianum' is resistant)	–	–			
G. hirsutum Acala 4-42 or DP	+		+	+	+	+	+			+
G. hirsutum 'Marie Galante'					+ (mild)	–				

Host	1	2	3	4	5	6	7	8
Hibiscus cannabinus	+		+	+				+
H. esculentus	+	+	+	−	+			−
H. sabdariffa				+	+			
Ipomoea batatas				−				
Lycopersicon esculentus	+							−
Malvaceous	+					+ (unknown species)		−
Malva parviflora	+			+				+
Malvastrum coromandelilanum								
Malaviscus arboreus		+			+		+	
Nicandra physaloides	+	+	+					
Nicotiana tabacum	+			−				
Petunia hybrida				−	+			
Phaseolus vulgaris	+		−	+				
Phyllanthus niruri			−		+			
Sida spp. (*alba, cordifolia, melongena, micrantha, rhombifolia*)	+ (*cordifolia, micrantha, rhombifolia*)	+ (*alba, cordifolia, melongena, micrantha, rhombifolia*)	+ (*micrantha, rhombifolia*)		+ (*alba, cordifolia*)			+ (*cordifolia, micrantha, rhombifolia*)
Solanum melongena	−			−				
S. tuberosum				−				
Vigna unguiculata				−				
Wissadula amplissima								−

* Infectious variegation or American mosaic.

a Host range: = suspected, not demonstrated.

References: b Costa, 1955; Costa and Carvalho, 1969. c Nour, 1959; Bink, 1973; Lourens *et al.*, 1972. d Costa and Saver, 1954; Waterhouse *et al.*, 1988. e Cauquil and Follin, 1983. f Dickson *et al.*, 1954; Brown and Nelson, 1986. g Farquharson, 1912; Kirkpatrick, 1931. h Rosberg, 1957; Halliwell *et al.*, 1980. i Mali, 1978a. k Granillo, *et al.*, 1974. l Schuster, 1964.

Fig. 9.2. Cotton mosaic disease symptoms in Brazil (A.S. Costa).

to discuss known aspects of the most prevalent and well-characterized diseases of cotton described to date. Further, an attempt is made to summarize, by disease name or description, the multitude of cotton disorders described in the literature, but for which minimal details are available. It is hoped that ongoing efforts by plant pathologists in cotton-growing regions of the world will yield more definitive information on these lesser-known problems, as well as on the better-characterized diseases. A concerted effort must be made to facilitate characterization of these viral pathogens in both biological and molecular terms in order to optimize cotton production strategies, and thus meet world demands for cotton fibre, oil and subsequent by-products.

African Cotton Mosaic

HISTORY AND DISTRIBUTION

African cotton mosaic (Cauquil and Follin, 1983) or cotton leaf mottle (Nour, 1959; Bink, 1975) has been reported throughout Africa, primarily in the regions south of the Sahara Desert (Ebbels, 1976; Fauquet and Thouvenel, 1987). The first outbreak of the disease, which was termed cotton leaf mottle, was reported in the Sudan in Egyptian cotton, *Gossypium barbadense* L., by Nour (1959). This disease was shown to be caused by a graft-transmissible agent, but attempts at insect transmission failed using the jassid, *Empoasca lybica*, or the cotton whitefly, *Bemisia tabaci* (Nour, 1959). In 1968, a major disease outbreak occurred in the susceptible *G. hirsutum* cultivar, BJA592, in Chad (Brader and Atger, 1971; Cateland, 1971), and subsequently further attention was focused on the problem. Because a diagnostic test is not yet available for identification of the African cotton mosaic disease agent, it is still debatable whether the leaf mottle (Nour, 1959) is identical to that which caused the outbreak of African cotton mosaic in Chad and, more recently, throughout Africa. Nevertheless, the cotton mosaic disease has now been reported in cotton in Benin, Cameroon, Central African Republic, Chad, Ivory Coast, Ghana, Mali, Nigeria, Sudan, Tanzania and Togo (Cauquil and Follin, 1983; Fauquet and Thouvenel, 1987; Brunt *et al.*, 1990). In all cases, the whitefly, *B. tabaci*, is associated with the disease.

Mosaic symptoms have been observed in cotton plantings elsewhere in the world, but there is no adequate information available to determine similarities or differences between the causal agents with respect to the African cotton mosaic disease. Whitefly-associated mosaics were reported in South America (Brazil and Colombia) (Fig. 9.2), Central America (Guatemala, Nicaragua and El Salvador) and the Caribbean Basin (Dominican Republic) (Fig. 9.3) (Costa and Forster, 1938; Costa *et al.*, 1954; Costa, 1955, 1966, 1976; Silberschmidt and Tommasi, 1955; Tarr, 1964; Kraemer, 1966; Costa and Carvalho, 1968, 1969; Delattre, 1970; Desmidts, 1970; Quant, 1977; Watkins, 1981; Cauquil and Follin, 1983; Brown *et al.*, 1991c) and in India (Mali, 1978a). Other mosaic diseases of unknown aetiology are associated with aphid infestations and affect cotton in the Philippines (Escober *et al.*, 1963), and more recently, in India (Mali, 1978a). No insect or other suspected vectors were reported for similar diseases occurring in Louisiana (USA) (Neal, 1946) and Texas (USA) (Schuster and Lambe, 1966). Generally in these latter cases, disease incidence and the impact on yields have been minimal, although in certain years the problems are worse than in others. There are no definitive updates on most of the above-described mosaic diseases. Additional investigations are needed to ascertain the level of losses that are currently sustained.

LOSSES

Yield losses attributed to cotton mosaic are a result of reduced flower production, boll shedding and stunting of susceptible plants (Nour, 1959; Bink, 1973; Cauquil and Follin, 1983). Five different symptom intensities or severities were described by Bink (1973) for use in screening for disease resistance in *G. barbadense* and *G. hirsutum* varieties, which can be defined by one of four categories. In type 1 symptom category, leaves exhibited chlorotic patches of discoloration 2–5 mm in diameter and vein-yellowing; in type 2

Fig. 9.3. Cotton yellow mottle disease symptoms in the Dominican Republic (J. Bird).

plants, leaves showed larger, more angular chlorotic patches 5–15 mm in diameter (considered the characteristic 'cotton mosaic' symptom); the type 3 symptom involved chlorosis, vein-thickening and deformation of leaf veins; and, in type 4, stunting of the actively growing tips plus severe mosaic and sterility is observed (Bink, 1973). Losses are particularly severe in BJA592 and significantly less in HG9 and several other resistant/tolerant varieties.

Symptoms

Infected plants exhibit an irregular mottle or yellow mosaic (see Plate 6A), stunting and yield losses due to a reduced canopy, fewer flowers and boll shedding. The severity of the symptoms varies with the cotton variety and with environmental conditions (soil, climate and fertilizer) (Bink, 1973; Cauquil and Follin, 1983; Fauquet and Thouvenel, 1987).

Causal agent

Based upon symptoms and whitefly transmissibility, the disease is believed to be incited by a plant virus; however, the aetiology is unknown. There are no reports describing ultrastructural studies, virus isolation or purification, or completion of Koch's postulates. Although the yellow mosaic symptoms associated with the disease are reminiscent of those incited by several other whitefly-transmitted geminiviruses, the nature of the causal agent will remain unknown until further studies are undertaken.

Host range

A definitive host range study has not been undertaken, but the common distribution of other mosaic-affected weed and garden species is considered evidence that some play a role as alternate hosts of the disease. Suspected (inferential) hosts are *Aspilia kotschyi* Oliv.(Compositae), *Hibiscus esculentus* (Malvaceae), *Sida micrantha* L. (Malvaceae) (Cauquil and Follin, 1983), *S. alba* L., *S. carpinifolia* L., *S. cordifolia* L. and *S. rhombifolia* L. (Nour and Nour, 1964; Lourens *et al.*, 1972; Bink, 1973).

Vector

The only known natural vector of the cotton mosaic agent is the cotton whitefly, *B. tabaci*. Among the three whitefly species present in cotton fields in Chad, *B. hancocki* Corbett, *B. tabaci* and *Trialeurodes desmondii* Corbett, only *B. tabaci* transmits the causal agent (Lourens *et al.*, 1972). In addition, *B. tabaci* makes up about 80–90% of the total whitefly population inhabiting cotton fields in the area, compared with 10–20% for *T. desmondii* and less than 5% for *B. hancocki*. The presence of *A. gossypii* and the jassid, *Empoasca lybica*, has been documented in affected cotton fields, but neither is known to vector the cotton mosaic agent (Nour, 1959; Bink, 1975).

B. tabaci has an apparently broad host range of plants (Mound, 1983; Cock, 1986) upon which it is capable of feeding and reproducing. The whitefly develops from egg to adulthood in about three weeks under tropical conditions. *B. tabaci* is capable of infesting, among others, plants within the Compositae, Cucurbitaceae, Euphorbiaceae, Fabaceae, Malvaceae and Solanaceae. Plant species within these families and many others provide food and shelter for whitefly populations during the dry season and when cotton is not available. Thus, representatives of these families may also serve as alternate hosts of plant viruses and virus-like agents. *B. tabaci* is capable of moving short distances from plant to plant, and of dispersing greater distances, by passive transport on the wind, in search of suitable food sources (Gerling, 1990). Whiteflies are homopteran insects which characteristically feed by way of a stylet which penetrates the phloem tissues through an intercellular path (Pollard, 1955). Because of these characteristics, polyphagous whiteflies like *B. tabaci* serve as ideal vectors of plant viruses. Likewise, many of the whitefly-transmitted plant viruses which have been described to date typically inhabit the phloem of their host plants. Hence, characteristic symptoms of infection by these viruses are yellow mosaics, leaf curling or yellowing, and are probably due in part to interference with normal phloem transport functions.

Transmission

African cotton mosaic agent is experimentally transmissible by grafting and *B. tabaci* (Cateland, 1971; Bink, 1973) and in nature by *B. tabaci*. Symptoms develop in 9–60 days in susceptible cultivars, with an average of 15 days (Bink, 1973). The agent is not transmissible by aphids or several other whitefly species tested (Bink, 1975), by mechanical means through sap or by seed (Cauquil and Follin, 1983).

Exacting studies on virus–vector relationships have not been carried out, but it is known that male and female *B. tabaci* adults and immature forms may become inoculative following a one-hour acquisition-access feed (AAF). A 24-hour latent period is required before the insect can transmit the agent and, once inoculative, whiteflies continue to transmit the disease agent for several days. The agent is not passaged through the egg (transovarial) because first instars hatching from eggs deposited by an inoculative female were shown to be incapable of transmitting the disease agent (Bink, 1973; Cauquil and Follin, 1983).

Epidemiology

Alternate host plants of the disease agent have not been identified to date. Bink (1973) lists plant species in 15 families which are present seasonally in the region, and which could serve as hosts of the whitefly vector and/or as alternate hosts of the disease agent. It is likely that one or more indigenous hosts of the agent exists, and that *B. tabaci* acquires the agent from those source(s) prior to dispersal to seedling cotton fields. Secondary spread from cotton to cotton would be facilitated following subsequent infestations and short-distance dispersal by the vector within the field. A hypothetical life cycle for *B. tabaci* is suggested by Bink (1973). *B. tabaci* populations are present in winter weed hosts in January and February, from which they disperse to colonize spring weeds in March, April and May. A second dispersal occurs to summer weeds and cotton (June planting) in July, where whiteflies increase to high population levels throughout July, August, September and November. About mid-December, following cotton harvest, populations begin to decline, and reinfestation of winter weeds occurs to complete the cycle. Typically, whitefly infestations must coincide with the natural succession of suitable host plants in an area since they are not thought to carry out true annual migrations in search of food (as occurs with insects in more temperature zones). However, dramatic population increases usually occur during the warmer, more humid (not rainy) times of the year when agricultural crops are growing rapidly. Though epidemiological evidence for the cotton mosaic disease is only circumstantial, it is most likely that the mosaic agent is introduced into cotton fields by a relatively low number of inoculative *B. tabaci* which have initially acquired the agent from locally occurring, infected, indigenous weeds, which serve as seasonal hosts of the whitefly and/or as reservoirs of the mosaic agent.

CONTROL

Varietal resistance

Several cotton varieties have been identified with tolerance or resistance to the cotton mosaic disease in Africa (Cateland, 1971; Fournier and Cateland, 1971; Cateland and Bink, 1974; Cauquil and Follin, 1983). Following the 1968–1971 epidemic in Chad, the resistant variety, HG9, replaced the susceptible BJA592, and the disease has been controlled since that time (Cauquil and Follin, 1983). According to Cateland (1973), susceptibility to cotton mosaic originated from Triumph, while resistance may be found in the N'Kourala lineage. Varieties originating from the HAR triple hybrids (*G. hirsutum* × *G. arboreum* × *G. raymondii*) are intermediate in susceptibility (Cauquil and Follin, 1983).

Knowledge that the disease agent is graft-transmissible has led to the development of a screening technique in which buds of the test lines are grafted on to infected 75–199-day-old stock plants to screen for resistance to the disease (Cauquil and Follin, 1983). From these tests, varieties may be classified into one of three groups based upon disease resistance. Susceptible varieties are BJA592, Acala del Cerro, Acala 15–17 BRI, Coker 417, Deltapine 16, Lockett B2 and MacNair 1032B; intermediate varieties are Acala 1517V, Acala 1517–70 and Coker 413A; and tolerant varieties include Allen, Reba BTK12, Coker 310, HG9, SRI-F4, Y1422 and N'Kourala (Cauquil and Follin, 1983). From earlier studies conducted by Bink (1973), very susceptible varieties were BJA592 and Y1422 × BJA592-D733. A high degree of tolerance occurred in EH9 and Y1422.

Chemical control of the vector

Insecticide programmes utilizing organic pesticides to control whiteflies have generally met with short-term success, followed by long-term failure. There is now evidence of insect resistance to many pesticides (DDT, methyl-parathion, endosulfan) used prior to the 1990s. Application of insecticides in these cases exacerbates the problem by contributing to population increases due to destruction of natural predators and parasites. There is currently a renewed interest in developing alternative pesticidal compounds to replace or use in combination with organic insecticides. These include products such as mineral and cottonseed oils, soaps, repellents such as neem extract and others (Broza *et al.*, 1988; Phadke *et al.*, 1988; Butler and Henneberry, 1989, 1990; Butler *et al.*, 1991). The future will see more of these approaches, combined with biological control strategies, since there is currently no effective chemical control for whiteflies.

Cultural control

Certain cultural practices influence the severity of the African cotton mosaic disease. Lack of proper moisture predisposes plants to poor growth, thus increasing the impact of the disease. Application of a complete fertilizer treatment compared with no fertilizer application resulted in a greater percentage infection rate of 13.4% and 3.5%, respectively (Cauquil and Follin, 1983). As with most insect-vectored plant viruses, disease pressures may be reduced by implementing routine sanitation practices. Removal of seasonal weeds, especially adjacent to cotton fields and ditch-banks in irrigated areas, and tilling under residues from infected or infested crops help to reduce potential levels of disease inoculum and insect vector populations.

Cotton Anthocyanosis Disease (Vermelhão or Reddening)

HISTORY AND DISTRIBUTION

The cotton anthocyanosis, or vermelhão (reddening), disease (Fig. 9.4) was first observed in *G. barbadense* plantings in Brazil (Costa and Sauer, 1954; Coury *et al.*, 1954). It now occurs throughout the country in both *G. barbadense* and *G. hirsutum* cotton varieties (Costa, 1957) and is considered the most serious virus-like disease of cotton in Brazil (Cauquil and Follin, 1983). The disease affects 5–100% of cotton plantings in all 25 counties surveyed in the states of São Paulo, Rio Grande Norte and Paráiba. Up to 10% of the total crop may be lost to the disease in severe years. Symptoms are consistently associated with aphid infestations.

Similar aphid-associated disease symptoms have also been reported in Thailand and the Philippines (Cauquil and Follin, 1983), India (Mali, 1978a) and the Sudan (Yassin and Defalls, 1982). These reports are based solely on symptomatologies and in some cases are associated with an aphid vector. However, the causal agents have been neither isolated nor characterized to any degree.

LOSSES

All commercial varieties of cotton grown in Brazil are susceptible to the disease. Damage is particularly severe in Campinas 817 and IAG 43/981, and somewhat mild in IA7111-028 and IA7387. The severity of the disease is directly proportional to the earliness of infection with regard to plant age. In experimentally induced infections of plants at the two- to four-leaf stage, up to 35% yield reductions occur (Costa, 1957). Plants infected at later growth

Fig. 9.4. Cotton anthocyanosis (vermelhão) disease symptoms in Brazil (A.S. Costa).

stages suffer fewer losses; typically field infections result in 10% losses (Costa, 1957).

SYMPTOMS

Cotton anthocyanosis is characterized by foliar chlorosis (see Plate 6B) followed by patchy purpling or reddening, and eventually leaves become entirely purple/red except for the veins (see Plate 6C). Symptoms are noticeable first on older leaves and progress to affect the developing leaves. Older affected leaves may be shed while newest leaves are asymptomatic; then newer leaves begin to discolour as described above. If plants are cut back (ratooned), regrowth of asymptomatic leaves occurs, and the progression of symptoms occurs again. Early in the season, affected plants may be observed in scattered areas in the field, and eventually symptoms become widespread as secondary infections occur (Costa, 1957; Cauquil and Follin, 1983).

CAUSAL AGENT

Although a plant virus is suspected, there is no direct evidence concerning the aetiology of the disease (Cauquil and Follin, 1983). Based on a tentative classification of disease agents with similar symptomatologies and biological characteristics, a luteovirus has been implicated as the causal virus of the disease (Waterhouse *et al.*, 1990; Brunt *et al.*, 1990).

HOST RANGE

The host range of the disease is limited to the Malvaceae. Naturally infected species which serve as sources of inoculum are *G. barbadense*, *G. hirsutum*, *Hibiscus cannabinus*, *H. esculentus* and *Sida rhombifolia*. In experimental transmission studies, these same species, plus *G. arboreum* L. and *S. micrantha*, were shown to be susceptible to the disease. *S. rhombifolia* L. is more susceptible, and easier to recover infectivity (by aphid transmission) from, than is *S. micrantha* (Costa, 1957).

VECTOR

Aphis gossypii Glover has been shown to vector the disease agent. A single insect is capable of transmission, but 100% transmission occurs with ten aphids. Aphid vectors which are reared on infected plants transmit the agent to healthy cotton in 1 h, but a 24–48 h acquisition feed followed by a 12 h or greater inoculation access feed (IAF) is optimum for experimental studies. Aphids are present in cotton throughout the growing season, and population levels as well as disease incidence increase as the season progresses (Costa, 1957).

TRANSMISSION

The disease agent appears to be transmitted in a persistent manner by *A. gossypii*, with an undetermined latent period. The agent may be transmitted for the life of the vector, but is not transmitted transovarially. Transmission by grafting, but not by inoculation with sap, has been demonstrated. The agent has not been shown to be transmitted through seed (Cauquil and Follin, 1983).

EPIDEMIOLOGY

Primary spread of the suspect virus occurs when migrating, inoculative *A. gossypii* colonize cotton seedlings early in the season. Because the agent is not seed-borne, migratory aphid vectors must acquire the agent from alternate host reservoirs. Suspected reservoirs are ratooned cotton, kenaf, okra and several malvaceous weeds, including *Sida micrantha* and *S. rhombifolia*. Inoculum overwinters in these and possibly other reservoirs, and is transported by the vector into cotton fields early in the growing season (November–January). Secondary spread within cotton occurs following population increases, which stem from founder aphid colonies. A second migration may occur in mid summer (February–March) which harbours additional inoculum and adds to the vector population levels in cotton fields. Thus, several repeated cycles of vector infestations occur, followed by an increase in disease incidence.

Control

Varietal resistance

Costa (1957) reported some degree of tolerance in germplasm available in Brazil, however, the resistance has not been integrated into the varieties which are most commonly cultivated (Cauquil and Follin, 1983). Additional effort is needed to develop resistant varieties for local use.

Chemical control

Routine pesticide application to reduce aphid populations is practised. Perhaps careful monitoring of insect vector flights in the early part of the season could aid in determining optimum timing of insecticide applications in order to reduce pesticide use, encourage the development of beneficial insects and reduce the chance for development of insect resistance to pesticides.

Cultural control

Eradication of sources of inoculum is the best way to control the early-season introduction of inoculative aphid vectors into the cotton crop. Thus, removal of infected cotton debris generated by the previous season's crop and rigorous control of weed hosts located near cotton fields are encouraged. The primary goal is to delay the onset of the disease, since plants infected during early development stages are most susceptible to infection.

Cotton Blue Disease and Leaf Roll Diseases

History and distribution

Cotton blue disease was first documented in 1949 in the Central African Republic (CAR) (Tarr, 1964; Delalande, 1970; Dyck, 1979; Cauquil and

Follin, 1983). The disease incidence remained relatively moderate until 1966–68 when severe outbreaks were experienced as a result of heavy infestations of the insect vector, *Aphis gossypii*. During the epidemic years of the mid- to late 1960s, disease incidence was routinely 25–30% in some fields and overall production was reduced by 3–6%. The cotton blue disease is now known to occur in Benin, Chad, Cameroon, Ivory Coast and Zaïre, but incidence and resulting damage are routinely low (Cauquil and Follin, 1983; Fauquet and Thouvenel, 1987). Although the impact of the disease in CAR has been reduced, since the 1966–68 epidemic, with the cultivation of tolerant/resistant varieties (Cauquil and Vaissayre, 1971) and rigorous insect vector control, the disease remains a consistent problem in cotton production areas in Africa (Cauquil and Follin, 1983).

Losses

The severity and thus the impact of the disease depend on plant age at the time of infection. An early infection (< 50 days post-emergence) results in no yield in susceptible cotton varieties whereas, for infections occurring about 100 days post-emergence, 15–20% losses are sustained. Late infections result in minimal to no loss probably because there is a minimum flush growth to support virus replication, and because the metabolism of more mature cotton plants renders them less susceptible to damage. Losses due to infection by the blue disease agent are manifest as a reduction in quality of harvested seed cotton with a low seed index, reduced fibre length (*c.* 10%), and a decrease in fibre resistance. In addition, fewer and smaller flowers are produced and boll shedding occurs (Cauquil and Follin, 1983).

Symptoms

The so-called blue disease is named for the dramatic dark green to bluish colour that the leaves take on when plants are affected (Cauquil, 1977). In addition, leaves roll under and become leathery in texture. Epinasty may be very severe, with reddened petioles and veins, and with early-season infection stunting of plants is pronounced. Approximately 3–4% of the affected plants die, whereas recovery from the symptoms is observed in about 6% of the plants (Cauquil and Follin, 1983).

These or similar leaf-rolling symptoms have been reported to occur in Brazil (Costa, 1976), and the *A. gossypii* populations infesting fields were presumed to serve as vectors. During the 1930s and 1940s in the USSR, leaf-rolling, accompanied by a brittle texture, brilliant red discoloration of foliage and stunting, was termed leaf roll or curliness disease (Moskovetz, 1971; Tarr, 1951, 1957, 1964). This disorder occurred in Azerbaijan, Turkmenistan and Armenia, and affected *G. barbadense, G. hirsutum, Hibiscus cannabinus* and *Solanum dulcamara* L. Although experimental transmission studies were not conducted, aphids were believed responsible for transmitting the disease agent (Cauquil and Follin, 1983; Moskovetz, 1941).

In the Philippines, symptoms similar to those described for the blue disease were also reported (Escober *et al.*, 1963). Subsequent studies demonstrated that one or possibly two agents were responsible for the resulting distinctive leaf curl or foliar mosaic symptoms, respectively. Symptoms were reproducible by aphid transmission, grafting and mechanical transmission experiments (Escober, 1963).

Another similar leaf roll disease has occurred in Thailand for over 25 years. The disease agent was shown to be aphid-transmissible with a 24 h IAF using inoculative *Aphis gossypii* (Cauquil and Follin, 1983). Blue disease leaf roll was also observed in Paraguay and Argentina, but there have been no additional studies to confirm the reports or to demonstrate transmissibility (Cauquil and Follin, 1983).

Causal agent

There has been no attempt to isolate or characterize the causal agent, although an aphid-transmitted virus is suspected. Another possibility is the involvement of an insect-related toxin.

Host range

Despite efforts to identify alternate host species by transmission studies using *A. gossypii*, cotton remains the only known host of the disease (Cauquil and Vaissayre, 1971). Plants in the Fabaceae (Leguminosae), Malvaceae and Solanaceae and several monocotyledonous species were inoculated, using *A. gossypii* which was previously exposed to blue disease-affected cotton, but symptoms were not observed in any plants tested (Cauquil and Follin, 1983). It is possible that symptomless hosts exist in those or other plant families, but additional experiments will be needed involving inoculation to and back-inoculation from a wider variety of test plants (using susceptible cotton as an indicator host).

Vector

The only known natural vector is *A. gossypii*, as was demonstrated in transmission studies (Cauquil and Vaissayre, 1971). Attempts were made to transmit the agent or induce the disease by feeding potentially inoculative *B. tabaci*, *Empoasca* spp. or *Hemitarsonemus latus* on susceptible cotton test plants, but no disease symptoms were observed (Cauquil and Vaissayre, 1971; Cauquil and Follin, 1983).

Transmission

The disease is associated only with feeding by *A. gossypii*. Symptoms develop within 9–28 days following an inoculation access feed, with an average of 18 days when plants are inoculated (exposed) at the two-true-leaf stage. The disease agent has been shown to be transmissible by grafting, but not through soil, seed or mechanical inoculation from cotton to cotton (Cauquil and Vaissayre, 1971; Watkins, 1981; Cauquil and Follin, 1983).

Epidemiology

Aphid infections occur throughout the season in cotton. Primary infections occur as a result of feeding by migrating alate *A. gossypii*, which arrive in June–July to infect the early-season vegetation and subsequently cotton seedlings as they emerge. The severity of the disease is determined by infestation and infection levels following the early-season migration of the aphid vector. A second vector migration generally occurs in September, and these aphids probably carry additional disease inoculum. Secondary spread results from transmission by founder female alates and apterous progeny, which are maintained at a low level in fields throughout the growing season (Vaissayre, 1971; Cauquil, 1977; Cauquil *et al.*, 1982).

Control

Varietal resistance

Triple hybrid cotton lines originating from the Ivory Coast and arising from introductions of germplasm material from the United States have shown good resistance. Crosses of *G. hirsutum* × *G. arboreum* × *G. raimondii* (HAR) have proved resistant to the disease, with *G. arboreum* serving as the primary source of immunity (Cauquil and Follin, 1983). Using these triple hybrids as parents, additional crosses have been made with different African and American varieties to produce HAR G198-9/BJA-610-1186, which yields well and has a high fibre quality (Mahama and Cauquil, 1976; Cauquil, 1981; Watkins, 1981). One recurrent selection from Chad, the cultivar SR1-F4, has been cultivated successfully since 1974 and is commonly used throughout Africa. Although it seems to have resistance under natural conditions, graft inoculation has shown that it is susceptible to the disease (Cauquil and Follin, 1983). Nevertheless, over half of the acreage once devoted to the cultivar BLA B2 is now (since 1983) cultivated with SR1-F4 in Central African Republic. There is some evidence that the tolerance level is diminishing; thus new varieties will probably be needed in the future to combat the disease (Cauquil and Follin, 1983).

Chemical control

Rigorous control of the aphid vector with insecticides early in the season is the best strategy for reducing disease incidence and severity (Cauquil and Vaissayre, 1971; Cauquil and Follin, 1983). In highly productive fields, a systemic insecticide applied as a seed-coating (disulphon at 3–4%), followed by one or two foliar applications of dimethoate within the first 60 days after emergence, has been used, but it is prohibitively expensive for general use. In addition, insecticide treatments, particularly applications of endosulphan-DDT, reduce valuable predator and parasite populations (which are just beginning to develop in fields), and result in resistance to pesticides in aphid populations (Cauquil and Follin, 1983). Conventional (or routine) aphid control is generally accomplished by three or four pesticide applications at two-week intervals during the first 70–75 days post-emergence (Cauquil and Follin, 1983).

Cultural control

The most important cultural control is the absolute eradication of infected cotton from fields at the end of the growing season or prior to potential early-season spring migrations of the aphid vector. In addition, avoidance of infection at early developmental stages (in cotton) may be accomplished with early planting. Good cultural practices involving tillage to control weeds, which might compete with cotton and/or serve as hosts of the aphid vector, and standard, routine fertilizer application to promote the growth of a vigorous plant reportedly helps combat the effects of the disease. High-density (70 000–80 000 plants ha^{-1}) planting of cotton is also recommended (Cauquil and Follin, 1983).

Cotton Leaf Crumple Disease

HISTORY AND DISTRIBUTION

The first report of what is now believed to have been cotton leaf crumple disease was made by Cook (1924) and termed acromania or crazytop. The disease was reportedly caused by a growth disorder which resulted from feeding damage inflicted by the cotton aphid, *A. gossypii*. A photographic record of the observed symptom suggests that the disease was in fact incited by cotton leaf crumple virus (CLCV), now known to be transmitted by the cotton whitefly, *B. tabaci*. Cook (1920, 1923) also described similar symptoms in cotton in China and Haiti, but aetiologies have not been investigated.

Cotton leaf crumple disease was widespread in Arizona and California (USA) cotton-growing areas adjacent to and along the Colorado River in the early 1950s, but it did not become economically limiting until 1954–1959, when some fields of seeded and stubbed (ratooned) cotton were nearly 100%

infected (Dickson *et al.*, 1954; Erwin, 1959; Allen *et al.*, 1960; Dickson and Laird, 1960). Leaf crumple disease affected nearly all *G. hirsutum* cultivar Acala 4-42 and *G. barbadense* fields in both states. Symptoms in *G. barbadense* were milder than those in *G. hirsutum*. The disease outbreak in the 1950s was believed to be related to the recently introduced practice of ratooning cotton to produce perennial plants, and thus an earlier-season harvest at a lower cost. Direct seed and perennial fields were grown in the same vicinities; thus ratooned plants infected the previous year probably served as the sources of inoculum. Early infection of seed cotton resulted in heavy yield losses in the first year of the crop, whereas ratooned cotton infected late the previous season did not suffer severe damage until the second growing season. The increased use of mechanization for planting and harvest and problems due to enhanced overwintering of insect pests in ratooned cotton eventually resulted in a shift to an annually seeded cotton crop in most areas by 1960–61. The disease incidence was subsequently reduced (Van Schaik *et al.*, 1962).

A second severe outbreak occurred, however, in Arizona, California and the Mexicali Valley of Mexico in 1981–82, and was believed to be associated with the reintroduced practice of ratooning cotton (Russell, 1982; Brown *et al.*, 1983). Another contributing factor was the unseasonably early (March–April) increase in populations of the cotton whitefly, *B. tabaci*, the insect vector of the virus, compared with the characteristically late-season (July and August) infestation (Butler *et al.*, 1986c).

Losses

In a study conducted in California in 1958, early-season infections resulted in a 71–85% yield reduction in Acala cultivars. Some infected plants produced no bolls (Van Schaik *et al.*, 1962). An earlier study indicated that at least eight varieties were affected, but Acala 124-68 and Acala 124-6 were most susceptible to damage (Allen *et al.*, 1960). Yield reductions of 20.6% and 16.8% in the seeded and stubbed crops, respectively, were reported (Allen *et al.*, 1960). Losses are due to reduced boll size (by weight) and number, reduction in seed index and a slight reduction in fibre length. Other fibre characteristics (lint weight, lint index) are not affected (Allen *et al.*, 1960; Van Schaik *et al.*, 1962; Brown *et al.*, 1987). The primary factor that determines the degree of damage and subsequent yield loss is the time of infection. Infection of plants in early growth stages (two- to three-leaf stage (Brown *et al.*, 1987); six- to eight-leaf stage (Butler *et al.*, 1986a)) results in the most significant reductions in yield, whereas infections later in the season (after flowering) have much less impact (Allen *et al.*, 1960; Van Schaik *et al.*, 1962; Russell, 1982; Brown *et al.*, 1987). Plant height in late June is a useful parameter with which to predict potential yield losses in CLCV-infected cotton plants (Butler *et al.*, 1986a).

In a study conducted in 1983 in the Salt River Valley of Arizona (USA), losses in first-year stub cotton affected by the disease were estimated at 23–55%, whereas estimates in seeded cotton planted adjacent to stubbed

Fig. 9.5. (a) Symptoms caused by cotton leaf crumple virus in Arizona (USA); and (b) healthy, virus-free cotton (J.K. Brown).

(a)

(b)

fields were 11–17%. Results of subsequent analysis of crop yields indicated however, that, although the seed index was reduced, the quality of the fibre produced was not affected (Russell, 1982).

Pairs of normal leaf and okra leaf cultivars (Auburn okra 310/Coker 310; Auburn okra 201/Coker 201; Auburn okra 149/TH-149; Auburn okra

Fig. 9.6. Symptoms caused by cotton leaf crumple virus on cotton flower petals (J.K. Brown).

56/Auburn 56; and Auburn okra 7A/Stoneville A) were assessed for damage by CLCV in greenhouse and field experiments after inoculation at the six- to eight-leaf stage (Butler *et al.*, 1985). All varieties were stunted (25%), produced fewer (47%) open bolls and experienced a 50% reduction in seed index compared with healthy control plants. Two okra leaf isolines exhibited more mild symptoms than their normal leaf counterparts, but there were no differences in yield parameters for those varieties (Butler *et al.*, 1985).

Symptoms

Infected cotton (*G. hirsutum*) exhibits mosaic or mottling, downward or upward curling of leaves and veinal hypertrophy which results in puckering or crumpling (Fig. 9.5a,b and Plate 6D), the most prominent symptom, from which the name of the disease was derived. Hypertrophy, small enations and ruffling are also apparent on petals of the flowers (Fig. 9.6 and Plate 6E), and bolls may be bumpy and irregular in size and shape. *G. barbadense* cultivars are susceptible but symptoms are relatively mild. Symptoms are most pronounced when plants become infected during early growth stages and stunting results when plants are inoculated at the two- to three- and eight- to ten-true-leaf stages (Erwin and Meyer, 1961; Brown *et al.*, 1987). Plants inoculated at later stages (12–14-true-leaf) develop symptoms primarily in rapidly growing leaves which emerge during the virus replication phase, but existing leaves remain asymptomatic (Erwin and Meyer, 1961; Van Schaik *et al.*, 1962; Brown *et al.*, 1987). Stubbing of infected but nearly asymptomatic plants results in the development of extremely severe symptoms on subsequently developing flush growth (Erwin and Meyer, 1961; Brown *et al.*, 1987). Thus, plants infected late in the growing season may appear virtually symptomless, although they are infected. These infected plants potentially serve as important sources of inoculum for spring-sown crops, and for infection of weeds in which the virus may overwinter. This further explains why past epidemics were associated with cultivation of ratoon cotton. Cotton plants which were defoliated for autumn harvest were asymptomatic and appeared healthy, but developed severe symptoms on spring regrowth.

Causal agent

Two different strains of the disease agent (Fig. 9.7 and Plate 6F) were initially reported by Erwin and Meyer (1961). The milder strain apparently protected against challenge inoculation by the severe strain when Acala 4-42 was inoculated as the experimental variety (Erwin and Meyer, 1961). No other reports occur, nor have similar observations been made by others working with the disease. It is possible that plants infected with the mild strain are

Fig. 9.7. Symptoms caused by mild (left) and severe (right) strains of cotton leaf crumple virus in California (USA) (D.C. Erwin).

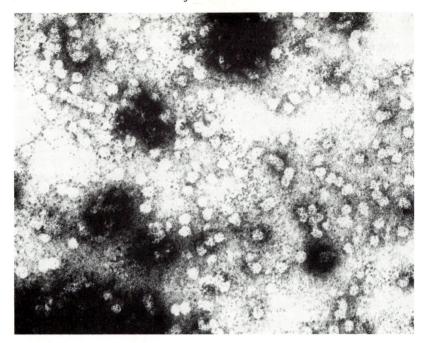

Fig. 9.8. Transmission electron micrograph showing negatively stained, labile particles of the cotton leaf crumple virus (*c.* 20 nm for monomers and 20 × 30 for dimers) as viewed in partially purified preparations from bean (J.K. Brown).

thought to be unaffected, and are thus overlooked, or that the mild strain has become uncommon and is not acknowledged by current observers.

The agent responsible for the CLC disease is a geminivirus (Brown and Nelson, 1984), provisionally assigned to a whitefly-transmitted subgroup (Brunt *et al.*, 1990). In retrospect, early evidence for geminivirus aetiology was provided following the visualization of amorphous or granular nuclear inclusion bodies in the nuclei of phloem parenchyma (vascular bundles and bundle sheath cells), but not in the cytoplasm or nucleoli of symptomatic, graft-inoculated *G. hirsutum* (Tsao, 1963). At this time, geminiviruses had not yet been discovered and were not established as a virus group until 1978. However, following partial purification of virions from CLCV-infected bean, *Phaseolus vulgaris* L., in 1983, single, paired and chains of geminivirus-like particles were observed by transmission electron microscopy (Brown and Nelson, 1984) (Fig. 9.8). The virus was subsequently assigned to the geminivirus group. Virus particles have not been observed in extracts from cotton due to difficulties associated with purification from a malvaceous host. To date, viral DNA has been isolated from both infected cotton and infected bean, and infectivity of DNA preparations was recently demonstrated using biolistic inoculation of cotton plants with DNA-coated microprojectiles (Brown and Ryan, 1991).

Two DNA-containing, ethidium bromide-staining bands were observed when nucleic acids were isolated from virus particles purified from bean and subsequently fractionated by agarose gel electrophoresis. The bands were judged to contain single-stranded DNA, based upon resistance to digestion by RNAse but susceptibility to treatment with DNAse or mungbean nuclease (J.K. Brown, unpublished data).

CLCV particles are generally unstable and easily degraded; thus attempts to obtain adequate amounts of viral DNA for further characterization have been only moderately successful. Additional evidence for the geminivirus aetiology of CLCV is the development of precipitin lines in gel double diffusion tests, using antisera raised to a well-characterized isolate of bean golden mosaic virus from Puerto Rico (BGMV-PR) (Goodman and Bird, 1978). Precipitin bands formed around wells containing extracts from BGMV-PR-infected bean and CLCV-inoculated cotton, bean and *Malva parviflora* L. (cheeseweed), but not with the respective healthy, uninoculated controls (J.K. Brown, unpublished data).

Likewise, geminivirus is detectable based upon results from DNA hybridization assays, using a panel of DNA (molecular) probes for several well-characterized whitefly-transmitted geminiviruses. Positive, but heterologous (non-identical), reactions were obtained when extracts of CLCV-infected cotton, bean and *M. parviflora* were analysed with several probes (Brown, 1989; Brown and Poulos, 1990). The strongest positive reactions were observed with DNA probes for several geminiviruses from the Americas. However, no reactions were observed in tests using DNA probes and virus extracts for whitefly-transmitted geminiviruses from Africa or the Middle East (Brown, 1989; Brown and Poulos, 1990; Brown *et al.*, 1990), or with DNA probes specific for several leafhopper-transmitted geminiviruses (J.K. Brown, unpublished data), members of the leafhopper-associated subgroup of geminiviruses.

HOST RANGE

Leaf crumple virus has a relatively narrow host range with respect to the number of plant families involved. Only species within the Malvaceae and the Fabaceae (Leguminosae) are recognized as experimental and/or natural hosts of the virus (Erwin and Meyer, 1961; Brown, 1984; Brown and Nelson, 1986). CLCV has been recovered from naturally infected *Abutilon theophrasti* Mill, *G. hirsutum*, *G. barbadense*, *Malva parviflora* (Brown and Nelson, 1986) and *Phaseolus vulgaris* (Brown *et al.*, 1986). In addition, experimental infection and recovery of the virus by back-inoculation to a susceptible cotton indicator host is reported for the following species in the Malvaceae: *Abutilon theophrasti*, *Althaea officinalis* L., *Althaea rosea* Cav., *Althaea* sp. 'Malavisco', *G. arboreum* L. (439, 440, 446, 449, CB2698, 2700), *G. barbadense* (CB3032 Domains Sakel), CB-3033 (Sakel L or X 17304, Monserrat Sea Island, Pima S-1, Pima S-2), *G. herbaceum* L. (CB 2939, CB 2940, CB 2941,

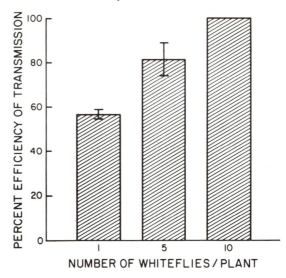

Fig. 9.9. Efficiency of transmission of cotton leaf crumple virus by *B. tabaci* following 48 h acquisition-access and 3-day inoculation-access times at 26 °C.

var. Wightianum/Texas 435), *G. hirsutum* (Acala 4-42, Delta Pine (DP) 'Glandless', DP 70, DP 90, R15 (experimental) Marie Galante/Texas 371), *G. thurberi* Tod. (which grows wild at elevations above 1200 m in some areas of southern Arizona), *Hibiscus cameronii* Knowles and Westc., *H. diversifolius* Jacq., *H. palustris* L., *H. sabdariffa* L. and *M. parviflora* (Erwin and Meyer, 1961; Brown and Nelson, 1986). Host species identified within the Fabaceae (Leguminosae) are *Castanospermum australe* Cunn and Fraser 'Delgado Bean', *Glycine max* L. Merr., *Phaseolus acutifolius* Grey var. *latifolius*, *P. angularis* Wight, *P. aureus* Roxb. and *P. vulgaris* (Brown and Nelson, 1986).

VECTOR

The only known natural vector of CLCV is the cotton whitefly, *B. tabaci* (Laird and Dickson, 1959), previously designated as *B. inconspicua* (Quaint.) or the sweet potato whitefly. The virus is not experimentally transmissible by the greenhouse whitefly, *Trialeurodes vaporariorum* (Westw.) (J.K. Brown, unpublished), *T. abutilonea* (Halde.) (Laird and Dickson, 1959) or the cotton aphid (*A. gossypii*) (J.K. Brown, unpublished). The cotton whitefly is capable of infesting a wide range of plant genera and species (Mound, 1969; Butler *et al.*, 1986b), and is characteristically present in all cotton-growing regions of the south-western USA and Mexico, where the disease has been reported. *B. tabaci* is a phloem-feeding homopteran insect (Pollard, 1955; Butler *et al.*, 1986b; Cock, 1986; Gerling, 1990), thus making it an ideal vector for a phloem-inhabiting plant virus like CLCV (Tsao, 1963).

TRANSMISSION

The CLCV is transmissible in nature by the cotton whitefly, *B. tabaci*, but not through seed or by sap inoculation (Erwin and Meyer, 1961; Brown and Nelson, 1986). Experimental transmission is by grafting and *B. tabaci*. The virus does not replicate in its whitefly vector, and is not transovarially transmitted to progeny. Recently, the virus has also been experimentally transmitted in the laboratory, using biolistic bombardment of cotton seedlings with tungsten microprojectiles coated with viral DNA (isolated from infected cotton plants) (Brown and Ryan, 1991). In experimental transmission studies, virus could be transmitted by a single whitefly (58% transmission), but efficiencies were increased to 82 and 100% with five and ten insects respectively, using a 24 h acquisition-access feed (AAF) and three-day inoculation-access feed (IAF) for experimental conditions (Fig. 9.9).

The CLCV is transmitted by *B. tabaci* in a persistent manner. The minimum AAF required with 20 *B. tabaci* plant^{-1} is 1 h (24% transmission). Using AAF times of greater than 2 h (up to 48 h) increases transmission efficiencies to 72–98%. The minimum IAF with 20 whiteflies plant^{-1} is 24 h, following a 2 h AAF (88% transmission) and 10 min (31% transmission), respectively. Transmission efficiencies increase to nearly 100% when AAF are increased to 24 and 48 h and with IAF of between 2 and 48 h (Figs 9.10 and 9.11). The transmission threshold is between 26 and 28 h, and there is a latent period of 24–28 h before virus transmission occurs, once the IAF has been initiated (Fig. 9.11) (Brown, 1984; Brown and Nelson, 1986).

B. tabaci continues to transmit the virus for 4–7 days following a 2 h AAF (Fig. 9.12) and from 5–9 days with a 48 h AAF (Fig. 9.13) under experimental conditions (Erwin and Meyer, 1961; Brown, 1984; Brown and Nelson, 1986), although it is possible that, in nature, whiteflies are capable of transmitting the virus for longer periods of time.

Temperature has an effect on the efficiency of virus transmission by *B. tabaci* (Figs 9.10, 9.12 and 9.13) (Brown, 1984). When experiments were conducted at 26, 32 and 37 °C, optimal transmission (100%) was at 32 °C using ten whiteflies plant^{-1}; transmission efficiencies were lower at 26 or 37 °C, and mortality of the insects was greatest at 37 °C (Brown, 1984).

EPIDEMIOLOGY

The disease is prevalent every year in cotton fields in the south-western USA and Mexico. The severity depends upon the age of the cotton plants at time of infection, and, in most years, winters are harsh enough to significantly reduce the overwintering whitefly populations such that they do not reach pest status until mid-season (June, July and August). When this is the case, there are minimal losses due to the disease, provided that only seeded, not stubbed cotton is grown. Due to increasing problems with CLCV and boll weevil in the early 1980s, the cultivation of stubbed cotton was prohibited in the area,

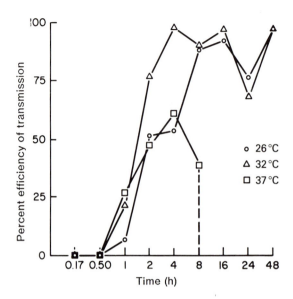

Fig. 9.10. Acquisition-access times of cotton leaf crumple virus by *B. tabaci* at three temperatures (26 °C, 32 °C and 37 °C) using a three-day inoculation-access feed.

and virus incidence has been correspondingly reduced, although late-season infections can still be documented on an annual basis. Periodically, mild winter temperatures, accompanied by sufficient rainfall, result in the early-spring proliferation of numerous weed species in the desert environment, and

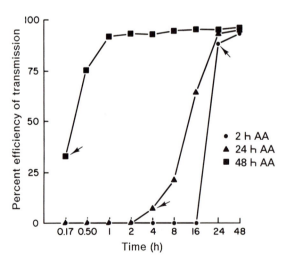

Fig. 9.11. Inoculation-access times and subsequent latent periods (arrows) following minimum (2 h), intermediate (24 h) and maximum (48 h) acquisition-access of *B. tabaci* to cotton leaf crumple-infected source plants at 26 °C.

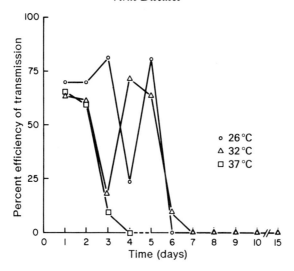

Fig. 9.12. Length of retention of cotton leaf crumple virus by *B. tabaci* through 24 h serial transfer of whiteflies following a minimum (2 h) acquisition-access to CLCV source plants at three temperatures (26 °C, 32 °C and 37 °C).

B. tabaci populations may reach unusually high levels by March or April. Host plants may become overcrowded or begin to senesce with the advent of even warmer temperatures, and whiteflies disperse to locate additional food sources. When this coincides with emergence of seedling cotton, the incidence

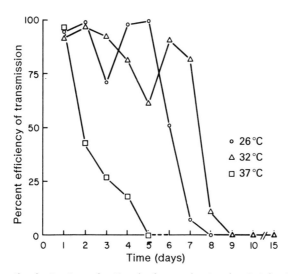

Fig. 9.13. Length of retention of cotton leaf crumple virus by *B. tabaci* through 24 h serial transfer of whiteflies following a maximum (48 h) acquisition-access to CLCV source plants at three temperatures (26 °C, 32 °C, 37 °C).

of CLCV can be as high as 100% in fields near sources of whitefly populations and sources of virus inoculum. Yield losses under these conditions may be as great as 50% for susceptible varieties such as Delta Pine 90. This situation has occurred only once since the 1981–82 outbreak (J.K. Brown, personal observation).

Regardless of the disease incidence and time of infection, *B. tabaci* populations overwinter (typically at very low levels) as immature instars on winter weed hosts, and characteristically cold winter temperatures delay development time. Whiteflies become active as adults and begin oviposition as spring daytime temperatures increase in February and March. However, as a result of characteristically cool night-time temperatures throughout the months of February, March and April, population levels remain low and are usually not even noticeable until cotton reaches the 10–12-leaf stage. Typically, populations do not peak until midsummer (July, August and September) and, by then, infection does little damage. It is believed that the combination of warmer temperatures and high humidity experienced during summer months, as a result of seasonal, summer rains in the region, contributes to the seemingly sudden annual population explosions of *B. tabaci* in cotton. In addition, there is some speculation that environmental factors plus certain physiological conditions of cotton plants during this stage of cotton development, work in concert to enhance whitefly population growth (Butler *et al.*, 1986b,c).

The CLCV is believed to overseason (winter/spring; summer/autumn) in hosts such as *M. parviflora*, which is a frutescent weed (an annual which grows as a perennial, in that in mild climates it has two growing seasons dictated by temperature and rainfall patterns). This weed and others are present along ditch banks and irrigated fields. In addition, there are many other weeds which do not serve as hosts of the CLCV, but which support large populations of reproducing *B. tabaci*. These include, among others, field bindweed (*Convolvulus*), ground-cherry (*Physalis*), jimson-weed (*Datura*), nightshade (*Solanum* spp.), sow-thistle (*Sonchus oleracea* (L.)), sunflower (*Helianthus annuus* (L.)), wild lettuce (*Lactuca serriola* (L.)) and wild morning-glory (*Ipomoea* spp.) (Brown *et al.*, 1991a,b). Cultivated crops, such as alfalfa, autumn- and spring-planted lettuce, melons and squashes, have in the past served as whitefly sources which infest winter and/or spring weeds. For the first time, in 1990–91, *B. tabaci* was documented reproducing and thus overwintering on cole crops grown in the Colorado River Valley (Brown *et al.*, 1991b). A new strain or biotype of *B. tabaci* which has a broader host range than the indigenous populations has recently been discovered in the USA and elsewhere in North America and the Caribbean Basin (Costa and Brown, 1990, 1991; Brown *et al.*, 1991a). The impact which this new strain may have on cotton production in Arizona and California is not currently known.

Overseasoning whiteflies or progeny of overwintered populations acquire the virus from infected weed sources and disperse to adjacent cotton fields, where primary infections are established. Under conditions which favour high populations, within several weeks fields may become 100% infected as whiteflies move short-range distances from plant to plant throughout the fields. However, in most years this occurs late in the growing season, and thus losses are probably less than 10% and are generally unnoticed. Recently, heavy *B. tabaci* infestations have contributed to a high incidence of 'sticky cotton' throughout the USA. The honeydew excreted by whiteflies contaminates cotton fibres, making ginning difficult, if not impossible. This most blatant and immediate problem has overshadowed concerns associated with infection by CLCV.

CONTROL

Leaf crumple has been controlled by elimination of stub cotton as a routine practice in the south-western USA since 1982. This practice has successfully reduced disease incidence by preventing early-season infection of cotton. Resistant varieties are needed, however, to control the disease and reduce losses in years when mild winters promote the unseasonably early infestation of the virus vector. Resistance or tolerance to CLCV was identified in several experimentally inoculated (grafting) species or varieties (Erwin and Meyer, 1961). Mild disease symptoms occur in inoculated *G. barbadense* Pima S-1, S-2, CB 3033 (Sakel LX 17304 and CB 3032 (Domains Sakel)). The *G. hirsutum* variety 'Marie Galante' (Texas 371) exhibited mild symptoms, as did *G. herbaceum* (L.) 'Wightianum', whereas other *G. hirsutum* lines, CB 2939 and CB 2940, were affected as severely as the susceptible Acala 4-42. Only *G. arboreum* (L.) lines 439, 440, 446, 449, CB 2698 and 2700 were either highly resistant or immune (Erwin and Meyer, 1961).

Some levels of resistance to CLCV have been identified in the *G. hirsutum* variety Cedix (FDW946 and FPW 1021) from El Salvador, which is asymptomatic and considered highly resistant, and in four Nicaraguan lines, NIC 71-23, 71-1655, 71-1728 and 71-1870, which exhibit mild symptoms (Wilson *et al.*, 1989). Cedix has been introduced into a breeding programme in Arizona (USA) as the source of CLCV resistance. The mechanism of inheritance was studied, using crosses made between Cedix and susceptible Delta Pine 90, and resulted in the first documentation of the inheritance of resistance to a virus disease (Wilson and Brown, 1991). In this study, factors controlling symptom expression (Fig. 9.14) were inherited as duplicate factors, and the symptomatic phenotype (genotype $c_1c_1c_2c_2$) was recessive to the asymptomatic (A) phenotype (genotypes C_1 and C_{2-}). These data are consistent with the duplicate factor hypothesis (Wilson and Brown, 1991), which has been documented in other cases (Endrizzi *et al.*, 1984). However, this mode of inheritance and the involvement of recessive genes are expected to complicate efforts to transfer CLCV resistance to elite US germplasm, since progeny testing required for identification of the true breeding 'A' phenotypes will be necessary (Wilson and Brown, 1991).

In addition to different levels of CLCV resistance available in these lines,

Fig. 9.14. Symptoms of CLCV infection in susceptible cv. Delta Pine 90 (left), compared with absence of symptoms in the resistant cv. Cedix (right). (J.K. Brown and D. Wilson.)

there is also evidence that whiteflies (banded wing) do not breed well on Nicaraguan lines NIC 71-2 and NIC 71-3, and on several other okra-leaf, super okra-leaf and smooth-leaved lines. Additional evidence was provided for a similar phenomenon observed for *B. tabaci* populations in the southwestern USA. Whitefly populations reproduced to lower levels on the smooth-leaved compared with hairy-leaved cottons tested (Butler *et al.*, 1986c). In another study, the open canopies of okra-leaf and super okra-leaf varieties tested discouraged the development of large populations of *B. tabaci* in the Sudan (Bindra, 1985). A combination of CLCV and insect vector resistance in the same lines would provide optimum protection for cotton varieties grown in areas where CLCV is a routine disease problem.

The fact that varieties reported resistant to leaf curl virus from the Sudan (Kirkpatrick, 1930, 1931; Tarr, 1964) are not resistant to CLCV indicates that the viruses are distinct and/or that environmental factors play an important role in resistance to viruses of cotton. The observation that some germplasm originating from the Americas has provided a source of resistance to CLCV (although *G. arboreum* lines obtained from Korea, but of unknown origin, are also tolerant), believed to be a virus which evolved in the Americas, is also of interest when studying the relationships and evolution of plant viruses and their hosts. Further evidence for the hypothesis that CLCV is distinct from leaf curl virus comes from hybridization analysis of CLCV with DNA probes which represent a panel of viruses originating from different geographical regions. Only DNA probes for other WFT geminiviruses of 'American' origin, but not probes for those viruses from the Far East or Africa, cross-hybridize or react at all to CLCV extracts under highly stringent experimental conditions (Brown and Poulos, 1990; J.K. Brown, unpublished). Similar analysis using extracts of cotton leaf curl-infected material from the Middle East is in progress (J.K. Brown and A.M. Idris, unpublished).

African Cotton Leaf Curl

HISTORY AND DISTRIBUTION

Leaf curl was first reported in Nigeria in 1912, and the native cotton species, *G. peruvianum* and *G. vitifolia*, were affected (Farquharson, 1912). A second outbreak occurred in Nigeria in 1924 (Jones and Mason, 1926), in which symptoms described were similar to those observed in 1912. Several years later (1927–28), leaf curl disease reached epidemic proportions in the Sudan (Golding, 1930; Kirkpatrick, 1930; Massay and Andrews, 1932; Prentice, 1972; Idris, 1990). The disease now occurs in Africa north of the equator, except in Egypt and the Maghreb, and in Benin, Chad, Ivory Coast, Togo and Upper Volta (Couteaux *et al.*, 1968; El-Nur and Abu Salih, 1970; Cauquil and Follin, 1983; Fauquet and Thouvenel, 1987). The reports by Golding (1930) and Kirkpatrick (1930) are often cited as the earliest recognition of whiteflies as potential vectors of disease agents, and, more specifically, as the first reports of a virus-like disease agent transmitted by *B. tabaci*.

LOSSES

The greatest damage and subsequent losses occur when cotton is infected at early growth stages; later-season infections result in only minimal damage. Up to 20% losses are reported from early-season inoculations as a result of fewer bolls set and a reduction in boll weight. *G. barbadense* var. 'Sakel' is seriously damaged in the Sudan (Andrews, 1936; Bedford, 1938; Boughey, 1947; Tarr, 1964; Ahmed, 1987).

SYMPTOMS

Two symptom types are distinguished in early reports by Kirkpatrick (1930, 1931), Tarr (1951, 1964) and Nour and Nour (1964). The most prevalent symptom (small vein-thickening) type is described in which mosaic, enations, vein-thickening and leaf curling are evident, but no stunting is observed (Kirkpatrick, 1930; Tarr, 1964; Bink, 1975; Cauquil and Follin, 1983). Losses are primarily experienced when plants become infected at an early growth stage. The second and more severe symptom type described involves severe leaf curling and foliar mosaic (Bink, 1975), stunting, blistering of leaves, shortened internodes (Babayan, 1954; Fauquet and Thouvenel, 1987) and main vein-thickening (Cauquil and Follin, 1983).

Another symptom type described in the literature, termed 'bunchy top', is characterized by shortened internodes, a reduction in size of leaves and flowers and vein-thickening, but no mosaic symptoms on *G. barbadense*, *G. peruvianum* and *G. vitifolia* (Bedford, 1938; Babayan, 1954; Tarr, 1964). It is possible that a distinct virus (or disease agent) is involved in all three of these situations, or that the variation in symptomatologies are related to genetic differences of the specified species. The latter symptom type is reportedly observed infrequently.

Discrepancies exist in the literature with respect to these leaf curl and/or mosaic symptomatologies in specific cotton species and suspect weed hosts (Tarr, 1951; Bink, 1975; Ebbels, 1976; Cauquil and Follin, 1983). It is possible there is more than one strain or causal agent involved, but no additional, more specific information is currently available. Another possibility is damage by herbicides, which induce leaf curling in affected plants and which is at times confused with biologically mediated diseases (Tarr, 1964).

CAUSAL AGENT

The causal agent is believed to be a plant virus transmitted by the cotton whitefly, *B. tabaci*. It has not been isolated or characterized to date. There may be more than one virus involved (El-Nur and Abu Salih, 1970).

HOST RANGE

Experimental transmission trials have been conducted with the leaf curl agent. Although leaf curl from cotton affects a large number of species within the Malvaceae, it nevertheless has a relatively narrow host range. Host species which have been identified include *Abutilon theophrasti* (Nill.), *Althaea rosea* (Cav.), *A. ficifolia*, *A. kurdica*, *A. nudiflora*, *A. pontica*, *A. sulphurea*, *G. barbadense*, *G. hirsutum*, *Hibiscus cannabinus* (L.), *H. esculentus* (L.), *H. ficulneus*, *H. huegelii*, *H. trionum*, *H. sabdariffa* (L.), *Lavatera cretica*, *Malva alcea* (L.), *M. silvestris* (L.), *M. moschata* (L.), *Malvaviscus arboreus* Car., *Pavonia hastata* (L.), *Sida acuta* (Burm.), *S. alba* (L.), *S. cordifolia* (L.) (Tarr, 1951, 1957; Bink, 1975; Cauquil and Follin, 1983; Fauquet and Thouvenel, 1987) and possibly *Nicotiana tabacum* L. (Tarr, 1951; Fauquet and Thouvenel, 1987). *G. arboreum* and *G. herbaceum* are resistant to leaf curl (Cauquil and Follin, 1983).

Similar leaf curl symptoms occur on a variety of species in other plant families in Africa, but whether the African leaf curl virus is responsible remains unknown. These include *Cochorus fascicularis* Lau. (Tiliaceae), *Phyllanthus niruri* L. (Euphorbiaceae), *Clitoria ternatea* (L.) and *Phaseolus vulgaris* (Fabaceae), *Sida urens* (Malvaceae), *Petunia* sp. (Solanaceae) and

Urena lobata (Tarr, 1951, 1957; Nour and Nour, 1964; El-Nur and Abu Salih, 1970).

Vector

The only known vector of the leaf curl agent is the cotton whitefly, *B. tabaci*, which is prevalent in all cotton-growing regions where the disease occurs. The agent is transmitted in a persistent manner by its whitefly vector (Kirkpatrick, 1931; Fauquet and Thouvenel, 1987).

Transmission

The African leaf curl causal agent is transmissible naturally only by its whitefly vector *B. tabaci* and is not passaged by transovarial means (Kirkpatrick, 1930, 1931; Nour and Nour, 1964; El-Nur and Abu Salih, 1970). It is not transmitted by aphids, flea beetles, thrips or jassids (Kirkpatrick, 1931). The agent is transmitted by grafting, but not by seed or by mechanical inoculation with sap (Tarr, 1951; Nour and Nour, 1964; Watkins, 1981; Cauquil and Follin, 1983; Fauquet and Thouvenel, 1987). Single *B. tabaci* are able to transmit the leaf curl agent (Yassin and El-Nur, 1970), but greater transmission efficiency is observed when additional *B. tabaci* are present (Cauquil and Follin, 1983). In experimental studies, a 3.5 h AAF and 30 min IAF were required for transmission and the transmission threshold is reported as 6.5 h (Kirkpatrick, 1930, 1931). Inoculative whiteflies can remain so for their entire life following a successful AAF; therefore a persistent, circulative relationship is postulated. Symptoms develop on inoculated plants within 15–30 days (Lagière and Ouattara, 1969).

Epidemiology

Climate conditions (rainfall, wind, temperature) affect the epidemiology of leaf curl in Africa. Periods of rainfall prior to seeding result in the development of a high population of the whitefly vector due to the abundance of food sources (Bink, 1975). Because cotton is grown for only part of the year (and without ratooning), alternate weed and cultivated hosts probably serve as virus reservoirs. Whitefly populations typically infect cotton fields and primary sites of infection are established. Secondary spread to other cotton plants in the field probably occurs from those sites, and from additional vectors which enter the fields throughout the growing season (Giha and Nour, 1969). The highest population levels of *B. tabaci*, and thus disease incidence occur early in the season (October-November) following August-September planting dates when cotton is most susceptible to damage (Giha and Nour, 1969; Idris, 1990). When four varieties with differential susceptibilities to the disease were field tested, disease progression rates were highest in the cotton variety GS(83)3, which was also the most susceptible. Rates were particularly high when GS(83)3 was interplanted with okra (Idris, 1990). Cultivation of okra by interplanting with cotton or okra grown adjacent to cotton increases disease incidence, suggesting that okra is an alternate host of the virus and/or a host plant favoured by the whitefly vector (Gifford, 1978). However, okra was reported as a non-host of the virus (Bink, 1975). Due to the lack of a reliable assay for the disease agent(s) involved, the particular plant species which play a direct role in the disease cycle are not known.

CONTROL

Resistant varieties

The susceptible cultivars Domains Sakel, Bar 14/25 and several others are no longer grown in the areas where the disease occurs. *G. barbadense* Sakel varieties selected in the 1930s by Lambert have been developed that are highly resistant to the disease and yet retain the desired agronomic traits (Tarr, 1964; Bink, 1975; Watkins, 1981). These include XLI, X1530, X1730A, X1030 and B6L. Currently, the most widely grown variety is B6L, known commercially as Barakat (Siddig, 1968), and this cultivar has been grown successfully for some years in the Sudan (Giha and Nour, 1969; Idris, 1990). Recently, however, virus-like symptoms were observed in promising extra-long-staple lines of the EB series (Idris, 1990). The mechanism of resistance appears to involve reduced multiplication of the disease agent, since plants with mild symptoms are observed. Further, it is difficult to experimentally recover (by transmission tests) symptoms in susceptible indicator hosts, thus supporting the hypothesis.

Chemical control

Pesticides which were used in the past to control thrips and jassids seemed to exacerbate the whitefly problem (Prentice, 1972). Some effort has been directed toward insecticide programmes, but reliance on resistant varieties and cultural control practices have been more fruitful.

Cultural control

Elimination of volunteer perennial cotton and suspected or known alternate hosts (weed reservoirs) is considered partially effective in reducing disease incidence (Tarr, 1964; Bink, 1975; Watkins, 1981; Idris, 1990). The cultivation of ratooned cotton has not been practised for many years and, as a result, in combination with the cultivation of resistant varieties, disease incidence has declined.

Cotton Yellow Vein (Texas Cotton Vein-clearing)

HISTORY AND DISTRIBUTION

The cotton yellow vein (CYV) disease was first described as a vein-clearing disorder of cotton in the lower Rio Grande Valley of Texas, USA, in 1957 (Rosberg, 1957). The disease, which causes vein-clearing and veinal chlorosis late in the growing season, was also documented as far south as Tampico in the state of Tamaulipas, Mexico (Halliwell *et al.*, 1980). It was first thought to be caused by a strain of the whitefly-transmitted cotton leaf crumple virus (CLCV) (Dickson *et al.*, 1954), but subsequent association with a leafhopper vector instead of *B. tabaci* suggests that a pathogen distinct from CLCV is involved (Halliwell and Rosberg, 1964).

LOSSES

The degree of severity of the disease is unclear due to the lack of yield loss data, but infected plants are stunted, indicating the potential for decreases in yield. Two cotton cultivars, *G. herbaceum* L. cv. Wightianum Tex. 435 and *G. hirstutum* L. cv. Tex. 371 were particularly susceptible to the disease (Halliwell *et al.*, 1980).

SYMPTOMS

Cotton yellow vein disease symptoms are bright yellow vein-clearing, vein-banding and eventually mosaic (Fig. 9.15). Crumpling and mottling of the leaves and mild to severe stunting are also observed. Compared with CLCV described in the south-western USA (Dickson and Laird, 1960; Erwin and Meyer, 1961; Brown *et al.*, 1983, 1986), CYV causes a brighter yellow mosaic and more severe stunting than does CLCV; however, leaf distortion and crumpling were more pronounced with CLCV when the same cotton var-ieties were inoculated with the two viruses (Halliwell and Rosberg, 1964).

CAUSAL AGENT

The disease agent is believed to be a plant virus, but Koch's postulates have not been completed for this disease. Electron micrographs taken of thin sections of CYV-infected cotton leaves indicate the presence of isometric virus-like particles (VLPs) (approximately 23 nm in diameter) in phloem cells (Halliwell *et al.*, 1980) (Fig. 9.16), which is suggestive of viral aetiology. Treatments with tetracycline (tetracycline HC1 and chlorotetracycline) by soil drench or infusion into stems at 500, 1000 and 1500 ppm, did not result in prevention of symptoms (despite phytotoxicity observed in some cases). Furthermore, MLOs were not found when infected leaves were examined using the electron microscope (Halliwell *et al.*, 1980).

Fig. 9.15. Symptoms caused by cotton yellow-vein virus in cotton, Texas (USA) (R. Halliwell and S. Lyda).

HOST RANGE

Information on the host range of CYV disease is limited. The disease affects *G. hirsutum* L. cultivars G + P 37-74, Acala 1517D, Western Stormproof, Lankart 57, Empire WR61, Stoneville 7A, Deltapine 15 and Texas 371, and *G. herbaceum* cv. Wightianum Texas 435. The latter two varieties were reported as resistant to CLCV. Unidentified malvaceous species were also experimentally infected by grafting (Halliwell *et al.*, 1980).

VECTOR

The only known natural vector of the disease is the leafhopper, *Scaphytopius albifrons* Hepner.

TRANSMISSION

Experimental transmission was achieved by grafting from cotton to cotton, and from cotton to malvaceous spp. Successful transmission to cotton (four-leaf stage) by the leafhopper vector was demonstrated using a 24 h acquisition access feed and a 72 h inoculation access feed. Exacting information is not

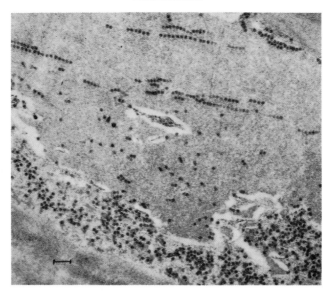

Fig. 9.16. Transmission electron micrograph of virus-like particles (~23 mm diam.) viewed in thin sections of cotton infected by cotton yellow-vein virus (bar = 100 nm) (R. Halliwell).

available on virus–vector interactions; thus, length of persistence in the leafhopper vector is not known. Efficiency of transmission is quite high, based upon experimental transmission rates in which 7/9 and 9/9 plants became infected following exposure of groups of three cotton test plants to ten potentially inoculative leafhoppers (Halliwell *et al.*, 1980). Seed transmission and transmission by mechanical or sap inoculation are reported as negative.

Epidemiology

Little information is available on CYV except that the disease occurs late in the growing season, and symptoms are associated with the presence of the leafhopper, *S. albifrons*. Severity of symptoms varies with the variety involved (Halliwell and Rosberg, 1964). Several other virus-like symptoms have been documented in cotton in the same region, but appear to be caused by different pathogens (Sleeth *et al.*, 1963; Schuster, 1964; Schuster and Lambe, 1966).

Control

No specific methods are cited; however, based upon the inference that the disease agent is transmitted by *S. albifrons* in a persistent manner, control of the insect vector and of sources of perennial disease inoculum would probably aid in reducing disease incidence.

Diseases with Symptoms of Mycoplasma-like Organisms (MLOs)

HISTORY AND DISTRIBUTION

Several diseases of cotton caused by suspect or demonstrated mycoplasma-like organisms (MLOs) have been described, but, for the most part, very little information is available. The diseases can be distinguished and have been designated based upon the characteristic symptoms which are produced in the cotton host plant. These include the flavescence, flower virescence, phyllody, psyllosis and small-leaf (stenosis) diseases.

The flavescence disease was described in Central African Republic and was later observed in Benin, the Ivory Coast, Mozambique (Cauquil and Follin, 1983) and Upper Volta (Delattre *et al.*, 1974; Delattre and Joly, 1981) and it is believed to be caused by an MLO.

Cotton flower virescence and phyllody diseases have been reported in Upper Volta and Mali since 1946 (Delattre, 1965, 1968; Follin, 1982; Cauquil and Follin, 1983) and for the first time in Ivory Coast in 1970. The flower virescence disease has been studied in more detail (Delattre, 1965; Lagière and Ouattara 1969; Cousin *et al.*, 1970; Cauquil and Follin, 1983).

Psyllosis was observed in the Belgian Congo in the 1940s (Dineur, 1957), but has not been reported elsewhere (Tarr, 1964), and small-leaf (stenosis) is the earliest virus-like disease reported on cotton in India and Pakistan (Kottur and Patel, 1920; Uppal *et al.*, 1944; Capcor *et al.*, 1972). Comparatively little is known about these diseases and other similarly symptomatic diseases of cotton described in Haiti and China (Cook, 1920, 1923, 1924) and in India (Gokhale, 1936). A report of shortened internodes of cotton occurring in Armenia in 1952 (Babayan, 1954) described a similar disorder, but no additional information is available (Tarr, 1964).

LOSSES

In general, the diseases incited by MLO-like agents result in a severe interruption of the plant metabolism. The outcome in most cases is that affected plants produce little to no fruit and no mature cotton fibres or seed. Infected plants do not recover, and yields are low to non-existent in most cases studied thus far.

SYMPTOMS

Cotton affected by flavescence disease exhibits chlorosis and reddening of leaves. Limbs are narrow and petioles are curved and longer than normal.

Plants are often bushy in appearance and are sterile (Delattre, 1968; Delattre *et al.*, 1974; Cauquil and Follin, 1983).

Symptoms of the flower virescence disease were described by Delattre (1965, 1968) and Lagière and Ouattara (1969). The disease is characterized by production of phyllody or transformation/replacement of flower petals and squares into leaf-like structures containing chlorophyll. Symptoms first occur when cotton plants are young (40–50 days old). Leaves become chlorotic and stunted, and subsequently phyllody symptoms become apparent as cotton plants enter the flowering stage. Stunting of the entire plant is pronounced, internodes are shortened and proliferation of tissues from the axillary buds is stimulated while primary meristem growth ceases. Floral parts become overgrown, flowers bloom prematurely while petals do not develop normally but become thickened. In some cases, foliation of the ovules occurs, stamens become green and the flower eventually becomes a vegetative structure (Cauquil and Follin, 1983). In minor contrast, in cotton plants affected by the phyllody disease, leaves become strap-like first and stunted, and eventually phyllody or flower virescence occurs as described for virescence disease (Watkins, 1981). These diseases are probably caused by similar or related disease agents.

Psyllosis-affected cotton exhibits scale-like leaves in rosettes and purple discoloration of older foliage and stems, and plants are sterile. Symptoms are more pronounced when plants are affected during early stages of rapid growth (Dineur, 1957). Exposure to sunlight apparently accelerates symptom development (Soyer, 1947). Irrespective of the purple discoloration observed, the symptoms are reminiscent of those termed 'bunchy top' and attributed to cotton leaf curl virus (Cauquil and Follin, 1983).

Plants affected by small-leaf disease are stunted and leaves have fewer lobes and are dramatically reduced in size (Cauquil and Follin, 1983). In addition, leaves develop in clusters, flower size is reduced and abortion of ovaries occurs. Flower-buds and young bolls are shed prematurely and little if any viable seed is produced. Apparently the root systems of affected plants are poorly developed and many secondary roots develop, which affect the plants' ability to remain anchored in the ground (Tarr, 1964). This and the observation that symptoms may occur on only portions of the above- or belowground plant parts suggest an MLO may be involved.

CAUSAL AGENTS

Flavescence disease is caused by a graft-transmissible MLO-like agent and may be linked to infested soils. A putative MLO has been cultured and is believed to be the causal agent (Delattre *et al.*, 1974).

Virescence disease is also believed to be caused by an MLO. Characteristic structures have been observed by ultrastructural examination of affected cotton tissues (Gourret and Maillet, 1969; Cousin *et al.*, 1970). The appearance of MLOs in the phloem of diseased plants has been correlated with symptom development (Cauquil and Follin, 1983), and one report describes the *in vitro* culturing of an MLO from virescence-diseased cotton, which is believed to be the causal agent (Giannotti and Delattre, 1972). There is no information on the causal agent of the similarly symptomatic disease termed phyllody; but an MLO is suggested as the causal agent based upon characteristic disease symptoms (Watkins, 1981).

Psyllosis disease is believed to be caused by an MLO-like agent transmitted by a psyllid vector (Soyer, 1947; Dineur, 1957; Tarr, 1964; Cadou, 1970; Desmidts and Rassel, 1974). There is no additional characterization of the causal agent reported.

The small-leaf disease is believed to be caused by a transmissible agent because symptoms may be reproduced by grafting (Tarr, 1964). An MLO pathogen is suspected, based upon a report in which MLO-like structures were observed in affected plants (Capcor *et al.*, 1972). Tetracycline treatments are also cited as a means of suppressing symptoms in affected cotton plants (Cauquil and Follin, 1983).

HOST RANGE

The only host reported for the flavescence agent is cotton (Delattre, 1980; Cauquil and Follin, 1983).

The flower virescence/phyllody MLO-like pathogen affects cotton and possibly kenaf and several weed species present in the area which exhibit similar symptoms. Possible alternate hosts listed are *Acacia albida* Del., *Combrettum glutinosum* Perr., *Crotalaria* sp., *Eragrostis* sp., *Miltracarpus scaber* Zucc., *Sida* sp., *Urena lobata* and *Vicia leptoclada* L. Experimental transmission studies have demonstrated leafhopper transmission from cotton to *Sida cordifolia* and *S. rhombifolia*. (Laboucheix *et al.*, 1973). The host range of the pathogen is probably restricted by the host range of its leafhopper vector, as with other better-characterized MLOs.

Although the only known host of the psyllosis disease is cotton (Tarr, 1964), more information is available concerning hosts of the small-leaf disease. The Indian varieties of *G. herbaceum* are quite susceptible to small-leaf but varietal differences are reported (Cauquil and Vaissayre, 1971). Other cotton varieties of Asiatic origin, including Rozi (*G. arboreum* var. typicum f. indicum) in western India and Mungari (mixture of *G. herbaceum* and *G. arboreum*) in Madras, are particularly affected. Although not a natural host of the disease agent, American cottons of *G. hirsutum* lineage can be experimentally infected, but Egyptian cotton, *G. barbadense*, is reported to be immune or resistant. Various hybrids containing American and Indian germplasm are also reportedly immune to the disease (Tarr, 1964). This information suggests that the small-leaf symptoms reported by Cook (1920, 1923) in *G. barbadense* in China and Haiti may be incited by a different pathogen(s).

VECTOR

The suspected vector of the cotton flavescence agent is the scale insect, *Margarodes* sp. The insect survives in the soil in brown, spherical cysts (1–2 mm in diameter), and cysts have been observed in soils where flavescence-affected cotton plants are grown. The vector of flower virescence/phyllody disease agent(s) is the cicadellid leafhopper, *Orosius cellulosus* (Lindberg), and transmission from cotton to cotton has been demonstrated (Laboucheix *et al.*, 1973).

In contrast, the psyllosis disease agent is believed to be transmitted by larvae of the psyllid, *Paurocephala gossypii* (Tarr, 1964), while vector studies with the small-leaf disease indicate that no insect or other natural vector(s) have been identified (Uppal *et al.*, 1944; Cauquil and Follin, 1983). As would be expected, all suspected or proved vectors of these MLO and MLO-like agents are phloem-feeding homopteran insects.

TRANSMISSION

The cotton flavescence agent is experimentally transmissible by grafting; evidence for natural transmission by the scale insect (*Margarodes* sp.) is at this point circumstantial. The flower virescence/phyllody disease agent(s) is also experimentally transmissible by grafting and by the leafhopper, *O. cellulosus*. Successful transmission (nearly 50%) is achieved using five leafhoppers per plant and long acquisition-access and inoculation-access feeding times (15 days) (Laboucheix *et al.*, 1973). Tests with numerous other potential vectors have not resulted in successful transmission (Cauquil and Follin, 1983).

The only information concerning transmission characteristics of the psyllosis disease agent indicates possible transmission by the larvae of *P. gossypii*, with symptom production following a two to eight-week incubation period. The greater the psyllid population, the more severe the disease symptoms observed (Tarr, 1964).

Little is known about transmission of the small-leaf disease agent except that the disease is transmissible by grafting, but not through soil or seed or by mechanical inoculation. The disease has been documented only sporadically, with up to 12 years lapsing between outbreaks (Uppal *et al.*, 1944; Tarr, 1964; Cauquil and Follin, 1983).

EPIDEMIOLOGY AND CONTROL

Various studies are reported concerning epidemiology and control of MLO-like diseases of cotton. In general, due to lack of knowledge about the pathogens, vector identities and/or relationships and the difficulties associated with manipulation, characterization and detection of MLO-like diseases, there is not much definitive information. Depending on the vector involved, symptoms tend to occur early in the growing season and are observed in scattered patches in the field. Because of the characteristically lengthy latent periods in the vector and/or long incubation periods in diseased plants prior to symptom expression, secondary infection may occur without obvious symptoms at the outset of the disease. Generally, standard practices should include adjustment of planting dates to avoid the suspect vector, control to minimize vector populations (especially early in the season) and eradication of infected plants at the end of the season. Control of weed species known to harbour the vector and/or causal agent in the immediate vicinity of cotton plantings may be of some value, although with migratory vectors, such as leafhoppers, this may not result in a noticeable decrease in primary infections.

Cultivation of resistant or tolerant varieties, if available, is also standard practice. For flower virescence disease, *G. hirsutum* is very susceptible, particularly smooth-leaf varieties. *G. herbaceum* (Budi) and *G. punctatum* are likewise susceptible, but *G. barbadense* cultivar 'Mono' was immune when tested in Togo. The lack of reliable inoculation techniques to assess resistance in cotton germplasm has been a major impediment in development of varieties which can tolerate the damage caused by these MLO-like diseases (Cauquil and Follin, 1983).

References

AHMAD, M. (1978) A whitefly-vectored yellow mosaic of jute. *FAO Plant Protection Bulletin* 26, 169–71.

AHMED, K.E. (1987) Field evaluation of selected strains of cotton, *Gossypium barbadense* L. for resistance to leaf curl virus disease and its effect on cotton yield and quality. MSc Thesis, University of Gezira.

ALLEN, R.M., TUCKER, H. AND NELSON, R.A. (1960) Leaf crumple disease of cotton in Arizona. *Plant Disease Reporter* 44, 246–50.

ANDREWS, F.W. (1936) The effect of leaf curl disease on the yield of the cotton plant. *Empire Cotton Growers Review* 13, 287–93.

BABAYAN, A. (1954) Short nodes – a new kind of cotton disease (translated). *Cotton Raising* 4, 57–60 (in Tarr, 1964).

BEDFORD, H.W. (1938) Entomological Section, Agricultural Research Service, Report 1936–37. *Report, Agricultural Research Service, Anglo Egyptian Sudan, 1937,* 50–65.

BINDRA, O.S. (1985) Relation of cotton cultivars to the cotton pest problems in the Sudan Gezira. *Euphytica* 34, 849–56.

BINK, F.A. (1973) A new contribution to the study of cotton mosaic in Chad. I. Symptoms, transmissions by *Bemisia tabaci* Genn. II. Observations on *B. tabaci*. III. Other various diseases on cotton and related plants. *Coton et Fibres Tropicales* 28, 365–78.

BINK, F.A. (1975) Leaf curl and mosaic diseases of cotton in Central Africa. *Empire Cotton Growing Review* 52, 133–41.

BOUGHEY, A.S. (1947) The causes of variation in the incidence of cotton leaf curl in the Sudan Gezira. *Imperial Mycological Institute Mycological Paper* 22, 1–9.

BRADER, L. AND ATGAR, P. (1971) Quelques réflexions sur une nouvelle maladie épidémique des cotonniers au Tchad. *Coton et Fibres Tropicales* 26, 225–8.

BROWN, J.K. (1984) Whitefly-transmitted viruses of the southwest. PhD Dissertation, University of Arizona, Tucson, Arizona.

BROWN, J.K. (1989) The development of nonradioactive sulfonated probes for the detection and identification of whitefly-transmitted geminiviruses. In: *Proceedings of the Fourth International Plant Virus Epidemiology Workshop*, Montpellier, France, 3–8 September 1989.

BROWN, J.K. AND NELSON, M.R. (1984) Geminate particles associated with cotton leaf crumple disease in Arizona. *Phytopathology* 74, 987–90.

BROWN, J.K. AND NELSON, M.R. (1986) Host range and vector relationships of cotton leaf crumple virus. *Plant Disease* 71, 522–4.

BROWN, J.K. AND POULOS, B.T. (1990) Semi-quantitative DNA hybridization analysis of whitefly-transmitted geminiviruses. *Phytopathology* 80, 887 (Abstr.).

BROWN, J.K. AND RYAN, R. (1991) High velocity microprojectile mediated transmission of whitefly-transmitted geminivirus DNA or purified virions to intact plants. *Phytopathology* 81, 1217 (Abstr.).

BROWN, J.K., BUTLER, G.D. AND NELSON, M.R. (1983) Occurrence of leaf crumple associated with severe whitefly infestation in Arizona. *Phytopathology* 73, 787 (Abstr.).

BROWN, J.K., NELSON, M.R. AND LAMBE, R.C, (1986) Cotton leaf crumple virus transmitted from naturally infected bean from Mexico. *Plant Disease* 70, 981.

BROWN, J.K., MIHAIL, J.D. AND NELSON, M.R. (1987) Effects of cotton leaf crumple virus on cotton inoculated at different growth stages. *Plant Disease* 71, 699–703.

BROWN, J.K., POULOS, B.T. AND BIRD, J. (1990) Differential detection of whitefly-transmitted geminiviruses in weed species from Puerto Rico by hybridization analysis with non-radioactive probes. Paper presented at American Phytopathological Society Carribean Divisional Meeting, Mayaguez. May 1990.

BROWN, J.K., COSTA, H.S. AND BIRD, J. (1991a) Differential esterase banding pattern and silverleaf production in *Cucurbita* sp. by three biotypes of *Bemisia tabaci* Genn. *Phytopathology* 81, 1157 (Abstr.).

BROWN, J.K., COSTA, H.S. AND LAEMMLEN, F. (1991b) First incidence of whitefly-associated squash silverleaf (SSL) of *Cucurbita* in Arizona, and of white-streaking (WST) disorder of cole crops in Arizona and California. *Plant Disease* 76, 426.

BROWN, J.K., LASTRA, R. AND BIRD, J. (1991c) First documentation of whitefly-transmitted geminiviruses causing widespread disease in cotton, tobacco, and tomato in Dominican Republic and in tomato in Puerto Rico. *Fitopatogia* 26, 47.

BROZA, M., BUTLER, G.D., Jr AND HENNEBERRY, T.J. (1988) Cottonseed oil for control of *Bemisia tabaci* on cotton. In: *Proceedings of the Beltwide Cotton Production Research Conference*, New Orleans, Louisiana. National Cotton Council, Nashville, Tennessee, p. 31.

BRUNT, A., CRABTREE, K. AND GIBBS, A. (1990) *Viruses of Tropical Plants: Descriptions and Lists from the VIDE Database*. CAB International, Wallingford, UK. 707 pp.

BUTLER, G.D., Jr AND HENNEBERRY, T.J. (1989) Sweetpotato whitefly migration, population increase, and control on lettuce with cottonseed oil sprays. *Southwestern Entomology* 14, 287–93.

BUTLER, G.D., Jr AND HENNEBERRY, T.J. (1990) Cottonseed oil and Safer Insecticidal Soap: effect on cotton and vegetable pests and phytotoxicity. *Southwestern Entomology* 15, 257–64.

BUTLER, G.D., Jr., BROWN, J.K. AND HENNEBERRY, T.J. (1986a) Effect of cotton seedling infection by cotton leaf crumple virus on subsequent growth and yield. *Journal of Economic Entomology* 79, 208–11.

BUTLER, G.D., Jr., HENNEBERRY, T.J. AND HUTCHISON, W.D. (1986b) Biology, sampling and population dynamics of *Bemisia tabaci*. *Agricultural Zoological Review* 1, 167–95.

BUTLER, G.D., Jr., HENNEBERRY, T.J. AND WILSON, F.D. (1986c) *Bemisia tabaci* (Homoptera: Aleyrodidae) on cotton: adult activity and cultivar oviposition preference. *Journal of Economic Entomology* 79, 350–4.

BUTLER, G.D., Jr., HENNEBERRY, T.J., STANSLY, P.A. AND SCHUSTER, D.J. (1991) Effect of selected soaps, oils, and detergents on the sweetpotato whitefly: Homoptera: Aleyrodidae. *Journal of Economic Entomology* (in press)

BUTLER, G.W., WILSON, F.D. AND HENNEBERRY, T.J. (1985) Cotton leaf crumple virus disease in okra-leaf and normal-leaf cottons. *Journal of Economic Entomology* 78, 1500–2.

CADOU, J. (1970) Note sur la présence du psylle du cotonnier *Paurocephala gossypii* R. en RCA. *Coton et Fibres Tropicales* 25, 405–7.

CAPCOR, S.P., DANDE, P.K. AND SHINHA, R.C. (1972) Mycoplasma like bodies found in cells of 'small leaf' affected cotton plants. *Hindustan Antibiotics Bulletin* 15, 40–1.

CATELAND, B. (1971) Transmission par greffage d'une nouvelle maladie épidemique des cotonniers au Tchad. *Coton et Fibres Tropicales* 26, 263–5.

CATELAND, B. (1973) Etude de la résistance variétale des cotonniers à la mosaïque du Tchad. I. Choix d'un crible de sélection. *Coton et Fibres Tropicales* 28, 301–5.

CATELAND, B. AND BINK, F.A. (1974) Etude de la résistance variétale des cotonniers à la mosaïque du Tchad. *Coton et Fibres Tropicales* 29, 207–13.

CAUQUIL, J. (1977) Etudes sur une maladie d'origine virale du cotonnier: la maladie bleue. *Coton et Fibres Tropicales* 32, 259–78.

CAUQUIL, J. (1981) Recent developments for the control of blue disease of cotton in Central Africa. *Coton et Fibres Tropicales* 36, 297–304.

CAUQUIL, J. AND FOLLIN, J.-C. (1983) Presumed virus and mycoplasma-like organism diseases in subsaharan Africa and the rest of the world. *Coton et Fibres Tropicales* 38, 293–317.

CAUQUIL, J. AND VAISSAYRE, M. (1971) La maladie bleue du cotonnier en Afrique: transmission de cotonnier à cotonnier par *Aphis gossypii* Glover. *Coton et Fibres Tropicales* 26, 463–6.

CAUQUIL, J., VINCENS, P., DENÉCHÈRE, M. AND MIANZE, T. (1982) Nouvelle contribution sur la lutte chimique contre *Aphis gossypii* Glover, ravageur du cotonnier en République Centrafricaine. *Coton et Fibres Tropicales* 37, 333–60.

COCK, M.J.W. (1986) Bemisia tabaci – *a Literature Survey on the Cotton Whitefly with an Annotated Bibliography*. CAB International Institute of Biological Control, Ascot, UK, 121 pp.

COOK, O.F. (1920) A disorder of cotton plants in China: club-leaf or crytosis. *Journal of Heredity* 11, 99–110.

COOK, O.F. (1923) Malformation of cotton plants in Haiti: a new disease named smalling or stenosis, causing abnormal growth and sterility. *Journal of Heredity* 14, 323–5.

COOK, O.F. (1924) Acromania or 'crazytop', a growth disorder of cotton. *Journal of Agricultural Research* 28, 803–28.

Costa, A.S. (1955) Studies on *Abutilon* mosaic in Brazil. *Phytopathologische Zeitschrift* 24, 97–112.

Costa, A.S. (1957) Anthocyanosis, a virus disease in cotton in Brazil. *Phytopathologische Zeitschrift*. 28, 167–86.

Costa, A.S. (1961) Studies on Brazilian tobacco streak. *Phytopathologische Zeitschrift*. 42, 113–38.

Costa, A.S. (1966) Moléstias de virus do algodoeiro. *Divulgação Agronomica* 21, 25–34.

Costa, A.S. (1976) Whitefly-transmitted plant diseases. *Annual Review of Phytopathology* 14, 429–49.

Costa, A.S. and Carvalho, A.M. (1968) Mechanical transmission and properties of the *Abutilon* mosaic virus. *Phytopathologische Zeitschrift*. 37, 259–72.

Costa, A.S. and Carvalho, A.M. (1969) Comparative studies between *Abutilon* and *Euphorbia* mosaic viruses. *Phytopathologische Zeitschrift*. 38, 129–52.

Costa, A.S. and Forster, R. (1938) Nota preliminar sobre une nova molestia de virus do algodoeiro: mosaico das nervuras. *Compinas Institute Agronomico Boletin Technico* 51.

Costa, A.S. and Sauer, H.F.G. (1954) Vermelhão do algodoeiro. *Bragantia* 13, 237–46.

Costa, A.S., Pinto, A.J.D. and Neves, O. (1954) Un mosaico do algodoeiro causado pelo virus da necrose branca do fumo. *Bragantia* 13, 111.

Costa, H.S. and Brown, J.K. (1990) Variability in biological characteristics, isozyme patterns, and virus transmission among populations of *Bemisia tabaci* Genn. in Arizona. *Phytopathology* 80, 888 (Abstr.).

Costa, H.S. and Brown, J.K. (1991) Variation in biological characteristics and in esterase patterns among populations of *Bemisia tabaci* Genn., and the association of one population with silverleaf symptom development. *Entomologia: Experimentia et Applicata* 61, 211–19.

Coury, T., Malavolta, E., Ranzani, G. and Brazil Sobrinho, M.O.C. (1954) Contribiuçao as estudos do 'vermelhao do algodoeiro.' *Annais Escota Superior de Agricultura* Luiz de Oueros Piracicaba 11, 41–68.

Cousin, M.R., Maillet, P.I. and Gourret, J.P. (1970) La virescence du cotonnier (*G. hirsutum*), nouvelle maladie à mycoplasmer. *Compte Rendu de l'Academie des Sciences, Paris* (Series D268) 19, 2382–4.

Couteaux, L., Lefort, P.I. and Kuakuvi, E. (1968) Quelques observations sur le 'leaf curl' du cotonnier chez *G. barbadense* à la station d'Anie-Mono (Togo). *Coton et Fibres Tropicales* 23, 506–7.

Delalande, P. (1970) *Quelques observations sur la maladie bleue des cotonniers. Campagne 1969–1970, rapport non publié.* Station centrale de Bambari, RCA, and IRCT, Paris.

Delattre, R. (1965) La virescence du cotonnier. I. Recherches préliminaires. *Coton et Fibres Tropicales* 20, 289–94.

Delattre, R. (1968) La virescence du cotonnier: deuxième note. *Coton et Fibres Tropicales* 23, 386–90.

Delattre, R. and Joly, A. (1981) Résultats des enquêtes sur virescence florale du cotonnier effectuées en Haute-Volta de 1970 à 1978. *Coton et Fibres Tropicales* 36, 167–84.

Delattre, R., Giannotti, J. and Czarneckyd, D. (1974) Maladies du cotonnier et de la vigne liées au sol et associées à des cochenelles endogées: presence de mycoplasmes et étude comparative des souches *in vitro*. *Compte Rendu de l'Academie des Sciences, Paris* 279, 315–18.

DESMIDTS, M. AND RASSEL, A. (1974) Mise en évidence d'organismes du type mycoplasme chez les espèces végétales atteintes de jaunisse en Haute-Volta: rélations avec la phyllodie du cotonnier. *FAO Phytopathology Bulletin* 22, 138–41.

DICKSON, R.C. AND LAIRD, E.F. (1960) Disease of cotton. *California Agriculture* 14, 14.

DICKSON, R.C., JOHNSON, M.McD. AND LAIRD, E.F. (1954) Leaf crumple, a virus disease of cotton. *Phytopathology* 44, 479–80.

DINEUR, P. (1957) Psyllosis disease of cotton in the Belgian Congo. *Empire Cotton Growing Review* 37, (1960) Abstr. 137.

DYCK, J.M. (1979) La maladie bleue du cotonnier au Tchad. *Coton et Fibres Tropicales* 34, 229–38.

EBBELS, D.L. (1976) Diseases of Upland cotton in Africa. *Reviews Plant Pathology* 55, 747–63.

EL-NUR, E. AND ABU SALIH, M.S. (1970) Cotton leaf and virus disease. *PANS* 16, 121–31.

ENDRIZZI, J.E., TURCOTTE, E.L. AND KOHEL, R.J. (1984) Qualitative genetics, cytology, and cytogenetics. In: Lewis, C.F. and Kohel, R.J. (eds), *Cotton*. Agronomy Series No. 24, American Society Agron., Madison, Wisconsin, pp. 81–129.

ERWIN, D.C. (1959) Leaf crumple–its cause, damage and control. *Cotton Gin and Oil Mill Press* 60, 51.

ERWIN, D.C. AND MEYER, R. (1961) Symptomatology of the leaf-crumple disease in several species and varieties of *Gossypium* and variation of the causal virus. *Phytopathology* 51, 472–7.

ESCOBER, J.T., AGATI, J.A. AND BERGONIA, H.T. (1963) A new virus disease of cotton in the Philippines. *FAO Plant Protection Bulletin* 11, 76–81.

FARQUHARSON, C.O. (1912) *Report of Mycologist*. Annual Report, Agricultural Department, Nigeria (in Tarr, 1951).

FAUQUET, C. AND THOUVENEL, J.C. (1987) *Plant Viral Diseases in the Ivory Coast*. Editions de l'ORSTOM, Institut Français de Recherche Scientifique pour le Développement en Coopération, Paris. 243 pp.

FOLLIN, J.C. (1982) La virescence florale (phyllodie) du cotonnier en Côte-d'Ivoire: possibilités de lutte. *Coton et Fibres Tropicales* 37, 179–81.

FOURNIER, J. AND CATELAND, B. (1971) Etude du comportement de deux variétés de cotonniers cultivés au Tchad en présence d'une mosaïque peut-être nouvelle. *Coton et Fibres Tropicales* 26, 229–333.

GERLING, D. (1990) *Whiteflies: Their Bionomics, Pest Status, and Management*. Intercept Ltd, Andover, UK. 348 pp.

GIANNOTTIE, J. AND DELATTRE, R. (1972) Une nouvelle approache de l'étude épidémiologique d'une phyllodie: la virescence florale due cotonnier. Culture sélective de mycoplasmes etraits de quelques plants et insects homoptères de la biocénose. *Coton et Fibres Tropicales* 27, 371–8.

GIHA, O.H. AND NOUR, M.A. (1969) Epidemiology of cotton leaf curl in the Sudan. *Cotton Growing Review*. 46, 105–18.

GIVORD, L. (1978) Alternate hosts of okra mosaic virus near plantings of okra in southern Ivory Coast. *Plant Disease Reporter* 62, 412–16.

GOKHALE, V.P. (1936) Preliminary observations on small-leaf disease of cotton. *Indian Journal of Agricultural Science* 6, 475–84 (in Tarr, 1964).

GOLDING, F.B. (1930) A vector of leaf curl of cotton in southern Nigeria. *Empire Cotton Growing Review* 7, 120–6.

GOODMAN, R.M. AND BIRD, J. (1978) Bean golden mosaic virus. *CMI/AAB Virus Descriptions* No. 192, Kew, England. 5 pp.

GOURRET, J.P. AND MAILLET, P.L. (1969) Ultrastructure des mycoplasmes dans le phloème du cotonnier atteint de virescence. *Coton et Fibres Tropicales* 24, 27–8.

GRANILLO, C.R., DIAZ, A. AND ANAYA, M. (1974) The mosaic virus of kenaf (*Hibiscus cannabinus*) in El Salvador. *Phytopathology* 64, 768 (Abstr.).

HALLIWELL, R.S. AND ROSBERG, D.W. (1964) Texas vein clearing virus compared with California cotton leaf crumple virus. *Phytopathology* 54, 623 (Abstr.).

HALLIWELL, R.S., LYDA, S.D. AND LUKEFAHR, M.J. (1980) A leafhopper-transmitted virus of cotton. *Texas Agricultural Experiment Station Bulletin* MP-1465.

IDRIS, A.M. (1990) Cotton leaf curl virus disease in the Sudan. *Mededelingen van de Faculteit Landbouwwetenschappen Rijksuniversiteit Gent* 55, 263–7.

ISELY, D. (1946) The cotton aphid. *Arkansas Agricultural Experiment Station Bulletin* No. 462. 29 pp.

JONES, G.H. AND MASON, T.G. (1926) On two obscure diseases of cotton. *Annals of Botany* 40, 759–73.

KHALID, R.A. AND AL-ZARARI, A.J. (1982) Estimation of the economic threshold of infestation for cotton aphid, *Aphis gossypii* Glover in cotton in Mosul, Iraq. *Mesopotamia Journal of Agriculture* 17, 71–8.

KIRKPATRICK, T.W. (1930) Leaf curl in cotton. *Nature* 125, 672.

KIRKPATRICK, T.W. (1931) Further studies on leaf-curl of cotton in the Sudan. *Bulletin of Entomological Research* 12, 323–63.

KOTTUR, G.L. AND PATEL, M.L. (1920) Malformation of the cotton plant leading to sterility. *Agricultural Journal of India* 15, 640–3.

KRAEMER, P. (1966) Serious increase of cotton whitefly and virus transmission in Central America. *Journal of Economic Entomology* 50, 15–31.

KRING, J.B. (1959) The life cycle of the melon aphid, *Aphis gossypii* Glov., as an example of facultative migration. *Annals of the Entomological Society of America* 52, 284–6.

LABOUCHEIX, J., VAN OFFEREN, A. AND DESMIDTS, N. (1973) Etude de la transmission par *Orosius celulosus* (Lindberg) (Homoptera, Cicadellidae) de la virescence florale du cotonnier et de *Sida* sp. *Coton et Fibres Tropicales* 28, 461–72.

LAGIÈRE, R. AND OUATTARA, S. (1969) Contribution à l'étude d'une nouvelle maladie due cotonnier: la virescence. III. Résultats d'essais de transmission de la maladie. *Coton et Fibres Tropicales* 24, 403–11.

LAIRD, E.F. AND DICKSON, R.C. (1959) Insect transmission of leaf crumple virus of cotton. *Phytopathology* 49, 324–7.

LOURENS, J.H., VAN der LAAN, P.A. AND BRADER, L. (1972) Contribution à l'étude d'une mosaïque due cotonnier au Tchad: distribution dans un champ: Aleurodidae communs: essais de transmission de cotonnier à cotonnier par les Aleurodidae. *Coton et Fibres Tropicales* 27, 225–30.

MAHAMA, A. AND CAUQUIL, J. (1976) La sélection de variétés résistantes à la maladie bleue du cotonnier dans l'Empire Centrafricain. *Coton et Fibres Tropicales* 71, 439–46.

MALI, V.R. (1977) Cotton leaf crumple virus disease – a new record for India. *Indian Phytopathology* 30, 326–9.

MALI, V.R. (1978a) Anthocyanosis virus disease of cotton, a new record for India. *Current Science* 47, 235–7.

MALI, V.R. (1978b) Infectious variegation – a new virus disease of cotton in India. *Current Science* 47, 304–5.

MALI, V.R. (1979) Viral wilt – a new disease hitherto unrecorded on cotton. *Current Science* 48, 687–8.

MASSAY, R. E. AND ANDREWS, F.W. (1932) The leaf curl disease of cotton in the Sudan. *Empire Cotton Growers Review* 9, 32–45.

MOSKOVETZ, S.N. (1941) Virus disease of cotton and its control. In: *Plant Virus Diseases and Their Control: Transactions of the Conference on Plant Virus Diseases*. Moscow 4–7 February 1940. Institut Mikrobiologie Akademi Nauk SSSR, Moscow.

MOUND, L.A. (1969) *Whitefly of the World*. British Museum, Kew, England. 343 pp.

MOUND, L.A. (1983) Biology and identity of whitefly vectors of plant pathogens. In: PLUMB, R.T. AND THRESH, J.M. (eds), *Plant Virus Epidemiology*. Blackwell Scientific Publications, Oxford, pp. 305–13.

NEAL, D.C. (1946) A possible mosaic disease of cotton observed in Louisiana in 1946. *Phytopathology* 37, 434–5.

NOUR, M.A. (1959) Cotton leaf mottle: a new virus disease of cotton. *Empire Cotton Growers Review* 36, 32–4.

NOUR, M.A. (1960) On leaf curl of cotton in the Philippines. *FAO Plant Protection Bulletin* 8, 55–6.

NOUR, M.A. AND NOUR, J.J. (1964) Identification, transmission, and host range of leaf curl viruses infecting cotton in Sudan. *Empire Cotton Growers Review* 41, 27–37.

PHADKE, A.D., KHANDAL, V.S., RAHALKOR, S.R. (1988) Use of neem product in insecticide resistance management IPM in cotton. *Pesticides* 22, 36–7.

POLLARD, D.G. (1955) Feeding habits of the cotton whitefly, *Bemisia tabaci* Genn. (Homoptera, Aleyrodidae). *Annals of Applied Biology* 43, 664–71.

PRENTICE, A.N. (1972) *Cotton – With Special Reference to Africa*. Tropical Agricultural Series, Longman Group Ltd, UK. 282 pp.

QUANT, G.L. (1977) Virosis del algodonero en Guatemala. *El Algodonero* 28, 4–7.

REYES, G.M., MARTINEZ, A.L. AND CHINTE, P.T. (1959) Three virus diseases of plants new to the Philippines. *FAO Plant Protection Bulletin* 7, 141–3.

ROSBERG, D.W. (1957) A new virus disease of cotton in Texas. *Plant Disease Reporter* 41, 726–9.

RUSSELL, T.E. (1982) Effect of cotton leaf crumple on stub and planted cotton. *Arizona Plant Pathology/Cooperative Extension Service Newsletter* 1, 3–5.

SCHUSTER, M.F. (1964) A whitefly-transmitted mosaic virus of *Wissadula amplissima*.

SCHUSTER, M.F. AND LAMBE, R.C. (1966) Incidence and importance of a cotton mottle disease in the lower Rio Grande Valley. *Plant Disease Reporter* 48, 955.

SIDDIG, M.A. (1968) Genetics of resistance to cotton leaf curl in the Sakel cotton *Gossypium barbadense*. *Journal of Agricultural Science* 70, 90–103.

SILBERSCHMIDT, K.M. AND TOMMASI, L.R. (1955) Observaçoés e estudos sobre espécies de plantas suscetiveis à chlorose infecciosa das malvaceas. *Anais da Academia Brasileira de Ciencias* 27, 195–214.

SLEETH, B., LAMBE, R.C. AND HUBBARD, J.L. (1963) Terminal stunt of cotton in South Texas. *Plant Disease Reporter* 47, 587–9.

SOYER, D. (1947) La psyllose provoquée par *Paurocephala gossypii* Russel. *Publication d'Institute National d'Etudes Agronomiques de la Congo Belge (INEAC), Série Scient.* No. 33.

TARR, S.A.J. (1951) *Leaf Curl Disease of Cotton*. Commonwealth Mycological Institute, Kew, England.

TARR, S.A.J. (1957) The host range of the cotton leaf curl virus in the Sudan. *Empire Cotton Growers Review* 34, 258–62.

TARR, S.A.J. (1964) *Virus Diseases of Cotton*. Miscellaneous Publication No. 18. Commonwealth Mycological Institute, Kew, Surrey, England. 23 pp.

TSAO, P.W. (1963) Intranuclear inclusion bodies in the leaves of cotton plants infected with leaf crumple virus. *Phytopathology* 52, 243–4.

UPPAL, B.N., CAPOOR, S.P. AND RAYCHAUDHURI, S.P. (1944) Small leaf disease of cotton. *Current Science* 11, 284–5.

VAISSAYRE, M. (1971) *Nouvelle contribution à l'étude de la 'maladie bleue' du cotonnier: recherche du vecteur et conditions économiques da la transmission, campagne 1970–71*. Station Centrale de Bambari (RCA) and IRCT, Paris.

VAN SCHAIK, P.H., ERWIN, D.C. AND GARBER, M.J. (1962) Effects of time of symptom expression of the leaf-crumple virus on yield and quality of fiber of cotton. *Crop Science* 2, 275–7.

WATERHOUSE, P.M., GILDOW, F.E. AND JOHNSTONE, G.R. (1988) *AAB/CMI Descriptions of Plant Viruses* No. 239. CMI, Kew, Surrey, England. 9 pp.

WATKINS, G.M. (1981) Virus and mycoplasma like organisms. In: Watkins, G.M. (ed.) *Compendium of Cotton Diseases*; American Phytopathology Society, St Paul, Minnesota, pp. 56–9.

WILSON, D.F. AND BROWN, J.K. (1991) Inheritance of response to cotton leaf crumple virus infection in cotton. *Journal of Heredity*, 82, 508–9.

WILSON, D.F., BROWN, J.K. AND BUTLER, G.D., Jr (1989) Reaction of cotton cultivars and lines to cotton leaf crumple virus. *Journal of the Arizona and Nevada Academy of Science* 23, 7–10.

YASSIN, A.M. AND DAFALLA, G.A. (1982) Cotton reddening in the Gezira. *Tropical Pest Management* 28, 312–13.

YASSIN, A.M. AND EL-NUR, E. (1970) Transmission of cotton leaf curl by a single insect *Bemisia tabaci*. *Plant Disease Reporter* 54, 528–31.

10 Nematodes

J. BRIDGE*

Introduction

Nematodes occur in a wide variety of morphological forms ranging from thread to globular shapes. All the plant nematodes known to be parasitic on cotton have a life stage in the soil. However, the majority of nematodes that can be found in any soils are non-plant-parasitic genera, the so-called 'free-living nematodes', and it is necessary to distinguish these from the true plant parasites. It is also the case that a proportion of the plant nematodes that feed or browse on plant tissue have not been shown to cause crop damage. Most of the nematodes discussed in this chapter are those which are known to cause plant damage and yield loss.

Most plant nematodes are microscopic, varying in size from 0.13 to 12 mm long, with the adult stages of many around 1 mm in length. Cotton nematodes all feed on and cause damage to roots as migratory endoparasites, sedentary endo- and semi-endoparasites, or ectoparasites.

It is often very difficult to determine immediately a clear association between foliar disease symptoms and plant nematodes. Usually symptoms of nematode damage are very similar or identical to symptoms caused by mineral deficiencies in the soil, inadequate or excessive water or generally poor soils. Many so-called 'sick soils' or 'tired soils' can be a result of the build-up of plant nematodes. Nematode soil populations, when they reach economic threshold levels, can cause considerable mechanical or physiological root damage which inhibits or prevents the uptake of water and nutrients. In addition, certain nematodes of cotton are known to have synergistic and other interrelationships with fungal pathogens in disease expression. To determine the associations between nematodes and crop damage or disease, it is generally

* International Institute of Parasitology, 395a Hatfield Road, St Albans AL4 0XU, UK.

331

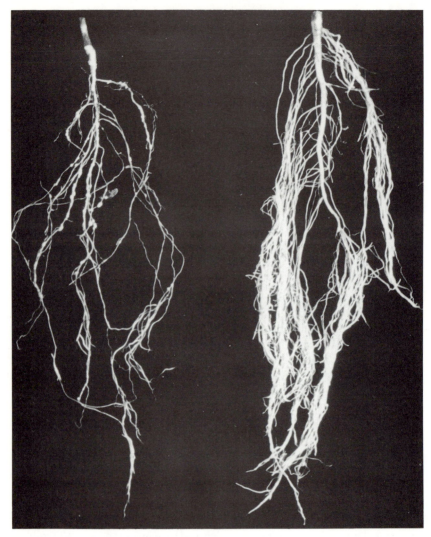

Fig. 10.1. Reduced root system with galled roots caused by *Meloidogyne incognita* (left) compared with healthy root system (right).

necessary to examine for characteristic root damage by uprooting the plants and to identify the nematodes microscopically, after extracting from soil and root material.

The plant nematode genera and species which are known to cause crop damage to cotton are *Meloidogyne* spp. (root-knot nematodes), *Rotylenchulus* spp. (reniform nematodes), *Pratylenchus* spp. (lesion nematodes), *Hoplolaimus* spp. (lance nematodes) and *Belonolaimus longicaudatus* (sting

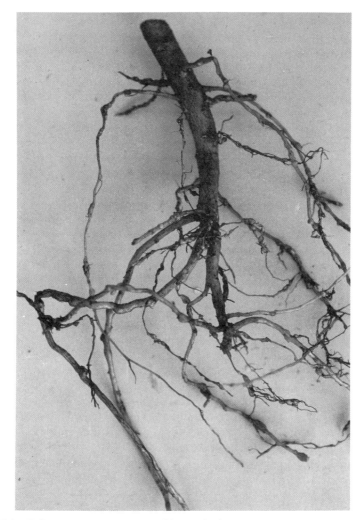

Fig. 10.2. Galls on cotton roots caused by *Meloidogyne incognita.*

nematode). In addition, other plant nematode genera suspected of causing injury to cotton are *Xiphinema, Longidorus, Paratrichodorus* and *Scutellonema.*

Meloidogyne spp., the Root-knot Nematodes

Root-knot nematodes are amongst the most damaging and economically important pests of sub-tropical and tropical crops throughout the world. Over 50 species of *Meloidogyne* have been identified but, of these, the only species which is both widespread and causes severe yield loss of cotton is

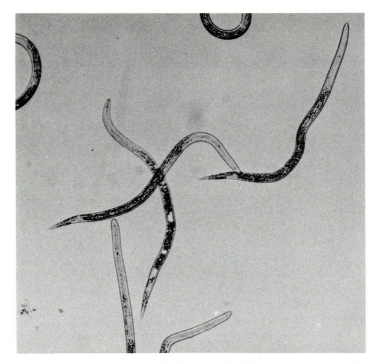

Fig. 10.3. Second-stage juveniles (infective stage) of *Meloidogyne incognita*.

M. incognita. Within this species different biological 'races' have been identified and two of these, races 3 and 4, are known to attack cotton.

The cotton races of *M. incognita* only occur, or have only been identified, in certain cotton-growing countries, although other races of *M. incognita* are common throughout the tropics. *M. incognita* causing damage to cotton has been found in every cotton-producing area of the USA. However, severity of damage varies among these areas (Heald *et al.*, 1981; Heald and Thames, 1982). It has also been found on cotton in the Central African Republic (Luc, 1968), Ethiopia (O'Bannon, 1975), Ghana (Hemeng, 1978), South Africa (Louw, 1982; Wyk *et al.*, 1987), Tanzania (Perry, 1962; Brown, 1968), Uganda (Wickens and Logan, 1960; Perry, 1963), Zimbabwe (Martin, 1954; Shepherd and Coombs, 1980), Brazil (Lordello *et al.*, 1984), El Salvador (*Meloidogyne* spp., probably *M. incognita*) (Pinochet and Guzman, 1987), Egypt (Shafshak *et al.*, 1985), Syria (Mamluk and Faust, 1975), Turkey (Yüksel, 1974, 1982), Pakistan (Maqbool and Saeed, 1981), China (Chengzhu and Pinson, 1986) and India (Thakar *et al.*, 1986). In these areas, *M. incognita* more often occurs in coarse textured or light sandy soils than in heavy soils.

One other species of *Meloidogyne* is known to seriously damage cotton. This is the African cotton root nematode, *M. acronea*, which has only been

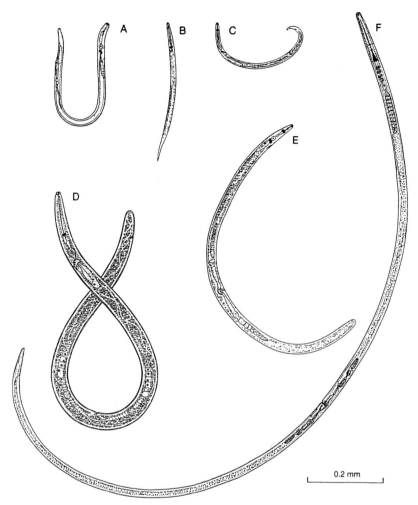

Fig. 10.4. Illustrations of different types of nematodes found in cotton soil or roots. A, *Pratylenchus* female. B, Second-stage juvenile of *Meloidogyne*. C, *Rotylenchulus* immature female. D, *Belonolaimus* female. E, *Hoplolaimus* female. F, *Xiphinema* female.

found on cotton in an area in the south of Malawi, but the nematode does also occur in South Africa (Bridge *et al.*, 1976; Starr and Page, 1990).

SYMPTOMS OF DAMAGE

Meloidogyne incognita is a sedentary endoparasite of cotton roots, feeding on the vascular elements and causing cellular modifications. Damage to the roots results in a smaller root system, normally with fewer lateral and feeder roots (Fig. 10.1). The nematode also induces increased cell division and

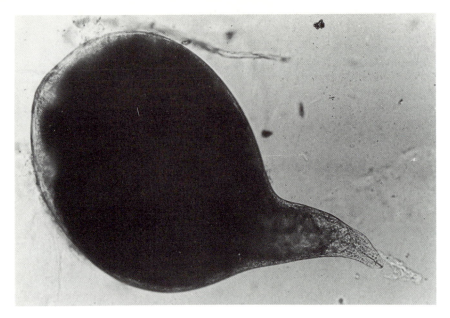

Fig. 10.5. Mature, swollen female of *Meloidogyne incognita*.

multiplication in the root cortex around the feeding sites, producing a characteristic knot or gall, mainly on the lateral roots. Unlike galls produced by *Meloidogyne* in many other plants, which tend to be large and prominent, the galls caused by *M. incognita* on cotton roots are more often inconspicuous swellings (Fig. 10.2).

Feeding by *M. incognita* inhibits or blocks the upward translocation of water and nutrients, which, together with a reduced root system, produces non-specific above-ground damage resembling nutrient or water deficiency symptoms. Plants are stunted, with smaller, yellowed leaves and smaller and fewer bolls. Infected plants are more prone to wilt in adverse conditions, such as drought and excessive temperatures. The field symptoms characteristically occur as circular or elongate-oval patches in a cotton field, depending on cultivation techniques, due to the gradual spread of the soil nematodes from a number of different and separate soil foci.

Field symptoms associated with *M. acronea* are similar to those caused by *M. incognita*. However, *M. acronea* does not produce galling of roots; instead, the symptoms of feeding are an increase in fine lateral root growth around the feeding site.

MORPHOLOGY AND BIOLOGY

Meloidogyne incognita, as with other plant parasitic nematodes, has four juvenile stages between the egg and the adult; moulting occurs between stages

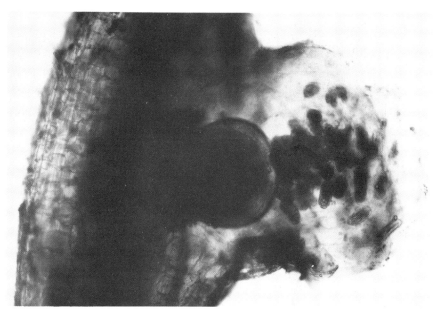

Fig. 10.6. Female of *Meloidogyne incognita* protruding from cotton root surrounded by egg sac containing eggs.

as the nematode increases in size. The first moult take place within the egg, and the second-stage juvenile (0.3–0.4 mm long) emerges as a vermiform, migratory nematode, which is the infective stage of this species (Figs 10.3 and 10.4B). The second-stage juveniles are normally found in the soil; all other stages developing towards the adult female are found in the root and are swollen and non-migratory. Infective juveniles penetrate roots, generally behind the root tips, and migrate to a permanent feeding site alongside the stele, where cells are transformed into a trophic system of transfer cells, which act as nutrient sinks for the nematodes. The mature female is a swollen, rounded nematode, 0.5–0.7 mm long, with a projecting neck and head (Fig. 10.5 and Plate 7A). The posterior of the female is normally positioned in the epidermal region of the root, and eggs are laid into the soil surrounded by a gelatinous matrix, the egg sac, which can be observed on the surface of the root (Fig. 10.6). Each female can lay 500–1000 eggs. During development, multiplication of cortical cells occurs, producing the characteristic gall. *M. incognita* reproduces by parthenogenesis, but the long vermiform males, although unnecessary for reproduction, can occasionally be found in soil extracts.

Meloidogyne acronea is morphologically similar to *M. incognita* in all life stages. Biologically it differs in that the females become semi-endoparasitic on the surface of roots, surrounded by egg sacs, when mature (Page, 1985).

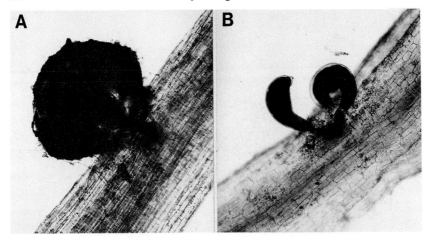

Fig. 10.7. A, Female of *Rotylenchulus reniformis* protruding from cotton root surrounded by egg sac. B, Mature females of *R. reniformis* on cotton roots with egg sacs removed.

Rotylenchulus reniformis, the Reniform Nematode

The reniform nematode is widely distributed in the tropics and sub-tropics and is known to be a serious pest of many crops. On cotton, it is known to cause severe yield losses in parts of the USA (Birchfield and Brister, 1963; Birchfield *et al.*, 1966; Heald *et al.*, 1981) and has also been found as a parasite of the crop in China (Chenzhu *et al.*, 1986), the Nile Valley, Egypt (Oteifa, 1970), India (Dasgupta and Seshadri, 1971; Palanisamy and Balasubramanian, 1983), Tanzania (J. Bridge and R.J. Hillocks, unpublished) and Ghana (Peacock, 1956). There is little doubt that it has a more extensive distribution on cotton, although it is unlikely to occur in all cotton-growing regions.

One other species of *Rotylenchulus*, *R. parvus*, is known to occur on cotton in southern Africa (Louw, 1982; Page *et al.*, 1985).

Symptoms of damage

Field symptoms of damage by *R. reniformis* resemble those described for root-knot nematodes. Uneven growth in patches, typical of root damage, can be observed; plants can be chlorotic as well as stunted, and wilting sometimes occurs.

R. reniformis does not cause galling or any other obvious symptoms on roots, but can cause reduction in root growth, which is particularly apparent in cotton seedlings (Birchfield, 1962).

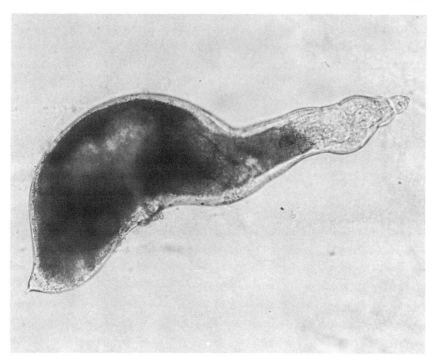

Fig. 10.8. Complete mature female of *Rotylenchulus reniformis*.

MORPHOLOGY AND BIOLOGY

Eggs of *R. reniformis* are deposited in egg sacs, which can be observed on the surface of the root at high magnification (Fig. 10.7A). The eggs develop into vermiform immature females and males in the soil. Immature females are small (0.23–0.64 mm) (Fig. 10.4C) and are the infective or invasive stage. They are sedentary semi-endoparasites and normally only the head and neck penetrate the cortex of the young roots, establishing a feeding site in the pericycle. The remainder of the body projects from the root becoming swollen and kidney-shaped (reniform) as feeding occurs (Figs 10.7B and 10.8 and Plate 7B). The gelatinous egg sac containing eggs is produced after fertilization. The life cycle on cotton from egg to egg-producing female is only 17–23 days (Birchfield, 1962), allowing many generations and large populations to develop during the growing season.

Hoplolaimus spp., the Lance Nematodes

Hoplolaimus spp., commonly known as lance nematodes, are migratory species feeding as endoparasites or ectoparasites on roots; all stages are vermiform and can be found in both roots and soil.

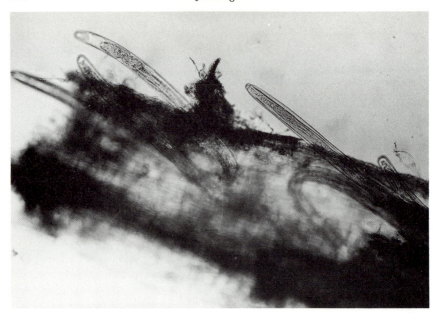

Fig. 10.9. *Hoplolaimus seinhorsti* females and juveniles emerging from a section of cotton root.

Five species of *Hoplolaimus* have been reported to be parasitic on cotton: *H. columbus* and *H. galeatus* in the USA (Krusberg and Sasser, 1956; Fassu-liotis *et al.*, 1968), *H. indicus* in India (Guar and Mishra, 1981), *H. aegypti* in Egypt (Shafie and Koura, 1969) and *H. seinhorsti* (syn. *H. sheri*) in Israel (Mor and Cohn, 1989), China (Chengzhu *et al.*, 1987) and Africa (Luc, 1958). *H. seinhorsti* has a much wider distribution on other crops throughout the tropics and sub-tropics (Van den Berg, 1976).

Symptoms of damage

During feeding, *Hoplolaimus* spp. cause cell destruction and disruption of the epidermal, cortical, endodermal (see Plate 7C) and sometimes phloem cells as they penetrate and migrate through the root tissues (Krusberg and Sasser, 1956; Lewis *et al.*, 1976). Large numbers of the nematodes can congregate in one area of the root (Fig. 10.9), causing root lesions.

Cotton plants infected with *Hoplolaimus* spp. can be stunted, with chlorotic or discoloured leaves and fewer squares and bolls.

Adult *Hoplolaimus* spp. are relatively long (1–2 mm) and the nematodes remain migratory and vermiform throughout all their life stages (Figs 10.4E and 10.9) and are found in both soil and roots. Juveniles are smaller versions of adults lacking sexual differentiation.

Nematodes can be found feeding on the epidermal cells as ectoparasites, partially embedded in root tissue as semi-endoparasites or completely endoparasitic in the root cortex. Eggs are laid free in both the roots and the surrounding soil.

Belonolaimus longicaudatus, the Sting Nematode

B. longicaudatus can be a devastating pest of many crops but the nematode has only been recorded in south-eastern and eastern USA from Connecticut to Texas (Orton Williams, 1974; Esser, 1976). It mainly occurs in very sandy soils (84–94% sand).

SYMPTOMS OF DAMAGE

Roots of cotton attacked by *B. longicaudatus* can be severely damaged or destroyed. Small, dark, sunken lesions on the secondary roots are the first signs of attack by the nematode, giving the root system a general unhealthy appearance. Root tips become shrivelled and blackened and damage leads to the loss of many of the lateral roots (Graham and Holdeman, 1953).

Field symptoms of damage by *B. longicaudatus* are marked stunting and chlorosis of plants and sometimes death. These symptoms occur in the field as variable size patches of reduced growth with the boundaries between healthy and stunted plants usually well marked (Graham and Holdeman, 1953).

MORPHOLOGY AND BIOLOGY

B. longicaudatus is longer than the average plant parasitic nematode (2.0–3.0 mm) with a very long feeding stylet (Fig. 10.4D). As with *Hoplolaimus*, all life stages are migratory and morphologically similar. The lance nematode is strictly an ectoparasitic nematode and is only found in the soil.

Other Nematodes of Cotton

Representatives of most of the plant-parasitic nematode families, in addition to those mentioned above, have been found in soils around cotton plants, but there is little evidence to show that all, or even many of these, are involved in damage to the crop (Birchfield *et al.*, 1966; Bird *et al.*, 1971; Bridge, 1979; Louw, 1982). It has been shown that cotton is extremely tolerant to one species, *Helicotylenchus dihystera*, commonly found on cotton (Bernard and Hussey, 1979). The other nematode genera and species which are suspected to damage cotton from the evidence available are as follows.

PRATYLENCHUS SPP., THE *LESION NEMATODES*

These are relatively small (less than 1 mm long), migratory and vermiform root-endoparasitic nematodes (Fig. 10.4A) that feed in the root cortex, causing cell breakdown and necrosis: *P. brachyurus* has been implicated in cotton disease in the USA (Graham, 1951) and Africa (Louw, 1982; Page *et al.*, 1985; Wyk *et al.*, 1987), but there is no proof that it causes direct damage. It has been suggested that the nematode is only important as part of a disease complex (Starr and Page, 1990). *P. sudanensis* has been found in the Sudan, where it is reported to reduce cotton yields (Yassin and Mohammed, 1980).

XIPHINEMA AND LONGIDORUS

These large (2–11 mm long) ectoparasitic nematodes (Fig. 10.4F), which only occur in soil, have been found associated with stunted cotton in Africa, causing root tip galling, the typical root symptoms of these nematodes (Bridge and Page, 1975; Starr and Page, 1990).

SCUTELLONEMA SPP.

A number of species of the genus *Scutellonema* have been found in considerable populations as endoparasites of cotton roots in the USA and Africa (Bridge and Page, 1975; Kraus-Schmidt and Lewis, 1979; Page *et al.*, 1985) and are possibly causing some yield loss of the crop.

PARATRICHODORUS SPP.

The stubby root nematodes, *Paratrichodorus* spp., are soil-inhabiting ectoparasites and have been found in large numbers in lands where cotton plants exhibited poor growth and 'stubby root systems', characteristic symptoms of damage by these nematodes (Louw, 1982).

Nematode Disease Complexes

Nematodes, as one of the groups of micro-organisms forming an integral part of the rhizosphere communities, are involved in complex interrelationships with other organisms, leading to disease expression. Details of these interrelationships are presented in Chapters 1 and 4.

The most well-known and well-documented nematode-fungus interrelationship on cotton is that between the root-knot nematode, *M. incognita*, and *Fusarium oxysporum* f.sp. *vasinfectum* in Fusarium wilt disease (see Plate 7D; Martin *et al.*, 1956; Starr *et al.*, 1989). There is reported evidence that other nematodes are also important factors in the incidence of Fusarium wilt of cotton; these include *Belonolaimus longicaudatus* (Holdeman and Graham,

1954) and *Rotylenchulus reniformis* (Neal, 1954) in the USA, *Pratylenchus sudanensis* in the Sudan (Yassin, 1974), and *Hoplolaimus seinhorsti* in India (Rajaram, 1979).

Both *M. incognita* and *R. reniformis* can also increase the incidence of Verticillium wilt of cotton (Khur and Alcorn, 1973; Prasad and Padeganur, 1980).

Root damage, resulting in poor growth and stress, caused by nematodes has been suggested to be the main factor in the nematode–fungi complexes of cotton seedling diseases. *M. incognita*, *R. reniformis* and *H. tylenchiformis* (probably *H. galeatus*) have been shown to increase the severity of seedling diseases caused by *Alternaria tenuis*, *Glomerella gossypii*, *Rhizoctonia solani* and *Pythium debaryanum*, as well as *F. oxysporum* f.sp. *vasinfectum* (Brodie and Cooper, 1964; Cauquil and Shepherd, 1970).

In another type of interrelationship, the vesicular arbuscular endomycorrhizal fungi have been shown to suppress plant parasitic nematodes on cotton (Saleh and Sikora, 1988; Starr and Page, 1990).

Economic Importance of Nematodes in Cotton

Accurate field assessments of yield losses due entirely to nematodes alone are difficult to obtain particularly when the nematodes are involved in disease complexes. The nematodes, and the crop damage they cause, can also be markedly affected by environmental conditions. Most of the reported yield losses are estimates, but are generally conservative in nature. Detailed studies and probably the most accurate estimates of cotton yield losses due to nematodes have been done mainly in the USA.

The annual cotton yield loss due to damage by plant parasitic nematodes on a world basis is estimated to be 10.7% (Sasser and Freckman, 1987). Reported yield losses are variable with the different nematode species. At individual field sites, the avoidable yield losses estimated after application of nematicides are reported to be 9.5–17.4% and 40–60% in India and Egypt respectively for *R. reniformis* (Oteifa, 1970; Palanisamy and Balasubramanian, 1983), and 10–25%, depending on cotton cultivar and environmental factors, for *H. columbus* plus *M. incognita* in the USA (Mueller and Sullivan, 1988). The most detailed estimates of yield losses due to nematodes on a regional basis have been done in different states of the USA, where nematicides are used regularly in a proportion of cotton farms. In Arkansas, yield losses are estimated at 1.5% annually, with *M. incognita* by far the most important nematode of cotton; in North Carolina, during 1987, nematode yield losses were estimated to be 6.8%, with *H. columbus* causing 5.2% and *R. reniformis* 1.1% loss, although overall losses in 1988 were reduced to 1%; in South Carolina, nematodes are major pests of cotton, accounting for an estimated 5% loss annually, with *H. columbus* and *M. incognita* being the most important (Kirkpatrick, 1988, 1989).

Control

CHEMICAL

Nematicides have been widely used as a means of controlling cotton nematodes in the USA. However, the choice of nematicides is becoming more and more restricted with the banning of some of the chemicals, particularly fumigants such as EDB and DBCP. The most effective nematicides for reducing nematode damage have been 1,3-dichloropropene (1,3-D) at 33.7–67.4 l ha^{-1} (or at 34 kg a.i. ha^{-1}), aldicarb at 0.84–1.68 kg a.i. ha^{-1}, and fenamiphos at 0.84–1.85 kg a.i. ha^{-1}. Combined treatments of 1,3-D and aldicarb are used in situations of high potential damage from nematodes (Kirkpatrick, 1988, 1989; Mueller and Sullivan, 1988).

CULTURAL

Crop rotation can be a practical and efficient means of controlling cotton nematodes in certain farming systems. It is more difficult to devise with those nematodes having a wide host range, especially *M. incognita* and *R. reniformis*. A weed-free fallow included in any rotation will most effectively reduce soil populations of nematodes, although this is not always possible. Sorghum and maize grown in rotation with cotton can reduce populations of *R. reniformis* and increase yields (Brathwaite, 1974; Thames and Heald, 1974). Groundnut (*Arachis hypogaea*) is a non-host for *M. incognita* and, when grown for two years preceding cotton, effectively decreases the amount of galling on the following cotton crop (Kirkpatrick and Sasser, 1984).

Control of *M. acronea* could be achieved by rotating cotton with poor or non-host crops, such as pearl millet (*Pennisetum typhoides*), finger millet (*Eleusine coracana*), guar bean (*Cyamopsis tetragonoloba*) or groundnut (Bridge and Page, 1977; Starr and Page, 1990).

Rotating rice with cotton is reported to control *H. seinhorsti* in China (Chengzhu *et al.*, 1987). Tobacco, watermelons and *Crotalaria* are non-hosts of the sting nematode, *B. longicaudatus*, and could be grown in rotation or as a cover crop with cotton to control the nematode. However, the rotation crop has to be kept free of weed hosts, such as crab grass (*Digitaria sanguinalis*), which would maintain nematode populations in the soil (Holdeman and Graham, 1953). Another difficulty of crop rotation is when multiple species of nematodes occur in cotton fields; crop rotation for control of one species may result in increased populations of and damage by another species (Kirkpatrick and Sasser, 1984).

Crotalaria spectabilis and *Tagetes minuta* grown as cover crops are very efficient for reducing a range of cotton nematodes, including *M. incognita*, *B. longicaudatus* and *Paratrichodorus* (Good *et al.*, 1965).

Annual subsoiling of cotton fields under the planting row to a depth of

35 cm increases yields of seed cotton in the presence of normally damaging population levels of *H. columbus* (Hussey, 1977).

RESISTANCE

Resistance in cotton has been found to both *M. incognita* and *R. reniformis*. Most cotton cultivars are susceptible to *M. incognita* races 3 and 4 but highly resistant to races 1 and 2. The cultivar Auburn 623 is known to be highly resistant to isolates of *M. incognita* cotton races (Sasser and Kirby, 1979) but some cultivars reported to be resistant are highly susceptible to the nematode (Starr and Page, 1990). Although no true commercial cultivars with high resistance are presently available to growers, breeding stocks with high resistance have been developed and the potential for producing acceptable resistant cultivars using these breeding stocks is great (Shepherd, 1988; Kirkpatrick and Shepherd, 1989; Starr and Page, 1990). Most breeding work has attempted to select for resistance to both root-knot and Fusarium wilt because of the relationship between the two diseases. Field resistance to Fusarium wilt does not necessarily mean that the genotype is resistant to the nematode (Starr and Page, 1990), but it has been thought possible to breed for Fusarium wilt resistance by breeding solely for high root-knot resistance (Shepherd, 1974b).

A number of screening techniques have been used for determining resistance to *M. incognita*. Shepherd (1974a) found that the most effective way of differentiating between resistant and susceptible entries was by using disinfected juveniles for inoculum of cotton grown in steam-sterilized soil. Juveniles were surface-disinfected by immersion in 0.001% 8-quinolinol sulphate for 30 minutes and then placed in tap-water. Eggs have also been used for inoculum in sterilized soil at population levels of 8000 eggs per plant (Shepherd, 1979). Field plots naturally infested with *M. incognita* can also be used for screening (Hyer *et al.*, 1979). Evaluation of cotton resistance to root-knot nematode can be based on severity of root galling, a plant response to the nematode, or egg production, a nematode response to the plant (Shepherd, 1979). Shepherd (1979) reported that egg counts appeared to be the most appropriate selection criteria rather than galling or root-knot indices.

High levels of resistance to the reniform nematode, *R. reniformis*, have been found in *G. arboreum* Nanking CB 1402, *G. barbadense* Texas 110, *G. somalense* and *G. stocksii*, with some resistance in other breeding lines of *G. arboreum* and *G. hirsutum*. Host resistance ratings are based on egg production per gram of root, female development and their fecundity (Carter, 1981; Yik and Birchfield, 1984).

Sampling, Extraction and Estimating Populations of Nematodes from Soil and Roots

Because of the subtle and non-specific nature of most nematode damage to cotton, normally the only sure means of determining the association between nematodes and crop loss is by microscopic examination of the roots themselves or of the nematodes after first extracting from soil and/or root material. The main exception to this is by assessing galling of roots caused by *Meloidogyne*.

FIELD SAMPLING

Assessment of the nematodes present in the field is done by collecting and analysing individual root and soil samples or by taking a number of samples from similar plants and bulking before extraction and processing. Most nematodes will occur in or around the roots at depths of 5–30 cm, with few occurring in the surface layers. Selective rather than random sampling will give the most pertinent information, with plants being selected on the basis of showing disease symptoms or being in yield decline. Sampling of healthy plants in close vicinity to the diseased plants will give very useful comparative information. Sampling time is important; the middle of the growing season will give the most representative picture of nematode populations.

Root galls

Although galls (root-knots) on cotton roots caused by *M. incognita* are not as obvious as those on other crops, the number of galls or extent of root galling can be used to give some estimate of both the populations of the nematode present and the amount of root damage. A root-knot index scale of 1 to 5 is generally used with cotton:

    1 = no galls (0%)
    2 = light galling
    3 = moderate galling
    4 = heavy galling
    5 = complete galling (100%)

EXTRACTION

Soil nematodes

At least one or more life stage of the different nematodes can be found in soils (juveniles and males of all species; immature females of *Rotylenchulus*; all stages of *Hoplolaimus, Belonolaimus, Pratylenchus, Paratrichodorus, Xiphinema, Paralongidorus* and *Scutellonema*).

Nematodes can be extracted from soil by many different methods, including the Baermann funnel and tray techniques, which use the activity of the nematodes to separate them from soil material, as well as flotation, elutriation, sieving and centrifugal flotation (Hooper, 1986, 1990).

The Baermann funnel method and its modifications extract most soil nematodes and are the simplest to use with the minimum of equipment. Rubber tubing is attached to the stem of an ordinary filter funnel and closed

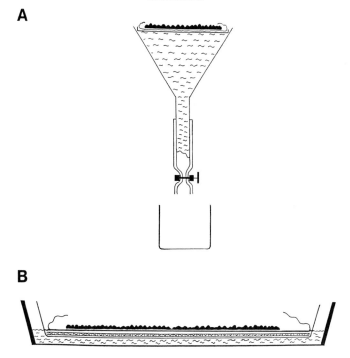

Fig. 10.10. Simple equipment for extracting live nematodes from soil and plant material. A, Baermann funnel. B, Tray modification.

with a clip. The funnel is supported above the work surface and partially filled with water. Soil (or chopped root tissue) is placed on a piece of muslin, tissue paper, nylon sieve, etc., supported in the funnel mouth submerged in the water (Fig. 10.10A). Nematodes pass into the water and the nematode suspension is run off after 24 hours or overnight. A simple modification is to use a tray or Petri dish instead of a funnel (Fig. 10.10B) (Hooper, 1986, 1990).

Root nematode

Active, migratory endoparasitic nematodes can be extracted from cotton roots using the Baermann funnel or tray method (Fig. 10.10A,B). Finely chopped roots are placed on the supporting tissue or sieve and left for 24–48 hours. More nematodes will be extracted if the roots are first macerated in an electric blender or liquidizer for 10–15 seconds (Hooper, 1986, 1990). This method will not extract the swollen females of *Meloidogyne* or *Rotylenchulus*, which have to be examined directly in root tissues.

Staining

All nematodes within root tissues can be clearly observed after staining in standard cotton blue or acid fuchsin stains. Lactophenol has been the normal staining medium but, to avoid using toxic phenol, the following method is recommended: roots are immersed for 3 minutes in a boiling solution of equal volumes of glycerol, lactic acid and distilled water, plus 0.05–0.1% acid fuchsin or cotton blue. After allowing to cool in the solution, roots are removed, washed in water and cleared for two to three days in a solution of equal volumes of glycerol and distilled water, acidified with a few drops of lactic acid (Hooper, 1986, 1990). This method will also stain egg sacs of both *Meloidogyne* and *Rotylenchulus* on the surface of roots to facilitate counting. When numbers of eggs are required, a further method can be used.

Extraction of eggs

To collect and count *Meloidogyne* eggs, a known weight of surface-dried, infested roots can be placed in a container with 1.05% NaOCl. Roots are shaken, preferably with a laboratory shaker. The NaOCl disperses eggs from the egg sacs and the resulting solution with eggs is poured through a bank of 75 μm and 25 μm-aperture sieves, and the eggs are washed off the latter (Shepherd, 1979). A similar method is used to extract *Rotylenchulus* eggs from egg sacs. Infested roots are placed in 0.5% NaOCl solution for 10 minutes to free the eggs. The roots are macerated for 10 seconds in a blender, and the resulting suspension is poured through a 45 μm-aperture sieve, and eggs are collected on an 18 μm-aperture sieve (Yik and Birchfield, 1984).

References

BERNARD, E.C. AND HUSSEY, R.S. (1979) Population development and effects of the spiral nematode, *Helicotylenchus dihystera*, on cotton in microplots. *Plant Disease Reporter* 63, 807–10.

BIRCHFIELD, W. (1962) Host–parasite relations of *Rotylenchulus reniformis* on *Gossypium hirsutum*. *Phytopathology* 52, 862–5.

BIRCHFIELD, W. AND BRISTER, L.R. (1963) Susceptibility of cotton and relatives to reniform nematode in Louisiana. *Plant Disease Reporter* 47, 990–2.

BIRCHFIELD, W., REYNOLDS, H. AND ORR, C. (1966) Plant-parasitic nematodes of cotton in the Lower Rio Grande valley of Texas. *Plant Disease Reporter* 50, 149–50.

BIRD, G.W., McCARTER, S.M. AND RONCADORI, R.W. (1971) Role of nematodes and soil-borne fungi in cotton stunt. *Journal of Nematology* 3, 17–22.

BRATHWAITE, C.W.D. (1974) Effect of crop sequence on populations of *Rotylenchulus reniformis* in fumigated and untreated soil. *Plant Disease Reporter* 58, 259–61.

BRIDGE, J. (1979) A study of the relationship between nematodes and Fusarium wilt of cotton. *Ministry of Overseas Development (ODM) Report on Cotton Research Project*, Ukiriguru, Tanzania. 17pp.

BRIDGE, J. AND PAGE, S.L.J. (1975) Plant parasitic nematodes associated with cotton in the lower Shire Valley and other cotton growing regions of Malawi. *Overseas Development Administration Technical Report*. London, UK. 40pp.

BRIDGE, J. AND PAGE, S.L.J. (1977) An assessment of the importance and control of plant parasitic nematodes in Malawi. *Ministry of Overseas Development Technical Report*, London, UK. 80pp.

BRIDGE, J., JONES, E. AND PAGE, S.L.J. (1976) *Meloidogyne acronea* associated with reduced growth of cotton in Malawi. *Plant Disease Reporter* 60, 5–7.

BRODIE, B.B. AND COOPER, W.E. (1964) Relation of parasitic nematodes to post-emergence damping-off of cotton. *Phytopathology* 54, 1023–7.

BROWN, A.G.P. (1968) Effects of fumigation to control nematodes on Fusarium wilt of cotton. *Cotton Growing Review* 45, 128–36.

CARTER, W.W. (1981) Resistance and resistant reaction of *Gossypium arboreum* to the reniform nematode, *Rotylenchulus reniformis*. *Journal of Nematology* 13, 368–74.

CAUQUIL, J. AND SHEPHERD, R.L. (1970) Effect of root-knot nematode–fungi combinations on cotton seedling disease. *Phytopathology* 60, 448–51.

CHENGZHU, M. AND PINSAN, C. (1986) The parasitic ability of southern root knot nematodes (*Meloidogyne incognita*) race 2 and 3 on cotton. *Acta Agriculturae Shanghai* 2, 81–8.

CHENGZHU, M., XUSHENG, X., DEYNAN, Z. AND MUONG, L. (1986) The species and genus distribution and population dynamics of plant parasitic nematodes in cotton fields of Shanghai. *Acta Agriculturae Shanghai* 2, 41–8.

CHENGZHU, M., ZHAOLIANG, L. AND GENGUI, N. (1987) Preliminary studies on the bionomics and control of *Hoplolaimus seinhorsti* on cotton. *Journal of Shanghai Agricultural College* 5, 124–8.

DASGUPTA, D.R. AND SESHADRI, A.R. (1971) Races of the reniform nematode, *Rotylenchulus reniformis* Linford & Oliveira, 1940. *Indian Journal of Nematology* 1, 21–4.

ESSER, R.P. (1976) Sting nematodes, devastating parasites of Florida crops. *Nematology Circular Florida Department of Agriculture and Consumer Services*, No. 18.

FASSULIOTIS, G., RAUN, G.J. AND SMITH, F.H. (1968) *Hoplolaimus columbus*, a nematode parasite associated with cotton and soybeans in South Carolina. *Plant Disease Reporter* 52, 571–2.

GOOD, J.M., MINTON, N.A. AND JAWORSKI, C.A. (1965) Relative susceptibility of selected cover crops and coastal Bermuda grass to plant nematodes. *Phytopathology* 55, 1026–30.

GRAHAM, T.W. (1951) Nematode root rot of tobacco and other plants. *South Carolina Agricultural Experiment Station Bulletin* 390.

GRAHAM, T.W. AND HOLDEMAN, Q.L. (1953) The sting nematode *Belonolaimus gracilis* Steiner: a parasite on cotton and other crops in South Carolina. *Phytopathology* 43, 434–9.

GUAR, H.S. AND MISHRA, S.D. (1981) Pathogenicity of the lance nematode, *Hoplolaimus indicus* to cotton, *Gossypium hirsutum*. *Indian Journal of Nematology* 11, 87–8.

HEALD, C.M. AND THAMES, W.H. (1982) The reniform nematode, *Rotylenchulus reniformis*. *Nematology in the Southern Region of the United States*. *USDA–CSRS Southern Cooperative Series, Bulletin* 276, 139–43.

HEALD, C.M., BIRCHFIELD, W., BLACKMON, C.W., HUSSEY, R.S., ORR, C.C., SHEP-HERD, R.L., VEECH, J. AND SMITH, F.H. (1981) Nematodes. In: Watkins, G.M. (ed.), *Compendium of Cotton Diseases*. American Phytopathological Society, St Paul, Minnesota, pp. 50–6.

HEMENG, B.M.S. (1978) The international *Meloidogyne* project. *Proceedings of the Second Research Planning Conference on Root-knot Nematodes*, Meloidogyne *spp.*, 20–24 February, Abidjan, Ivory Coast, pp. 21–3.

HOLDEMAN, Q.L. AND GRAHAM, T.W. (1953) The effect of different plant species on the population trends of the sting nematode. *Plant Disease Reporter* 37, 497–500.

HOLDEMAN, Q.L. AND GRAHAM, T.W. (1954) Effect of the sting nematode on expression of Fusarium wilt in cotton. *Phytopathology* 44, 683–5.

HOOPER, D.J. (1986) Extraction of free-living stages from soil. Extraction of nematodes from plant material. In: Southey, F.F. (ed.), *Laboratory Methods for Work with Plant and Soil Nematodes*. Ministry of Agriculture, Fisheries and Food, Her Majesty's Stationery Office, London, pp. 5–30, 51–8.

HOOPER, D.J. (1990) Extraction and processing of plant and soil nematodes. In: Luc, M., Sikora, R.A. and Bridge, J. (eds), *Plant Parasitic Nematodes in Subtropical and Tropical Agriculture*. CAB International, Wallingford, UK, pp. 45–68.

HUSSEY, R.S. (1977) Effects of subsoiling and nematicides on *Hoplolaimus columbus* populations and cotton yield. *Journal of Nematology* 9, 83–6.

HYER, A.H., JORGENSON, E.C., GARBER, R.H. AND SMITH, S. (1979) Resistance to root-knot nematode in control of root-knot nematode–Fusarium wilt disease complex in cotton. *Crop Science* 19, 898–901.

KHOURY, F.Y. AND ALCORN, S.M. (1973) Effect of *Meloidogyne incognita acrita* on the susceptibility of cotton plants to *Verticillium albo-atrum*. *Phytopathology* 63, 485–90.

KIRKPATRICK, T.L. (1988) Report of nematode management committee – 1987. *Proceedings of the Beltwide Cotton Production Research Conference*, Memphis, Tennessee, pp. 7–8.

KIRKPATRICK, T.L. (1989) Report of nematode management committee – 1988. *Proceedings of the Beltwide Cotton Production Research Conference*. Memphis, Tennessee, Book 1, pp. 7–10.

KIRKPATRICK, T.L. AND SASSER, J.N. (1984) Crop rotation and races of *Meloidogyne incognita* in cotton root-knot management. *Journal of Nematology* 16, 323–8.

KIRKPATRICK, T.L. AND SHEPHERD, R.L. (1989) Response of four root-knot nematode/Fusarium wilt resistant cotton breeding lines when grown in a field infested with both *Meloidogyne incognita* and *Fusarium oxysporum* f.sp. *vasinfectum*. *Proceedings of the Beltwide Cotton Production Research Conference*, 2–7 January, Book 1, Memphis, Tennessee, pp. 41.

KRAUS-SCHMIDT, H. AND LEWIS, S.A. (1979) Seasonal fluctuations of various nematode populations in cotton fields in South Carolina. *Plant Disease Reporter* 63, 859–63.

KRUSBERG, L.R. AND SASSER, J.N. (1956) Host–parasite relationships of the lance nematode in cotton roots. *Phytopathology* 46, 505–10.

LEWIS, S.A., SMITH, F.H. AND POWELL, W.M. (1976) Host–parasite relationships of *Hoplolaimus columbus* on cotton and soybean. *Journal of Nematology* 8, 141–5.

LORDELLO, A.I.L., LORDELLO, R.R.A., CIA, E. AND FUZATTO, M.G. (1984) Resultados parciais do levantamento de nematoides do genero *Meloidogyne* parasitos do algodoeiro em Sao Paulo. *Nematologia Brasileira* 8, 185–92.

LOUW, I.W. (1982) Nematode pests of cotton. In: Keetch, D.P. and Heyns, J. (eds) *Nematology in Southern Africa*, Republic of South Africa Department of Agriculture and Fisheries, Pretoria, pp. 73–9.

LUC, M. (1958) Les nématodes et le flétrissement des cotonniers dans le Sud-Ouest de Madagascar. *Coton et Fibres Tropicales* 13, 239–56.

Luc, M. (1968) Nematological problems in the former French African tropical territories and Madagascar. In: Smart, G.C. and Perry, V.G. (eds), *Tropical Nematology*, University of Florida Press, pp. 93–112.

Mamluk, O.F. and Faust, E.W. (1975) Distribution and identity of root-knot nematodes (*Meloidogyne* spp.) on cotton and sugar beet in the Syria Arab Republic and Republic of Lebanon. *Zeitschrift für Pflanzenkrankheiten und Pflanzenschutz* 11, 717–21.

Maqbool, M.A. and Saeed, M. (1981) Studies on root-knot nematodes in Pakistan. *Proceedings of the Third Research and Planning Conference on Root-knot Nematodes*, Meloidogyne *spp.*, 20–24 July, Jakarta, Indonesia, Review VI, pp. 115–19.

Martin, W.J. (1954) Parasitic races of *Meloidogyne incognita* and *M. incognita acrita*. *Plant Disease Reporter Suppl.* 227, 86.

Martin, W.J., Newsom, L.D. and Jones, J.E. (1956) Relationship of nematodes to the development of Fusarium wilt in cotton. *Phytopathology* 46, 285–9.

Mor, M. and Cohn, E. (1989) New nematode pathogens in Israel: *Meloidogyne* on wheat and *Hoplolaimus* on cotton. *Phytoparasitica* 17, 221.

Mueller, J.D. and Sullivan, M.J. (1988) Response of cotton to infection by *Hoplolaimus columbus*. *Annals of Applied Nematology* 2, 86–9.

Neal, D.C. (1954) The reniform nematode and its relationship to the incidence of Fusarium wilt in cotton at Baton Rouge, Louisiana. *Phytopathology* 44, 447–50.

O'Bannon, J.H. (1975) *Nematode Survey: Report to Institute of Agricultural Research Ethiopia*. FAO, Rome. 29pp.

Orton Williams, K.J. (1974) *Belonolaimus longicaudatus. CIH Descriptions of Plant-parasitic Nematodes*, Set 3, No. 40. Commonwealth Agricultural Bureaux, Farnham Royal, Bucks, UK.

Oteifa, B.A. (1970) The reniform nematode problem of Egyptian cotton production. *Journal of Parasitology* 56, 255.

Page, S.L.J. (1985) *Meloidogyne acronea. CIH Descriptions of Plant-parasitic Nematodes*. Set 8, No. 114. CAB International, Wallingford, UK.

Page, S.L.J., Mguni, G.M. and Sithole, S.Z. (1985) *Pests and Diseases of Crops in Communal Areas of Zimbabwe*. Overseas Development Administration Technical Report for the Plant Protection Research Institute, Harare, Zimbabwe. 203pp.

Palanisamy, S. and Balasubramanian, P. (1983) Assessment of avoidable yield loss in cotton (*Gossypium barbadense* L.) by fumigation with metham sodium. *Nematologia Mediterranea* 11, 201–2.

Peacock, F.C. (1956) The reniform nematode in the Gold Coast. *Nematologica* 1, 307–10.

Perry, D.A. (1962) *Fusarium* wilt of cotton in the Lake Province of Tanganyika. *Cotton Growing Review* 39, 14–21.

Perry, D.A. (1963) Interaction of root knot and Fusarium wilt of cotton. *Cotton Growing Review* 40, 41–7.

Pinochet, J. and Guzman, R. (1987) Nematodos asociados a cultivos agricolas en El Salvador: su importancia y manejo. *Turrialba* 37, 137–45.

Prasad, K.S. and Padeganur, G.M. (1980) Observations on the association of *Rotylenchulus reniformis* with Verticillium wilt of cotton. *Indian Journal of Nematology* 10, 91–2.

Rajaram, B. (1979) Biochemical association between infection of *Hoplolaimus seinhorsti* Luc, 1958 and development of *Fusarium oxysporum* f.sp. *vasinfectum* (Atk) Snyder & Hansen on cotton. *The Indian Zoologist* 3, 135–8.

SALEH, H.M. AND SIKORA, R.A. (1988) Effect of quintozen, benomyl and carbendazim on the interaction between the endomycorrhizal fungus *Glomus fasciculatum* and the root-knot nematode *Meloidogyne incognita* on cotton. *Nematologica* 34, 432–42.

SASSER, J.N. AND FRECKMAN, D.W. (1987) A world perspective on nematology. In: Veech, J.A. and Dickson, D.W. (eds), *Vistas on Nematology: a Commemoration of the Twenty-fifth Anniversary of the Society of Nematologists*. Society of Nematologists, Hyattsville, USA, pp. 7–14.

SASSER, J.N. AND KIRBY, M.F. (1979) Crop cultivars resistant to root-knot nematodes, *Meloidogyne* species. *Cooperative Publication of the Department of Plant Pathology, North Carolina State University and US Agency for International Development. International* Meloidogyne *Project.*

SHAFIE, M.F. AND KOURA, F.H. (1969) *Hoplolaimus aegypti* n.sp. (Hoplolaimidae: Tylenchida: Nematoda) from UAR. *Zoological Society of Egypt Bulletin* 22, 117–20.

SHAFSHAK, S.E., SHOKR, E., SALEM, F.M., EL-HOSARY, A.A. AND BARAKAT, A. (1985) Susceptibility of some field crops to the infection of certain nematode genera in Egypt. *Annals of Agricultural Science, Moshtohor* 23, 1003–11.

SHEPHERD, J.A. AND COOMBS, R.F. (1980) The interaction between four *Meloidogyne* species (Nematoda: Meloidogynidae) and seven cultivars of *Gossypium hirsutum*. *Zimbabwe Journal of Agricultural Research* 18, 123–4.

SHEPHERD, R.L. (1974a) Breeding root-knot resistant *Gossypium hirsutum* L. using a resistant wild *G. barbadense* L. *Crop Science* 14, 687–91.

SHEPHERD, R.L. (1974b) Transgressive segregation for root-knot nematode resistance in cotton. *Crop Science* 14, 872–5.

SHEPHERD, R.L. (1979) A quantitative technique for evaluating cotton for root-knot nematode resistance. *Phytopathology* 69, 427–30.

SHEPHERD, R.L. (1988) Effects of root-knot nematodes alone and in combination with fungi on cotton seedling disease. *Proceedings of the Beltwide Cotton Production Research Conference*, Memphis, Tennessee, p. 14.

STARR, J.L. AND PAGE, S.L.J. (1990) Nematode parasites of cotton and other tropical fibre crops. In: Luc, M., Sikora, R.A. and Bridge, J. (eds), *Plant Parasitic Nematodes in Subtropical and Tropical Agriculture*. CAB International, Wallingford, UK, pp. 539–56.

STARR, J.L., JEGER, M.J., MARTYN, R.D. AND SCHILLING, K. (1989) Effects of *Meloidogyne incognita* and *Fusarium oxysporum* f.sp. *vasinfectum* on plant mortality and yield of cotton. *Phytopathology* 79, 640–6.

THAKAR, N.A., PATEL, H.R. AND PATEL, C.C. (1986) Screening of cotton varieties/hybrids against root-knot nematodes, *Meloidogyne* spp. *Indian Journal of Nematology* 16, 265–6.

THAMES, W.H. AND HEALD, C.M. (1974) Chemical and cultural control of *Rotylenchulus reniformis* on cotton. *Plant Disease Reporter* 58, 337–41.

VAN DEN BERG, E. (1976) *Hoplolaimus seinhorsti. CIH Descriptions of Plant-parasitic Nematodes* Set 6, No. 76. Commonwealth Agricultural Bureaux, Farnham Royal, Bucks, UK.

WICKENS, G.M. AND LOGAN, C. (1960) Fusarium wilt and root knot of cotton in Uganda. *Cotton Growing Review* 37, 15–25.

WYK, R.J.-VAN, PRINSLOO, G.C., VILLIERS, D.A. DE AND MCCLURE, M.A. (1987) A preliminary survey of plant-parasitic nematode genera in the cotton-producing areas of South Africa. *Phytophylactica* 19, 259–60.

YASSIN, A.M. (1974) Role of *Pratylenchus sudanensis* in the syndrome of cotton wilt with reference to its vertical distribution. *Sudan Agricultural Journal* 9, 48–52.

YASSIN, A.M. AND MOHAMED, Z.E. (1980) Studies on the biology and chemical control of root-lesion nematode, *Pratylenchus sudanensis* Loof and Yassin 1970, from the Gezira. *Zeitschrift für Angewandte Zoologie* 67, 225–31.

Yik, C.-P. and Birchfield, W. (1984) Resistant germplasm in *Gossypium* species and related plants to *Rotylenchulus reniformis*. *Journal of Nematology* 16, 146–53.

YÜKSEL, H. (1974) Kök-Ur nemetodlarinin (*Meloidogyne* spp.) Türkiyedeki durumu ve populasyon problemleri. *Atatürk Universiti Zi Fak., Dergisi*, C. 4, No. 1, S., pp. 53–71.

YÜKSEL, H.S. (1982) Root-knot nematodes in Turkey. *Proceedings of the Third Research and Planning Conference on Root-knot Nematodes*, Meloidogyne *spp*. 13–17 September 1982. Region VII, pp. 103–11.

11 | Nutrient Deficiency Disorders

S.C. HODGES*

Introduction

Many of the common disorders of cotton in the field are related to nutrition of the cotton plant. The cotton plant is particularly sensitive to such disorders during the seedling stage, when plant root systems are small, and during the period following peak flowering. The indeterminate growth habit of cotton requires that the plant be supplied with sufficient nutrients for both vegetative growth and boll development for much of the season. This places a tremendous demand on the soil reserves, especially during the peak flowering period of the season when a high percentage of nutrient uptake occurs over a relatively short period. Problems may arise from nutrient interactions, nutrient imbalances and excessive nutrient levels as well as nutrient deficiencies. The goal of this chapter is to describe not only nutrient deficiency disorders but ways in which they may be prevented or corrected. This requires a knowledge of nutrient requirements and some means of assessing the soils' ability to supply nutrients. Cultural practices, methods of application and differences between cultivars can all affect plant nutrient utilization. Finally, plant nutrition can have profound effects on disease susceptibility. These topics will be reviewed in the context of field applications.

Nutrient Requirements

Cotton is adaptable to a wide range of soils provided they are not subject to waterlogging and rooting depth is not limited by salinity, acidity, stoniness, shallowness, low fertility or low moisture-holding capacity. Provided these

* Extension Agronomist, Soils and Fertilizers, University of Georgia, P.O. Box 1209, Tifton, Georgia 31793, USA.

conditions are met, the crop can be grown on soils ranging from deep quartzitic sands to rich heavy clays, from the deep alluvial soils of California and China to the takyr soils of Russia, and the cracking black clays of India, Egypt and Australia to the highly weathered ultisols of the south-eastern USA and Uganda and the Oxisols in Nigeria and the Brazilian Cerrado. With such a wide range of soil types, it should not be surprising to find differences in fertilizer requirements throughout the world. Unless a long history of soil amendment and fertility enrichment precedes cotton planting, observed nutritional problems in cotton are strongly related to organic matter, parent material and the extent of weathering and leaching of the soil.

Of the mineral nutrients now recognized as essential for plant growth and development, three (carbon, oxygen and hydrogen) are supplied by air and water. The major or macronutrient elements are nitrogen (N), phosphorus (P) and potassium (K); the secondary elements are calcium (Ca), magnesium (Mg) and sulphur (S); and the micronutrients are boron (B), manganese (Mn), zinc (Zn), copper (Cu), iron (Fe), molybdenum (Mo) and chlorine (Cl). Soils supply many of these nutrients in amounts sufficient to meet the needs of cotton.

The form of nutrients applied is relatively unimportant in cotton nutrition as long as adequate supplies are available and soil properties are not adversely affected. Cotton tolerates relatively high levels of most mineral elements, but does appear more sensitive to Al and Mn toxicity. Cotton is less efficient than most agronomic crops in removing P and K from the soil, and is more sensitive to low levels of K, B and S (Ergle and Eaton, 1951; Russell, 1973; Cope, 1981; Kerby and Adams, 1985).

Uptake of N, P and K

Studies on the accumulation of nutrients by plant dry matter as a function of time provide a logical starting-place for determining nutrient requirements of cotton. Such data provide an indication of the variation in nutrient concentrations with time, the amount of nutrients required to produce a crop, the nutrient uptake rates at various stages of growth and the nutrients removed with the crop. Such studies in the south-eastern USA date back to 1891 (McBryde, 1891; Anderson, 1894; McBryde and Beal, 1896). McBryde and Beal (1896) and Fraps (1919) used the nutrient requirement index, the amount of nutrient uptake required to produce 100 kg of lint, as a means of comparison between cultivars and locations. Such a comparison for several studies (Table 11.1) reveals considerable variation in the nutrient requirement index for N, P and K with yield, location, cultivar and time. Studies under irrigated conditions result in the lowest nutrient requirement indices (Bassett et al., 1970; Halevy, 1976; Halevy et al., 1987), while studies under rain-fed conditions typically have much larger indices. This usually results from a lower fruiting efficiency and higher ratio of vegetative matter to reproductive matter in rain-fed experiments. Uptake for studies typical of actual field values range

Table 11.1. Nutrient requirement index (kg 100 kg⁻¹ lint) for several studies reporting nutrient uptake and removal.

Location	Soil	Irrigated	Lint/yield (kg ha⁻¹)		N	P	K	Source
						kg 100 kg⁻¹ lint		
Upland	South-eastern USA	Udult	No	500	Uptake	18.7	3.3	11.5	Christidis and Harrison, 1955
					Removal	7.2	1.3	2.5	
Upland	Texas, USA	–	No	300	Uptake	25.0	3.0	14.0	Fraps, 1919
					Removal	6.6	1.0	2.5	
Upland	India	Vertisol	Yes	560	Uptake	13.5	4.7	8.0	Arakeri et al., 1959
					Removal	8.0	1.4	2.7	
Upland-Compact	India	–	Yes	1254	Uptake	10.4	3.5	10.8	Bhatt and Appukuttan, 1971
					Removal	5.6	2.1	5.0	
Upland-Bushy	India	–	Yes	1051	Uptake	20.7	6.9	21.6	Bhatt and Appukuttan, 1971
					Removal	9.2	3.2	7.0	
Acala	California, USA	Orthent	Yes	1400	Uptake	10.0	1.4	9.0	Bassett et al., 1970
					Removal	5.1	0.8	1.5	
Acala		Xeralf	Yes	1700	Uptake	13.5	2.6	10.2	Halevy, 1976
					Removal	6.1	1.2	2.7	
Acala		Xerert	Yes	1670	Uptake	6.6	1.9	7.2	Halevy et al., 1987
					Removal	4.0	1.0	2.1	
Acala		Xerert	Yes	2440	Uptake	10.9	1.9	8.5	Halevy et al., 1987
					Removal	6.1	1.2	2.4	
Acala		Xerert	Yes	2700	Uptake	11.2	1.4	8.5	Halevy et al., 1987
					Removal	6.2	0.9	2.4	
Upland	Mississippi, USA	–	No	336	Uptake	15.5	3.0	14.5	McHargue, 1926
					Removal	7.6	1.8	2.1	
Upland	USA	–	No	560	Uptake	15.2	2.6	8.3	Donald, 1964
					Removal	8.0	1.7	2.5	
Upland		–	No	560	Uptake	17.9	3.9	11.9	Berger, 1969
					Removal	7.4	1.3	2.6	
Upland	Georgia, USA	Udult	No	740	Uptake	20.3	4.1	15.1	Olson and Bledsoe, 1942
					Removal	11.7	3.2	6.8	
Upland & Acala	Alabama, USA	Udult	No	643	Uptake	19.9	2.5	15.3	Mullins and Burmester, 1990
					Removal	9.0	1.4	3.1	

Table 11.2. Uptake and removal of secondary nutrients by cotton.

	Ca	Mg	S	Source
		kg 100 kg⁻¹ lint		
Uptake	6.2–13.1	2.5–7.5	–	Bassett et al., 1970
Removal	–	1.2	–	
Uptake	14.8	2.5	6.8	McHargue, 1926
Removal	0.3	0.7	0.4	
Uptake	6.0	2.5	–	Donald, 1964
Removal	0.4	0.8	0.4	
Uptake	17.5	5.5	–	Olson and Bledsoe, 1942
Removal	–	–	–	
Uptake	–	–	2.1–3.1	Stanford and Jordan, 1966
Removal	–	–	0.5	
Uptake	10.2	2.9	–	Christidis and Harrison, 1955
Removal	0.5	0.8	–	

from 10 to 38.5, 1.4 to 4.7 and 8 to 24 kg 100 kg⁻¹ lint for N, P and K, respectively. Across the range of sites studied, uptake of N, P and K averaged 18.4, 2.8 and 13.5 kg ha⁻¹ 100 kg⁻¹ of lint produced. It is apparent in reviewing these figures that excess uptake of N, P and K commonly occurs. In a K rate study, Bennett et al. (1965) showed that uptake in excess of 13 kg 100 kg⁻¹ lint resulted in luxury consumption of K. Nutrient requirement indices of 15 for N, 2.6 for P and 13 for K appear indicative of efficient cotton production systems.

UPTAKE OF OTHER NUTRIENTS

Relatively few studies report on uptake and removal of secondary and micro-nutrients, so ranges and variability as a function of location and production system are less well known (Table 11.2). Since the majority of the Ca taken up by the plant remains in the leaves, total uptake varies with the amount of dry matter produced and the ratio of vegetative to reproductive growth. Uptake ranges from 6.2 to 17.5 kg Ca 100 kg⁻¹ (Christidis and Harrison, 1955; Bassett et al. 1970). Compounds containing Mg and S are involved in oil production, resulting in higher concentrations of these nutrients in the seed than of Ca. Uptake of Mg is also quite variable, ranging from 2 to 6.6 kg 100 kg⁻¹ lint, and often unrelated to yields (Christidis and Harrison, 1955; Bassett et al., 1970). Values of S uptake are reported from 7 to 33 kg S ha⁻¹ (Kamprath et al., 1957; Jordan and Ensminger, 1958; Hearn, 1981). In terms of nutrient requirement uptake index, about 1.5–2.4 kg S 100 kg⁻¹ lint are required by the plant.

There are few reports involving uptake of micronutrients which also include yields (Table 11.3). McHargue (1926) found copper concentrated in seed kernels and leaves, iron in leaves, manganese in hulls and leaves and zinc in the kernels. The values reported by Donald (1964) apparently do not include micronutrients in the vegetative portion of the plant. In a survey of 35 fields in Australia, Constable *et al.* (1988) found that Mn was taken up in the highest amounts, followed by Fe, Zn, B, Cu and Mo. They concluded that Fe uptake was complete by flowering, while Zn uptake must continue throughout the season. Anderson and Harrison (1970) reported tissue concentrations in Georgia consistent with these findings, but did not measure total uptake.

REMOVAL IN LINT AND SEED

The values reported above would suggest that cotton removes relatively high amounts of nutrients from the soil, and this would be true if all plant parts were removed from the soil, as is common practice in some parts of the world. A majority of cotton-producing areas, however, remove only the lint and seed.

Expressed in terms of kg of nutrient removed per 100 kg lint, reported values range from 4 to 8 for N, 0.8 to 1.76 for P and 1.5 to 3.1 for K. The best data on removal of secondary nutrients indicate 0.14 kg Ca, 0.4 kg Mg (Bassett *et al.*, 1970) and 0.55 kg S (Kamprath *et al.*, 1957) per 100 kg lint. Removal of micronutrients in lint and seed seldom exceeds 30 g 100 kg^{-1} lint (Table 11.3).

Table 11.3. Uptake and removal of micronutrients by cotton.

Nutrient		McHargue	Donald	Constable	Alimov (g ha^{-1})
		— g 100 kg^{-1} lint —			
Manganese	Uptake	30.0	–	450.0 (g ha^{-1})	85–162
	Removal	13.8	24.0	29.3 (g 100 kg^{-1} lint)	–
Zinc	Uptake	55.6	–	60.0 (g ha^{-1})	25–38
	Removal	32.4	64.4	5.8 (g 100 kg^{-1} lint	–
Iron	Uptake	244.4	–	600 (g ha^{-1})	–
	Removal	56.0	–	29.3 (g 100 kg^{-1} lint)	–
Copper	Uptake	12.4	12.0	20 (g ha^{-1})	19–26
	Removal	6.2	12.0	1.3 (g 100 kg^{-1} lint)	–
Boron	Uptake	–	–	200 (g ha^{-1})	66–107
	Removal	–	–	9.3 (g 100 kg^{-1} lint)	–
Molybdenum	Uptake	–	–	–	2–4
	Removal	–	–	–	–
Cobalt	Uptake	–	–	–	2–4
	Removal	–	–	–	–

McHargue = McHargue, 1926; Donald = Donald, 1964; Constable = Constable *et al.*, 1988; Alimov = Alimov and Ibragimou, 1976.

Table 11.4. Applied N, P and K for different cotton-growing regions.

Soils	Conditions	N (kg ha⁻¹)	P (kg ha⁻¹)	K (kg ha⁻¹)	Source
Udults – sandy	Rain-fed – CP	40–80	0–35	0–81	[a]
Udults – loams and clays	Rain-fed – Piedmont	40–60	0–50	0–81	[a]
Inceptisol, Alfisol, Mollisol	Rain-fed – Delta	50–100	0–27	0–80	[a]
Xeralfs, Xerolls	Dryland – High plains	20–60	0–10	0–30	[a]
Alfisols, Mollisols, Inceptisols	Irrigated – Delta and plains	50–100	13–35	0–50	[a]
Argids, Xerolls	Irrigated – SW USA, low elev.	50–300	0–35	0–30	[a]
Argids, Xeralfs, Xerolls	Irrigated – SW USA, high elev.	40–120	0–35	0–200 +	[a]
Lower Texas	Rain-fed + supplemental	60–120	0–30	0–50	[a]
Inceptisols	Northern China – irrigated	97–142	26–38	32–27	China Cooperative Research Group, 1989
Inceptisols	Central China – irrigated	112–164	30–44	37–55	As above
Inceptisols	Southern China – irrigated	120–172	32–46	48–55	As above
Argids, Xerolls, Inceptisols	USSR – irrigated	150	100	100	Mamedov and Ismailov, 1989
Argids, Xerolls, Inceptisols	USSR – irrigated	250	53	104	Bat'kaev et al., 1989
Argids, Xerolls, Inceptisols	USSR– irrigated	250	78	62	Meredov, 1987
Argids, Xerolls, Inceptisols	USSR – irrigated	150–250	67–111	0–58	Porter et al., 1974
Vertisols	India – irrigated	60–140	20–50	0	Bhole and Varade, 1988
Vertisols	India – irrigated	30–60	30	15	Innes, 1971
Vertisols	Sudan Gezira – irrigated	80–120	0	0	Burhan, 1971
Inceptisols, Ultisols, Oxisol	Latin America	46	25	31	FAO, 1987, p.43

[a] Compiled from Jones and Bardsley, 1968, Kamprath and Welch, 1968; Tucker and Tucker, 1968.

Fertilizer requirements

Average world use of N, P and K for cotton is in the order of 48, 8 and 28 kg ha^{-1}, respectively (FAO, 1987), but amounts vary greatly for different cotton-growing areas (Table 11.4). Fertilizer requirements are a function of both the yield potential for a given site and the ability of the farmer to manage inputs to meet that potential. Cotton yields vary greatly because of climatic limitations, unproductive soils or economic and technical inability to supply nutrients, water and pest control measures in a timely and effective manner. Fertilizer requirements must be determined on the basis of an attainable, economically profitable and environmentally acceptable yield goal. Where factors other than nutrient availability limit yields, this should be recognized and fertilizer rates adjusted accordingly.

While most soils will show considerable response to N additions, fewer show response to P and fewer still to K additions (Hearn, 1981). Responses to other nutrients are very dependent upon soil properties. This is a function of degree of weathering and the minerals remaining in the soil, as well as past fertilization history. Soil testing, as described later, is widely used as a basis for making fertilizer recommendations. The following overview will describe the range in fertilizer requirements for different soil and environmental conditions in major cotton-growing areas.

Nitrogen

Nitrogen rates are determined in response to uncontrollable yield-limiting factors over a period of years. While N deficiency limits yield, excessive N can extend the season, cause rank growth, increase disease and insect pressure and complicate defoliation (Hearn, 1981; Miley, 1982). Even though more bolls may be produced, fewer may be harvested. Because of this, N rates tend to reflect the yield potential of a particular cotton-growing area. For irrigated desert soils with a long growing season, yields of 7.5–10 bales ha^{-1} are not unreasonable with excellent management. In these areas, growers typically use 12–15 kg N for each 100 kg of lint they expect to produce, resulting in applications of 250 kg N ha^{-1} or more. Lowest N recommendations are typically found in areas where water stress, disease or pests place unpredictable limits on yield potential (Hearn, 1981; Kerby and Adams, 1985). Fertilizer trials in these regions seldom show yield increases with heavy N applications. Responses to N applications may reach a maximum at relatively low levels: 40 kg ha^{-1} in Tamil Nadu (Perichiappan *et al.*, 1987), 44–88 kg ha^{-1} in Nigeria (Smithson and Heathcote, 1977), 50 kg ha^{-1} in Pakistan (Malik *et al.*, 1987), 60 kg ha^{-1} in Jiangsu province of China (Yan and Zhu, 1986), 60–70 kg ha^{-1} in Georgia (Hodges *et al.*, 1989) and 60–100 kg ha^{-1} in the Mississippi Delta (Miley, 1982). Where supplementary irrigation is available, these values may increase by 30–60 kg N ha^{-1}. Similar increases may be justified with improved pest management or improved cultivars (Burhan, 1973; Burhan and Jackson, 1973).

Residual N following land clearing, bare fallow or legumes such as clover or alfalfa may provide all or a part of the needs of the crop. In the southeastern USA, recommended N rates are reduced by 25–30 kg ha^{-1} when cotton follows a legume crop such as peanuts or soyabeans. In tropical areas, 40 kg N ha^{-1} may be available following fallow or a leguminous crop (Hearn, 1981).

Phosphorus

Highly weathered, acidic red soils (Ultisols and Oxisols) may have a high P-fixing capacity. Fixation capacity increases with increasing clay and Fe oxide content. Deficiencies can be so severe that banded applications are inferior to broadcast for the first crop, simply because roots grow only in the soil volume where P is applied (Sanchez, 1977). Using a one-time broadcast application at a high P rate (320–600 kg P ha^{-1}) in conjunction with liming to neutralize exchangeable Al, Brazilian growers have developed large areas of the Cerrado region for agriculture. Banded applications of 20–35 kg P ha^{-1} in subsequent years are adequate for good production. Typical application rates in soils with moderate P availability range from 20 to 70 kg P ha^{-1} depending on soil pH and target yield (Hearn, 1981).

Phosphorus also reacts strongly with Ca and carbonates in calcareous soils to form relatively insoluble compounds. Availability to plants is greatly decreased above pH 7.5. Band applications of P are recommended to increase availability. Rather large amounts of P are applied in the cotton-growing areas of Russia and China (Table 11.4). This may be related to high pH or the emphasis placed on earliness at these far-northern latitudes.

Many alluvial soils contain adequate P to meet the needs of cotton. Burhan and Mansi (1970) reported erratic responses to P and no response to K addition for *G. barbadense* in the Sudan Gezira. In southern Texas, a maintenance application of 30 kg P ha^{-1} may be made once every five years. Where seasons are short, rates of 10–30 kg ha^{-1} have been applied to promote earliness of the crop (Jones and Bardsley, 1968).

Potassium

Many cotton-growing areas have high soil reserves of soil K, resulting in relatively low fertilizer requirements for an otherwise important plant nutrient (Kamprath and Welch, 1968). Where deficiencies are a problem, maximum yields were obtained with K applications of 22–75 kg ha^{-1} in Ultisols and 28 to 56 kg ha^{-1} in alluvial Entisols and Inceptisols (Kerby and Adams, 1985). In Nigeria, Lombin and Mustafa (1981) found 25 kg K ha^{-1} resulted in maximum yields for Oxisols, but two Inceptisols required no K. Application rates as high as 480 kg K ha^{-1} have been required in K-fixing soils of

California (Kerby and Adams, 1985; Cassman *et al.*, 1990). Irrigation water may supply more K than is taken off in the lint and seed in some areas.

Other nutrients

Responses to B and S are common on acidic, highly weathered soils low in organic matter (Hinkle and Brown, 1968; Hearn, 1981) Overliming of these soils can induce or aggravate existing deficiencies of B, Zn, Mn and perhaps Fe. Rates of 0.5–1 kg B ha^{-1} are usually recommended on an annual basis, although Honisch (1975) reports that 2–4 kg ha^{-1} were required to overcome deficiencies in fine-textured soils of Zambia. The need for S depends on local atmospheric inputs. Where such inputs are less than crop requirements, S applications of 10–20 kg S ha^{-1} will prevent deficiencies. Rates as high as 64 kg ha^{-1} applied as gypsum have lasted for several years (Hearn, 1981). Liming of acid soils provides sufficient Mg if dolomitic lime is used; otherwise applications of a water-soluble form of Mg (usually potassium–magnesium sulphate) should be applied at a minimum rate of 25 kg ha^{-1}.

Deficiencies of Zn, Mn and Fe are most common in fine-textured calcareous and alkaline soils. Soils with adequate organic matter will usually provide adequate levels of these nutrients unless pH exceeds 8.0. Above pH 7.0, these nutrients react quickly to form relatively insoluble compounds in the soil. Foliar applications at 0.2–1 kg ha^{-1} are usually most effective, although repeated treatments may be required. Zinc may be soil-applied at a rate of 3 to 20 kg ha^{-1} if soil pH is less than 7.5. The lower rates should be used on coarser-textured soils.

Fertilizer recommendations are best based on fertilizer trials using soils and adapted cultivars of a particular region. Application rates should be adjusted for variations in soil, climate, yield potential and management considerations. Continuous applications of P and K in excess of crop removal and other losses will build soil nutrient reserves and tend to increase nutrient availability over time. When this occurs, continued additions will have no effect on yields. Soil analyses can be of great financial benefit in this situation.

Determining Fertilizer Requirement

Fertilizer requirements for a given location usually evolve over a period of time. In newly developed areas, fertilizer requirements may initially be quite low. With continued cropping and depletion of organic matter, deficiencies begin to appear. Preliminary fertilizer trials may be used to determine elements that are most deficient. Eventually, more elaborate factorial studies are carried out to establish response curves and nutrient interactions. The primary objective of such testing is establishment of lime and fertilizer rates which result in the greatest profit. But fertilizer requirements must be based on the management level and cultural practices employed within the area to which they are applied. Requirements for highly mechanized, irrigated cotton production on a fertile Aridosol will be very different from those for hand-tended, rain-fed cotton produced on a nutrient-depleted Ultisol with heavy insect and disease pressure. In areas where it is practised, stalk removal greatly increases nutrient removal and should be figured into the fertilizer requirement. Ideally, fertilizer trials could be used to determine nutrient requirements. But fertilizer trials reflect results of only one season, and will vary from place to place in the field and from year to year with climate and with changing soil nutrient levels. Variability will be even greater over a district or region. Before firm recommendations can be made, a large number of trials are necessary.

Where sufficient information is available to combine soils into similar management groups, soil and tissue testing have proved valuable as a means of assessing fertilizer requirements. Intensive local soil research is still required to 'calibrate' or correlate the measured nutrient levels with fertilizer responses in order to provide a firm recommendation. Even in the absence of local research, soil and plant analyses can identify major deficiencies and indicate nutrients which should be used in trials.

Since many factors contribute to the yield potential and thus the economic response, fertilizer requirements depend not only on a measure of available nutrients, but on such factors as soil properties, climate, irrigation, weed control, crop rotations, interactions of fertilization with disease and insect infestations, and potential for environmental degradation.

Soil analysis

Soil testing has come to mean the rapid chemical analysis of a representative soil sample to assess the available nutrient status of a field or sampling unit. The goal of a soil-testing programme is to transfer the results of detailed fertilizer trials, greenhouse experiments, plant analysis and laboratory studies across a range of soils to a particular field. The final results should provide the producer with a measure of the nutrient status of the soil for cotton production and recommend appropriate fertilizer use. Thus, soil testing is composed of four components: sample collection, chemical analysis of the sample, interpretation of the chemical results and recommendation of lime and fertilizer. Individual components are discussed more fully in Walsh and Beaton (1973) and Whitney *et al.* (1985).

Sampling

Soil sample collection is often the weakest link in the process of soil analysis. On top of inherent soil variability, fertilizer and tillage practices complicate the collection of a representative sample. With proper attention to these sources of variability, samples can be extremely useful in determining lime and fertilizer requirements.

A farm or field should be divided into sampling areas of uniform soils and uniform past management. Sample areas should be 2–4 ha where the variability is great and no more than 8 ha for the most uniform areas. Practically speaking, the sampling area should be no smaller than the smallest area the farmer can expect to fertilize as a unit (Cline, 1945).

Although systematic sampling schemes may be used, a single composite sample composed of 10 to 30 random subsamples is typically used for soil analysis. The actual number of subsamples required depends on the soil variability, sampling depth and past management. Statistical methods are available to determine the optimum number of subsamples for various sampling schemes (Cline, 1944; Peterson and Calvin, 1965). Thomas and Hanway (1968) suggested the use of two or more composite samples to evaluate the adequacy of testing within a sample area. If results show unacceptable variation, the sample area should be subdivided or more subsamples should be taken.

Sampling depth should be determined by the depth of tillage. This is typically 15–30 cm. Where reduced or no-till methods are used, the upper 7.5 cm may better represent the soil pH and P status (Whitney *et al.*, 1985), but the mobility of N requires sampling to depths of 30–100 cm where this test is used. This test is useful only where soil nitrates are unlikely to be leached from the rooting zone of cotton during the season. Samples should be collected as near to planting as possible.

A uniform cross-section of the soil should be sampled at each location. Where micronutrients are to be analysed, all tools coming in contact with the sample should be constructed of stainless steel, plastic or wood. Subsamples should be thoroughly mixed in a clean (plastic) bucket. Samples to be analysed for nitrate-N should be immediately dried at a temperature of less than 40 °C. For other nutrients, samples may be submitted as mixed in the field.

Chemical analysis

Primary test elements include pH, P, K, Ca and Mg. Where deficiencies are likely, tests for B, Zn and Mn are often available. In alkaline soil areas, salinity measurements are common. There are many different extractants in use throughout the world. In acid, low-CEC (cation exchange capacity) soils, acidic extractants such as dilute double acid ($0.05 \text{ M HCl} + 0.0125 \text{ M H}_2\text{SO}_4$) and similar extractants are commonly used as a universal extractant for P, K, Ca, Mg and micronutrients. As CEC and soil pH increase, double acid underestimates exchangeable forms and may overestimate availability of acid-soluble minerals such as phosphates and carbonates. In calcareous and alkaline areas, $1 \text{ M NH}_4\text{OAc}$ (K, Ca, Mg) and 0.05 M NaHCO_3 (P) are more commonly used. Extraction with 2 M KCl or distilled water is most often used for nitrate-N determinations.

Even though crop response to lime is related more to yield-limiting factors, such as toxic levels of Al or Mn or perhaps Ca deficiency, than to pH,

Table 11.5. Soil fertility status based on soil tests

Element	Method	Low[a] (ppm)	High[a] (ppm)	Source
N	NO$_3$ 0.3 m depth	<10	>20	Hearn, 1981
	NO$_3$ incubation	<30	>40	Hearn, 1981
P	0.05 M HCl + 0.0125 M H$_2$SO$_4$			
	Sands	<15	>30	Plank *et al.*, 1989
	Loams and clays	<10	>20	
	0.05 M NaHCO$_3$	<5	>9	Reisenauer *et al.*, 1978
K	0.05 M HCl + 0.0125 M H$_2$SO$_4$			
	Sands	<35	>85	Plank *et al.*, 1989
	Loams and clays	<60	>125	
	Exchangeable			
	Loams	<40	>80	Reisenauer *et al.*, 1978
	Clays	<60	>100	
Mg	0.05 M HCl + 0.0125 M H$_2$SO$_4$			
	Sands	<15	>30	Plank *et al.*, 1989
	Loams and clays	<30	>60	
S	Sulphate ions	<15		Hearn, 1981
B	Hot water	<0.15		Hearn, 1981
Zn	0.05 M HCl + 0.0125 M H$_2$SO$_4$	<1.0	>4.0	Plank *et al.*, 1989
	Dithizane	<0.5		Hearn, 1981
Zn, Fe, Mn, Cu	DTPA	<0.4	>0.7	Reisenauer *et al.*, 1978

[a] Low status indicates that response to added nutrient is likely. High status indicates that little response is expected.

the strong correlation between pH and these factors allow its use as the primary criterion to determine the need for lime. Actual lime requirement is most accurately determined by a Ca(OH)$_2$ titration procedure, but this method is not suited to large numbers of samples. Using calibrated buffer solutions, many quick titrations of soil acidity can be made each day (Adams, 1984). Many suitable buffered solutions are available, each of which requires correlation with actual lime response in the field. Alternatively, exchangeable Al has been effectively used in highly weathered tropical soils of south America (Sanchez, 1977).

Although chemical analysis is typically the most accurate component of a soil analysis programme, small errors can destroy the credibility of the whole programme. Good quality control cannot be overstressed. Detailed analytical methods are described in Walsh and Beaton (1973), Kamprath and Watson (1980) and Page *et al.* (1982).

Interpretation and recommendations

While the actual extractants and methodology should allow accurate determination of a significant portion of the available nutrients, the extracted nutrients must be correlated with fertilizer trials on responsive soils to be meaningful. A good extractant will show a definable mathematical relationship with decreased crop response as the available nutrient extracted increases.

Where possible, recommendations should also consider target yield for a particular site in determining final fertilizer rates.

Where climatic conditions are reasonably constant and cotton is irrigated, nitrate-N tests can be used to measure the N-supplying power of the soil. While such tests are not particularly useful in determining the actual amount of fertilizer required, they have been related to the yield of the unfertilized soil (Gardner and Tucker, 1967). Response levels are shown in Table 11.5.

Extractants for P and K are adaptable over a wide range of soils when calibration data are available. Recommended rates of P vary from 0 at high soil test levels to 44 kg ha^{-1} at low values (Hearn, 1981; Plank *et al.*, 1989). An additional 50% P may be required if soil pH is excessively acid or alkaline. Potassium rates are much more dependent upon the cation exchange capacity (CEC) of the soil (Kerby and Adams, 1985). Recommended K rates range from 0 to 85 kg ha^{-1} where fixation is of little concern. In soils of the San Joaquin Valley of California, very high rates of K may be needed to overcome K deficiencies, even though current testing methods indicate that adequate amounts are available. Cassman *et al.* (1990) found that readily available K (water extraction) was better correlated with plant response than extractants which measure the soil K reserves.

Depending on the degree of deficiency and the exchange capacity of the soil, Mg rates range from 20 to 67 kg ha^{-1} (Hearn, 1981). Where lime is recommended, dolomitic lime should be used to supply Mg. Alternatively, rates from 4 to 63 kg S ha^{-1} have been used to correct deficiencies; rates in excess of 10–20 kg S ha^{-1} may have considerable residual effect (Hearn, 1981).

Where soil Zn levels are low, soil-applied rates of 3–20 kg ha^{-1} have been used if pH is not too high. Lower rates should be used on acid, sandy soils to avoid potential problems with zinc toxicity. Boron is required in very small amounts. Rates for soil application range from 0.5 to 4 kg ha^{-1} with the highest rates on finer-textured soils. Both Zn and B can be applied as foliar sprays at a rate of 0.25 kg ha^{-1} in 200 l of water.

FOLIAR ANALYSIS

Plant analysis can be of great value to the farmer, the extension agent or the researcher. It measures the intensity of nutrient supply at a given point in time. It lets the grower know if the soil is keeping up with plant demand. Plant analysis can be used to confirm the diagnosis of a visible symptom, or to identify problems which have no readily distinguishable symptoms other than a lack of vigour. In many cases, it is necessary to confirm that applied nutrients are actually entering the plant. If not, lack of response may be explained by nutrient–soil interactions, root damage or placement problems. Nutrient imbalances or multi-element deficiencies are more readily detected using plant analysis than through visual symptoms or fertilizer trials. There are problems with plant analysis that must be considered in its use. A correct interpretation depends on correctly sampling the appropriate plant part at the proper time and then maintaining the integrity of the sample by careful handling. In some cases, a soil sample provides essential information which is difficult or impossible to ascertain through plant analysis. Even when problems are correctly diagnosed through plant analysis, very often nothing can be done to correct the problem for the current crop. Finally, plant analysis results must be integrated with other information for correct interpretation and application. Drought and waterlogging (Hocking et al., 1987) profoundly affect nutrient uptake by cotton in otherwise well-supplied soils. Similar effects are noted for cloudy weather, soil compaction and inappropriate soil pH (Sabbe and Mackenzie, 1973). The concentration of nutrients in the cotton plant changes dramatically with time, with stage of growth and with boll load. Varietal differences and nutrient interactions within the plant can further obscure the interpretation of plant analysis data. In spite of these problems, plant analysis has proved to be a very useful tool not only in research, but in the diagnosis of field problems and preventing nutrient deficiencies.

Whole-plant, leaf or petiole samples may be used to assess the nutrient status of cotton. Whole-plant and leaf samples more accurately reflect levels of stored nutrients within the plant and are less affected by short-term climatic fluctuations than petioles (Sabbe and Mackenzie, 1973). Petiole analysis is better correlated with current uptake rates of nitrate-N, P, K, Ca and Mg in the soil (Joham, 1951). Analysis of whole plants and leaf blades is most commonly used in humid regions, where day-to-day climatic variation can limit the usefulness of a single petiole analysis. The prevalence of shorter seasons and use of determinate cultivars further limit the time available to obtain responses to fertilizer inputs, even if a deficiency is detected. The longer growing seasons and more predictable environment of irrigated, semi-arid regions allow adequate time for detection and correction of deficiencies, usually with little or no yield loss (Tucker and Tucker, 1968).

Whole-plant and leaf analysis

Whole-plant samples should be sampled near the appearance of the first square (budding). Leaf samples are taken from the uppermost fully expanded leaf, usually three or four nodes from the terminal. Youngest mature leaves are better indicators of plant-immobile nutrients, while older leaves and petioles are best for assessing K status (Kerby and Adams, 1985). A random sample of 30 leaves should be collected from within relatively uniform areas of the field. For problem-solving, soil and tissue samples should be taken from good and bad areas of the field.

Nutrient concentrations within the plant are affected by the stage of growth. Concentrations of N, P and K in the leaf tend to decrease with age, while Ca and Mg tend to accumulate. Thus, critical nutrient levels have been established for a number of locations and at various times of the season (Table 11.6). In spite of differences in nutrient uptake and sensitivity to nutrient stress, these levels have generally proved acceptable for identifying deficiencies in different cotton-growing areas (Sabbe and Mackenzie, 1973). Interpretation of plant analysis results is not always straightforward. For example, S deficiency results in accumulation of N and P in the leaf while having little effect on K (Ergle and Eaton, 1951; Joham, 1951; Ergle, 1953; Braud, 1974). Some deficiencies can be corrected if detected early. Corrective treatments for soil or foliar applications may be recommended for deficiencies of N, Mg, B and Zn if detected prior to flowering (Plank, 1989). Additions of

Table 11.6. Interpretations for nutrient concentrations in whole plant and leaf analysis of cotton.

Nutrient	Plant part[a]	Growth stage[b]	Critical	Adequate	Toxic
N (%)	YMLB	≤1st bloom		3.75–4.5	
		Early bloom		3.0–4.3	
P (%)	YMLB	≤1st bloom		0.3–0.5	
		Early bloom		0.3–0.65	
		Early fruiting	0.31		
		Late fruiting	0.33		
		Late maturity	0.24		
K (%)	WS	90 DAS	2.75–3.25		
		120 DAS	2.0–2.75		
	YMLB	45 DAS		>3.2	
		≤1st bloom		2.0–3.0	
		Early fruiting		0.9–1.96	
	Old LB + petiole	120 DAS	2.5		
S (%)	YMLB	≤1st bloom		0.25–0.80	
		Mid-season		0.59–0.99	
Ca (%)	YMLB	≤1st bloom		2.0–3.0	
		Early bloom		1.9–3.5	
Mg (%)	YMLB	≤1st bloom		0.5–0.9	
		Early bloom		0.3–0.75	
Cu (ppm)	WS	35 DAS		13	
	YMLB	≤1st bloom		5–25	
Zn (ppm)	WS	35 DAS			
	YMLB	≤1st bloom		20–60	
		37–43 DAS	11–14	17–48	200
Mn (ppm)	WS	35 DAS			494
	YMLB	Early bloom		25–350	
Fe (ppm)	YMLB	≤1st bloom		50–250	
	YMLB	Early bloom		30–300	
B (ppm)	YMLB	≤1st bloom	15	20–60	1000
Mo (ppm)	YMLB	Maturity	1.5–1.9	3–9	
Al (ppm)	WS	35–42 DAS			169–200
	YMLB	≤1st bloom		<200	

[a] YMLB = youngest mature leaf blade; WS = whole shoot; [b] DAS = days after sowing.
Compiled from Reuter, 1986; Plank, 1989.

other nutrients will seldom correct a deficiency, and deficiencies frequently appear late in the season. Plant analysis results can, however, be useful in planning fertilizer programmes for subsequent crops.

Based on extensive tests in seven tropical African countries, Braud (1974, 1987) developed relationships between key leaf nutrient levels to define critical values of N, P, K and S (Table 11.7). A sample of 30 leaves (without petioles) taken from the main-stem position subtended to the latest flowering branch is used for this analysis. The strength of this approach lies in the linkage of total nutrient content, nutrient interactions and stage of growth.

Petiole analysis

A petiole sample consists of 25–30 randomly sampled petioles taken from the uppermost fully expanded leaf. The collection area should be uniform in soils and management. Petioles should be removed from the leaf as soon after sampling as possible.

Petiole analysis indicates the current rate of nutrient uptake by the plant. This test has long been used in semi-arid conditions as a diagnostic tool for N, P and K (Tucker and Tucker, 1968; Bassett and MacKenzie, 1978). In these areas, plant and soil nutrient levels decline in a more predictable fashion, allowing the test to indicate adequacy at a given stage of growth (Table 11.8). Less than adequate concentrations do not indicate an immediate deficiency, but do indicate that insufficient accumulation is occurring. Under normal circumstances, petiole analysis indicates a need for N about two weeks before actual deficiency occurs (Ray and Tucker, 1962). As such, correction can be made before actual deficiencies occur. In the Sudan Gezira, Burhan and Babikir (1968) found that yield reductions were related to the number of days petiole N remained below the critical level.

The unpredictable nature of residual N and recurring problems related to excess N in rain-fed areas have prompted increased interest in petiole analysis as a means of reducing early-season N rates without risking a deficiency (Maples *et al.*, 1977; Miley, 1982; Lutrick, 1986). Monitoring begins two to three weeks after first squares appear, and continues weekly for eight to ten weeks. In order to deal with a more variable climate and heavier insect pressure, these programmes require weekly assessments of soil moisture and

Table 11.7. Equations for calculating critical levels for N, P, K and S according to Braud (1974).

N critical % = (91.65− 3.84/P− 1.32p)/14.55
P critical % = 5.87/(1.45p + 12.44F− 7.11)
K critical % = 71.03(3.11 + 2.99F + 2.25/S)
S critical % = 6.58/(0.6+1.13/P + 1.24F)

P = % phosphorus in leaf; S = % sulphur in leaf; p = dry weight of 30 leaves; F = number of branches flowering.

Table 11.8. Critical petiole concentrations for California, Arkansas and Georgia.

| Growth stage | Petiole nitrate concentration (ppm) | | | | | |
| | California[a] | | Arkansas[b] | | Georgia[c] | |
	Low	High	Low	High	Low	High
First bloom	12 000	18 000	8 500	28 000	4 500	12 000
Peak bloom (bloom + 30 d)	3 000	7 000	3 000	10 000	1 000	6 000
First open boll (bloom + 60 d)	1 500	3 500	1 000	2 500	500	4 000
Late season (bloom + 70 d)	1 000	2 000	800	1 000	500	4 000

[a] Bassett and MacKenzie, 1978; [b] Miley, 1982. Estimated from graph; [c] M. E. Walker (personal communication) and Lutrick *et al.*, 1986.

fruiting rate and a record of nutrient additions. With adequate soil moisture, a decrease in N is usually accompanied by increased P. When moisture is limiting, concentrations of both nutrients decrease. This knowledge, combined with the farmer's data is used to interpret the petiole analysis results and make appropriate recommendations. Both programmes will recommend appropriate amounts of soil-applied N (or K) through the third week of flowering. If nitrate-N levels drop from the adequate zone later in the season, foliar applications of urea are recommended at a rate of 8–10 kg N ha^{-1}. Positive responses to foliar urea have been reported when petiole analysis indicates a deficiency (Miley, 1982; M.E. Walker, unpublished data). Apparently the required petiole N levels vary with soils and location. M.E. Walker (unpublished data) and Lutrick *et al.* (1986) found that nitrate levels required to produce 1300 kg lint ha^{-1} in the Ultisols of Georgia and Florida were lower than in Arkansas (Table 11.8).

Visual Deficiency Symptoms

For generations, the appearance of plants has been used as a primary indicator of plant health. Man has observed the plant's appearance and attempted to interpret visual symptoms. Correct diagnosis requires a knowledge of what healthy plants look like, as well as how the various nutrients are translocated within the plant. It often helps to know that certain soil classes are predisposed to particular nutrient deficiencies. Proper interpretation often requires going beyond visual symptoms, which can be misleading. Very often visual symptoms are not evident until deficiencies are severe. This usually limits the chances of a successful remedy for the current crop. Environmental conditions, undetected diseases, insects or mechanical damage can also cause problems with plant nutrient utilization, even though adequate nutrients are available. Nevertheless, visual symptoms are usually the first indicator that a problem exists and, as such, deserve special attention. The following descriptions were compiled primarily from the works of Donald (1964) and Crawford (1982).

Nitrogen

Young cotton plants deficient in N produce small, pale yellow leaves (see Plate 8A). Deficiencies beginning early in the growing season may result in reduced stalk height, missing vegetative branches and fewer, shorter fruiting branches than on normal plants. If deficiency occurs later in the season, N is remobilized. This results in yellowing, premature reddening and early leaf shed, beginning with the older, lower leaves. The premature colouring and shedding are sometimes confused with early maturity. Deficient plants show extensive abortion of squares (flower-buds) and young bolls. Few mature bolls occur beyond the second lateral position of fruiting branches or at the top of the plant.

Nitrogen deficiencies may occur anywhere cotton is grown, although deficiencies are most likely to occur in soils low in organic matter.

Phosphorus

Symptoms of P deficiency are not strongly expressed. Growth rates may be affected if deficiency is severe, resulting in stunted, dark green foliage. As the condition persists, the dark green colour may move up the stalk. If other nutrients are well supplied, occasionally more distinctive symptoms are observed. These include stunting of leaves, stalks and fruiting branches, delayed fruiting, reduced flowering and boll set and delayed maturity of set bolls. In severe cases, the foliage becomes reddish purple in colour. Phosphorus translocates from older tissue to new, actively growing tissue quite readily, so symptoms should appear on older tissue first. Since the primary functions of P involve energy and growth regulation, deficiencies affect yield more than fibre or seed quality.

Potassium

Potassium deficiency occurs first on the older, lower leaves, since this element is readily translocated within the plant. Symptoms have typically begun as a yellowish white mottling in the interveinal area and at leaf margins. Leaves become a light yellowish green with yellow specks appearing between leaf veins, and the leaf margins may appear bronzed (see plate 8B). The specks become necrotic, causing the leaves to appear rusted or dotted with brown specks at leaf tip, margin and interveinal areas. As breakdown progresses, the margins and leaf tip shrivel, giving a ragged appearance. Eventually, the entire leaf is affected and drops prematurely. Premature shedding of leaves prevents boll development, resulting in small, immature bolls, many of which may fail to open. Symptoms normally proceed from the bottom to the top of the plant, eventually ending in death of the terminal in very severe cases.

Symptoms of the late-season K deficiency syndrome are similar to those described, but the symptoms first appear on the youngest mature leaves instead of the oldest (see Plate 8C and 8D). Symptoms have been observed on nearly 150 000 ha in California (Weir *et al.*, 1989; Cassman *et al.*, 1990), and are reported in Arkanas, Texas, Alabama, Israel and Africa. Associated soils usually have considerable K-fixing capacity, resulting from the presence of vermiculite in the clay fraction. Cassman *et al.* (1990) reviewed recent work concerning these new symptoms and attributed them to increased partitioning of nutrients to reproductive tissues in newer, faster-fruiting cultivars. In these cultivars, maturing bolls represent a much stronger nutrient sink than vegetative tissues. Coupling a reduced vegetative reservoir of nutrients and a heavier boll load within a shorter time period results in such a high demand that newly developing leaves are unable to compete with maturing bolls for K.

CALCIUM

There are no well-defined foliage symptoms for Ca deficiency. Seedlings grown in low-Ca solutions exhibit collapse of leaf petioles. Ca deficiency can result in poor resistance to damping-off organisms (Wiles, 1959). Poor fruiting of surviving plants, resulting in excessively large plants, has been reported from culture studies (Donald, 1964). Increased P, applied close to the seed as ammonium phosphate P, has resulted in seedling mortality through Ca deficiency (Hood and Ensminger, 1964). Low moisture levels increase sensitivity of seedlings to calcium deficiency (Presley and Leonard, 1948).

MAGNESIUM

Lower leaves become purplish-red with distinct green veins (see Plate 8E). Upper leaves develop symptoms as the older leaves senesce and shed prematurely. Normal ageing late in the season causes an orange-red colour on the older leaves, which is easily mistaken for magnesium deficiency.

SULPHUR

Symptoms include pale green to yellow leaves in the upper part of the plant (see Plate 8F), in contrast to N deficiency, which occurs first on the lower leaves. Plants may appear small and spindly with short, slender stalks. Unlike N deficiency symptoms, the veins of the leaf generally remain green. If the deficiency is severe, the plants will have fewer vegetative and fruiting branches. Fruiting is about normal for the plant size, but boll size is smaller than normal.

BORON

Symptoms in cotton vary widely with the stage of growth and severity of the deficiency. Growth and development of young tissue throughout the plant is strongly affected. The most consistent symptom is the irregular thickening of young petioles, accompanied by dark bands and necrotic pith. In Africa, the first visible sign was shortened flower corollas with ends of the petals folded inward. Corollas appeared water-soaked. The change from cream to pink colour was irregular and took longer than the normal 24 hours. This was followed by petiole thickening and necrosis. Incomplete fertilization of the seeds also resulted in hooked bolls (Rothwell *et al.*, 1967).

If deficiency is mild and other nutrients are adequate, poor square retention and shedding of young bolls result in excessive vegetative growth. As the deficiency becomes more severe, stunting, short internodes, splitting of petioles and possibly main stems, deformation of leaves, abnormal terminals and excessive branching may be observed. Discoloured or ruptured nectaries at the base of the flower and early bolls are common. When cut with a knife, lint near the base will be discoloured. New leaves are smaller than normal, with margins and main veins chlorotic and cupped upwards.

Rapid increases of soil pH and Ca levels caused by liming can induce B deficiencies in acid sandy soils. Induced deficiencies can also occur during periods of drought on non-irrigated soils and as a result of excessive leaching in sandier soils.

In most areas, deficiencies can be corrected by adding 0.5 to 1.0 kg B ha^{-1} applied with mixed fertilizer prior to planting in heavier soils, or as foliar sprays after squaring begins. Applications in soils are subject to leaching. Foliar sprays can be mixed with most insecticides. Honisch (1975) found that 2–4 kg ha^{-1} were required to correct deficiencies in Zambia. There was no difference between broadcast, drilled soil or foliar applications.

ZINC

Young plants have abnormally small, thickened and brittle leaves with interveinal chlorosis. Upward cupping of the leaf edges is common. Small dots of tissue in the chlorotic areas become necrotic, producing a bronzing effect. If deficiency continues, growth may essentially cease for a period, accompanied by shedding of any squares and flowers. Parallel elongation of the leaf tips may occur, resulting in a finger-like appearance. The distance between leaf nodes shortens, giving a bushy appearance to the top of the plant. Flowers often fall off before the bolls form, while bolls are small and sometimes blunt-nosed. If the deficiency is mild, yields may be unaffected, but maturity will be delayed and fibre quality will be lower.

Zinc deficiency is usually associated with soil pH values above 6.5. If plant analysis indicates a low plant Zn level, symptoms can be corrected with a foliar application of 0.25 kg Zn ha^{-1} in 200 l of water. A wetting agent will improve effectiveness (Hearn, 1981).

MANGANESE

Symptoms include leaf cupping and interveinal chlorosis of the youngest leaves with the veins remaining green. Chlorosis is yellow or reddish grey. Severe deficiency may result in shortening of internodes and stunting, eventually resulting in terminal death. Symptoms may be quite similar to those of zinc deficiency, but leaf distortion is much less than for zinc deficiency and its occurrence is quite rare. The author has observed Mn deficiency symptoms in somewhat poorly drained sands (Aquult) on the Coastal Plain of Georgia with a pH of 6.3. Plant analysis should be used to confirm deficiencies unless the situation clearly indicates a potential Mn problem.

Where deficiencies occur during the season, apply 0.57 kg elemental Mn ha^{-1} as manganese sulphate or 0.17–0.25 kg elemental Mn ha^{-1} as Mn chelate in 200 l of water to the foliage. Multiple applications at two-week intervals may be required.

COPPER

Unlike most micronutrients, copper deficiency results in interveinal chlorosis of lower leaves. Reduced growth is also reported. Observation of symptoms in the field is very rare (Hearn, 1981).

IRON

Although field observations are somewhat rare, iron deficiency can be readily recognized by the strong interveinal chlorosis of young leaves. As deficiency continues, chlorosis becomes more pronounced with each new leaf. Veins remain green, in sharp contrast to the chlorotic areas, which may become yellowish white. Leaf margins may curl, although cupping is not reported.

MOLYBDENUM

Kallinis and Vretta-Kouskoleka (1967) report that symptoms begin as interveinal chlorosis, followed by development of a greasy leaf surface and leaf thickening. Leaf cupping and necrotic spots and margins eventually develop. Bolls may develop, but will be abnormal and apparently do not mature or open normally. McClung *et al.* (1961) and Mikkelsen *et al.* (1963) report yield responses on Oxisols of the Brazilian Cerrado region, but deficiency symptoms have not been observed in the field.

Manganese Toxicity

Typical symptoms of Mn toxicity or 'crinkle leaf' include abnormally distorted or puckered leaves with dead spots along and between the veins. As this continues, the leaves become thickened and brittle, with ragged leaf margins. The phytotoxic effects of Mn occur primarily in the aerial portion of the plant, with little evidence of root damage. Symptoms must be quite severe before plant growth is limited solely by Mn toxicity (Hiatt and Ragland, 1963).

Manganese availability is a function of total soil Mn, easily reducible Mn and soil pH. Absorption by cotton depends on the soil solution concentration of the reduced, divalent form. This form increases with decreasing pH. Toxicity symptoms are possible when easily reducible soil Mn exceeds 50–100 mg kg^{-1} and pH is less than 5.5 (Adams, 1984). Leaf Mn levels of 3 g kg^{-1} are associated with visible symptoms in cotton (Adams and Wear, 1957).

Conditions leading to Mn toxicity are identical to those associated with Al toxicity. Toxic effects from Al and Mn may occur simultaneously where these conditions occur. Undue credit for reduced plant growth is often given to Mn toxicity, since its symptoms are highly visible (Adams, 1984). While Al toxicity is much more damaging, it primarily affects roots and is not as readily detected.

Interactions Between Nutrients

Many nutrients display antagonistic effects with one or more other nutrients when a deficiency or excess occurs. With increasing N rates, K concentration in the plant declines (Kamprath and Welch, 1968). Based on their study in highly weathered Ultisols, Bennett et al. (1965) recommended increased K rates to maintain plant K concentrations and increase yields when very high rates of N are applied.

Maples et al. (1977) found that N applications increased concentrations of nitrate and decreased concentrations of P in the petiole when adequate soil water was available. Significant depression of both N and P occurred when plants were water-stressed. This interaction was later used in their petiole monitoring programme to determine if petiole N reductions were indicative of actual nutrient stress or induced by drought.

Mutually antagonistic effects between K and Mg were reported in pot studies (Page and Bingham, 1965) and in fields of the Mississippi Delta (Pettiet, 1988). Miley et al. (1969) found that N and K applications significantly depressed B levels in plant tissue, but did not reduce yields in this particular study. In this responsive loessial soil, additions of B decreased total N concentrations in the petiole and led to better utilization of applied N. Liu et al. (1986) found that K applications to B-deficient soils resulted in decreased B content of cotton leaves and aggravated deficiency symptoms. Uptake of K was reduced at very low and very high levels of B.

Cotton can take up considerable amounts of Na, even though it is not an essential element. Where soils are responsive to K applications, Na can substitute for a portion of the K and in some cases delay or prevent the onset of deficiency symptoms (Christidis and Harrison, 1955; Hearn, 1981). Additions of Na have no effect where K levels are adequate. Evidence suggests that Na may substitute for stored Ca and Mg in older tissues, mobilizing these nutrients for use in new growth. Uptake of K, Ca and Mg may be synergistically increased by Na additions (Amin and Joham, 1968; Thenabadu, 1968).

Excessive applications of P to soils low in Zn can induce Zn deficiencies through increased growth, through interference of Zn uptake by cations added with the P or through P-enhanced adsorption of Zn to soil oxides and free lime (Marschner, 1986). Uptake of Zn by cotton is not suppressed by high P levels; rather, it appears that P uptake and translocation are enhanced by low Zn. The imbalance results in P toxicity symptoms and unusually high P : Zn ratios in the tissue (Cakmak and Marschner, 1986). Deficiency is not directly related to Zn–P reactions in the soil (Kissel *et al.*, 1985).

Le Mare (1977) found that low applications of P reduced cotton yields, while larger applications resulted in large yield increases. In the high-Mn soils of Uganda, addition of P in the presence of low Ca resulted in excess Mn uptake. With higher P rates, Ca added with the P prevented excess Mn uptake.

Effect of pH

Cotton grows well from pH 5.5 to 8.0 provided adequate nutrients are present. The primary effects of pH are related to solubility of metals and phosphorus compounds in the soil rather that to actual H ion concentrations. Metal solubility is high at low pH and decreases with increasing pH. Phosphorus availability is affected at both high and low pH by the concentrations of metals and by the availability of Ca and free carbonates. Generally, P availability is greatest in the range of pH 6.0–6.5, while the best balance between excess and deficiency of micronutrients is obtained at 5.5–6.0 (Tisdale *et al.*, 1985).

Cotton is one of the least tolerant crops to acid soils in spite of the fact that it will produce some yield even at low pH. In ten Ultisols with pH ranging from 4.9 to 5.4, liming increased seed cotton yields by 70–1010 kg ha^{-1} and averaged 433 kg ha^{-1} (Adams, 1984). Even greater effects of liming are reported in Tanzania (Le Mare, 1972). Acid-soil infertility is primarily related to Al and Mn toxicities, and Ca, Mg and Mo deficiencies. Excess Al is often the major root-limiting factor in acid soils. Cotton roots can be inhibited when Al saturation of the exchange complex exceeds 5–25% (Adams, 1981). Adams *et al.* (1967) found that yields were reduced by one-third when cotton roots were unable to penetrate into a strongly acid subsoil, while peanut roots and yields were unaffected by subsoil acidity.

Absorption of Mn depends on the amount of divalent Mn in the soil solution. As pH decreases from alkaline and neutral pH to the acidic range, the amount of divalent Mn increases. When easily reducible Mn in the soil exceeds 50–100 mg kg^{-1}, the potential exists for Mn toxicity as pH declines below 5.5. Solution Mn levels in excess of 10 mg l^{-1} have proved toxic to cotton (Adams and Wear, 1957).

Deficiencies of Ca, Mg and Mo may also limit growth of cotton in highly weathered, acid soils with low CEC and low organic matter. True Ca deficiencies in cotton are rare. On sandy soils with extractable Mg levels below 15 mg kg^{-1}, Adams (1975) reported that Mg additions increased seed cotton yields by 200 to 400 kg ha^{-1}. Additions of Mg-free liming materials and heavy applications of K increased incidence of the deficiency. Deficiencies of Mo would be expected on similar soils with high iron oxide contents (Adams, 1984). Although rarely observed, McClung et al. (1961) reported an increase in cotton yield when Mo was applied to an Oxisol of the Brazilian Cerrado region.

When P is added to highly acidic soils, adsorption and precipitation reactions with soil Al and Fe compounds tie up available P. Liming to a pH of 5.5–6.5 decreases available Al and maximizes P availability. McClung et al. (1961), Mikkelsen et al. (1963) and Sanchez and Uehara (1980) reported that less P was required when soils were limed.

When pH exceeds 7.0, added P can react with soil Ca and carbonate minerals, again limiting availability to plants (Kissel et al., 1985; Tisdale et al., 1985). Deficiency symptoms are rarely reported, but use of 10–40 kg P ha^{-1} is common where these soils are intensively cultivated and the growing season is short (Table 11.4).

Next to N and P, Zn is the most commonly deficient nutrient on high-pH soils (Kissel et al., 1985). Formation of relatively insoluble hydroxides, oxides and carbonates can limit availability of Zn, Mn, Cu and Fe when pH exceeds 7.0–7.5. Deficiencies of Zn related to high pH are most common in alkaline and calcareous soils where organic layers have been removed by erosion or during land levelling or shaping. Organic complexes with these metals help maintain adequate solution concentrations. Complexation tends to increase with increasing pH (Tisdale et al., 1985).

Effect of Soil Type

Cotton is grown on a wide range of soils, each presenting unique obstacles to cotton culture. The following descriptions are of the major soil orders on which cotton is grown, their advantages and their limitations are very generalized. More detailed information on soil formation, classification and properties is reviewed by Buol et al. (1989). Many other excellent texts are available which deal with soils of specific regions.

OXISOLS

Oxisols represent the ultimate step in soil weathering. These dark red soils have few weatherable minerals and low cation exchange capacity. Typically they are on nearly level to gently sloping landscape positions. Clay content varies little throughout the profile. They are generally characterized by low pH, low nutrient reserves and moderate to very high P fixation capacity, depending on the clay content. Physical properties of Oxisols are generally excellent, but may deteriorate somewhat as organic matter becomes depleted. Water-holding capacity is low because of the excellent stability of the soil structure. Inherent fertility is derived almost entirely from soil organic matter. Deficiencies of N, P, K, Ca, Mg, B, S and perhaps Mo have been reported (McClung *et al.*, 1961; Mikkelsen *et al.*, 1963; Sanchez, 1977; Lombin and Mustafa, 1981). With proper management, these soils can be highly productive. Oxisols occur extensively in tropical areas of South America and Africa.

ULTISOLS

Ultisols are acid soils with few bases remaining on the exchange complex, particularly in the subsoil. These soils have loamy to sandy surfaces over red or yellow subsoils. Clay content increases with depth. Ultisols occur on stable land surfaces over acid crystalline rocks or weathered coastal plain sediments. These soils occur in climates with long frost-free seasons and ample rainfall for rain-fed cotton production. Ultisols are used extensively for cotton production in the southern USA, Africa (south of the Sahara), southern China, South America and India. Continuous cultivation of these soils for cotton and rotation crops requires the addition of fertilizer and lime. Deficiencies of N, P, K, B and S are common. On sandier members of the order, Mg and Zn may be deficient.

ENTISOLS

These are very young soils which have little profile development resulting from soil-forming factors. The wetter members (Aquents or Fluvisols) occur as soils of varying depth deposited by rivers along their present or former courses. These soils may show layers of sand, silt and clay as a result of repeated deposition. Generally the alluvial soils are quite fertile, but this depends on the soils from which they have come. Sandy members of the order occur on more stable surfaces, but are typically infertile and prone to moisture deficit. Entisols often respond to N and P, but usually contain adequate K. Neutral to alkaline reactions are common, but some require lime. Flooding is a major problem within alluvial soils of this order. Many are drained to enable reclamation from salts or to prevent flooding during the cropping season. These soils occur extensively in many of the major cotton-producing areas of the world, often in association with soils of other orders.

INCEPTISOLS

These soils are weakly developed, young soils with few horizons. Horizon development indicates more stable land surfaces than typical in Entisols. Those horizons which are present occur near the surface. These soils typically develop on wind-blown or alluvial deposits in deltas and flood plains of major rivers in non-arid climates. Inceptisols are important cotton-producing soils in the USA, China, Pakistan and India. In terms of fertility, Inceptisols in cotton-growing areas tend to contain adequate levels of all nutrients except N and P. With intensive cropping, additions of K may be required. Some areas of acidic Inceptisols with lower fertility status also occur.

VERTISOLS

These self-inverting soils contain high amounts of swelling clays and develop deep, wide cracks when dry. Base saturation is usually high, with Ca and Mg dominating the exchange complex. Vertisols often respond to additions of N and perhaps P, but other nutrients are adequate (Sawhney and Sikka, 1960; Burhan and Mansi, 1970). At high pH, P and Zn deficiencies can be found. Poor drainage and sealing of the surface when moist make these soils difficult to cultivate and manage for irrigation. Widely known as 'black cotton soils', Vertisols are extensively used for cotton production in India, Egypt, Sudan, Australia and Texas.

ARIDOSOLS

These soils are formed in arid and semi-arid regions. Irrigation is essential for profitable cotton production. Problems associated with salinity, alkalinity, free $CaCO_3$ and high pH must be dealt with in growing cotton on Aridisols. Cotton tolerates many of these conditions much better than other cultivated crops and, with proper management, enormous yields are possible. In some cases, reclamation to remove excess Na and other salts is required. This usually involves replacement with Ca and leaching with good-quality irrigation water. Concentrations of B, fluorine, lithium, selenium and molybdenum can be high enough to inhibit growth (Dargan *et al.*, 1982). Major cotton-producing areas with Aridisols include the reclaimed desert areas of Arizona, California, Russia, India, Pakistan and Egypt.

MOLLISOLS

Mollisols are characterized by dark surface horizons rich in organic matter, and high to medium base status. Mollisols develop under relatively dry conditions, which prevent rapid decay of organic materials. Where carbonate is high near the surface (Ustolls or Kastanozems), P availability may be reduced. Subsoil salinity or alkalinity may also present problems. Moisture stress typically limits yields unless cotton is irrigated. In somewhat wetter climates (Udolls or Phaeozems), carbonate accumulations near the surface are less common. Mollisols are among the most inherently fertile soils in the world. They often show little response to N until cropped for some time. Rendolls (Rendzinas) have free carbonates near the surface and high pH, although salinity is seldom a problem. Mollisols occur in the central and western areas of the USA, Turkey, Russia, India and portions of South America.

ALFISOLS

Alfisols (Luvisols) range considerably in properties. Most have a high base status, and reasonable levels of weatherable minerals. Horizon development is strong, with subsurface accumulations of clay. In the native state, physical conditions are good. Dark red members (Udalfs or Nitosols) are among the most responsive to management. Nitrogen, P and perhaps K fertilization may be required, but can result in very high yields. Magnesium and B may be required for cotton. Wetter members of the Udalfs or Planosols formed in depressions and subject to occasional surface waterlogging are often drained and used for cotton production. Alfisols occur throughout cotton-growing areas as the more developed soils of flood plains and deltas. The drier members of this order (Xeralfs) are important cotton soils in the south-western USA, Russia, India, Egypt, Brazil, West Africa and China.

Effect of Cropping Practice

IRRIGATED VERSUS RAIN-FED

Cotton in rain-fed areas is subject to prolonged periods of drought, which can have marked effects on nutrient uptake and utilization. Drought-induced deficiencies can markedly reduce squaring and fruit set as concentrations of immobile nutrients such as Ca and B become limiting. If severe enough, drought can result in 'cut-out' of the plant, resulting in a delay of 10–14 days before fruiting resumes. When rains eventually come, light fruit loads can result in rank growth. While supplementary irrigation can overcome such problems, it cannot overcome extended periods of rainfall, cloudy weather and high humidity. These conditions can disrupt pollination, decrease photo-synthate production at critical growth stages and provide ideal conditions for disease, all leading to fruit-shedding and rank growth. Excessive rainfall can cause waterlogging and move leachable nutrients out of the rooting zone. Waterlogging causes rapid changes in nutrient uptake, although yields may not suffer from this condition alone (Hocking *et al.*, 1987). The potential for leaching, particularly of N, requires split applications in most rain-fed areas, with 25–50% of the total N requirement applied at planting. Excessive N rates under these conditions favour even greater production of vegetative growth over reproductive growth, adding potential for a prolonged season, increased boll rot, greater insect pressure (Tucker and Tucker, 1968) and confounding defoliation problems. To avoid these problems, rain-fed cotton generally receives less N than irrigated cotton. In irrigated regions, vegetative growth rate can be restricted by fruit load and regulation of water rather than nutrient regulation, allowing much higher rates of N to be applied. This in turn means that higher rates of other nutrients may be required (Tables 11.1 and 11.4) in proportion to the increased yield. Since a higher percentage of the crop may be set in a shorter amount of time, rates of release from the soil become critical under irrigated conditions. Higher inherent fertility may be required under irrigated conditions to supply demands at the peak uptake period (Cassman *et al.*, 1990).

Land-levelling operations for furrow irrigation in arid climates have exposed calcareous subsoils with low organic matter contents. In many cases, such areas have been implicated as the source of P, K and Zn deficiencies (Kerby and Adams, 1985; Kissel *et al.*, 1985).

ROTATION

Unlike many other annual crops, cotton is less responsive to crop rotation unless disease pressure is high. Continuous plots have been maintained in parts of Africa for as much as 50 years with few ill effects (Munro, 1987). The use of annual or perennial cover crops, deep-rooted legumes and, in some cases, fallow can maintain or increase organic matter and residual N in the soil. This improves the soil tilth, water-holding capacity and non-capillary porosity and may bring plant nutrients from deep in the soil profile to the surface (Christidis and Harrison, 1955).

On Vertisols of Australia, a fallow or soyabean rotation greatly reduced N requirements and improved yields at equal N rates (Standley *et al.*, 1988). In an intensive cropping system in the Sudan Gezira, long-term trials without fertilization showed the benefits of fallow and legumes preceding cotton and the detrimental effects of sorghum. In more recent trials, fertilizers decreased but did not eliminate these rotational effects (Burhan, 1969). Similar declines in cotton yields following sorghum have been reported in the USA and India (Christidis and Harrison, 1955). In the south-eastern USA recommended N rates are reduced by 25–30 kg ha^{-1} when cotton follows a legume crop such as peanuts or soyabeans. In tropical areas, 40 kg N ha^{-1} may be assumed to be available following fallow or a leguminous crop (Hearn, 1981).

Deficiency of K was more common following peanuts where the hay was removed, than in continuous cotton in the south-eastern USA (Skinner *et al.*,

1946; Volk 1946). Even though peanut yields were unaffected by K applications, cotton yields increased in four of the five experiments. In addition, K requirements for the rotation were less when K was applied to cotton rather than to peanuts. Rouse (1960) found that 45 kg ha^{-1} gave maximum cotton yields in continuous cotton, but 90 kg ha^{-1} were required for maximum cotton yields in a cotton-peanut rotation. Requirements were also increased following deep–rooted perennial legumes such as alfalfa, sericia and annual lespedeza.

In spite of a general consensus that inclusion of legumes and green manures in rotation with cotton may raise soil fertility levels and increase yields, these practices are not widely adopted (Sawhney and Sikka, 1960; Prentice, 1972). Fertilization and other cultural practices provide many of the same effects more economically and efficiently (Prentice, 1972). Where land values are high, cotton production is intensely mechanized or population pressure requires intensive use of the land, rotational crops must generally yield an economic return in their own right. In general, if the soil has adequate fertility and pests can be controlled, several studies indicate that cotton can be grown continuously for a number of years (Christidis and Harrison 1955; Munro, 1987).

Where land use is less intense, shifting cultivation has been used to negate the effects of declining fertility (Jones 1976; Munro 1987). Cotton is considered a good crop on newly cleared land in Africa (Munro, 1987), and could be grown as a first crop in the Brazilian Cerrado region to pay initial liming and fertilization costs (Mikkelsen *et al.*, 1963). Under shifting cultivation, the time required to re-establish soil fertility varies with soil and climatic conditions; in excess of 12 years may be required in the African Congo. In East Africa, agricultural departments have long recommended a rotation of three years of cropping followed by three years of grass. In such a rotation, fertilizers gave little response and failed to improve yields in continuous cropping situations. Jones (1972, 1976) found that soil fertility was strongly tied to soil organic matter and that this rotation allowed adequate build-up to maintain soil fertility. However, continuous cropping could be achieved using a mixed fertilizer containing N, P, K, Ca and S.

Methods of Fertilizer Application

The goal of a particular method of fertilizer placement is to maximize plant uptake efficiency using a convenient and economical operation while preventing fertilizer injury. Soils and fertilizers have inherent physical and chemical properties which begin to interact the moment a fertilizer is added to the soil. These interactions determine the concentration and form of nutrients available to plants in the soil solution. The effectiveness of a particular method of fertilizer placement depends on how well the method performs in balancing soil–fertilizer interactions to provide adequate levels of required nutrients under the prevailing conditions.

Nutrients may be broadcast or band-applied to the soil surface or under the soil surface or sprayed directly on the plant foliage. Proper choice of a method depends on the form of the nutrient and how it reacts with the soil or plant. The following discussion briefly describes the primary chemical properties and reactions of nutrients in the soil, and the implications for fertilizer placement. For a more complete discussion, the reader should consult more detailed reviews (Russell, 1973; Randall *et al.*, 1985).

Ammonia (NH_3) is normally applied as a pressurized gas, and must be applied in a subsurface band to prevent volatile losses. It readily transforms to the ammonium (NH_4^+) form upon reaction with soil water. Application must be deep enough (20–30 cm) to prevent seedling damage and be made several days to weeks ahead of planting. The urease enzyme readily converts urea to the ammonium form in warm moist soils. Band applications of urea can accumulate significant amounts of ammonia in all but very acid soils, which may be toxic to germinating seeds and young seedlings (Randall *et al.*, 1985). Above pH 7.0, ammonium can be converted to ammonia, and substantial amounts of N can be lost to the atmosphere when surface-applied. Some form of incorporation or subsurface application in neutral to alkaline soils should decrease losses. Ammonium is held in an exchangeable form by charged sites on soil clays and organic matter or broken down by soil microbes to the nitrate (NO_3^-) form. As ammonium levels decrease in the soil solution, additional increments are released from the soil exchange sites.

Nitrate-N, B and Cl are not strongly reactive with most soil clays and are readily leached. In acid soils, S and Mo may be retained on soil clays but are relatively mobile above pH 5.5 (Kamprath and Foy, 1985).

Phosphorus reacts strongly with the soil and moves little from the point of placement. The intensity of this reaction depends on the presence of excess Al and Fe (low pH) or Ca (high pH). Band application of granular, water-soluble P sources can greatly increase uptake efficiency on low P soils. As P levels increase, localized placement shows less effect (Barber, 1958). Availability of less soluble sources, such as rock phosphate, increases with increasing contact with the soil. These sources should be finely ground, broadcast and incorporated (Sanchez and Uehara, 1980; Randall *et al.*, 1985). In acid tropical soils very low in P and having a high P fixation capacity, initial broadcast applications were more effective than band applications. Deficiency was so severe that roots developed only in areas where P was applied. A combination of a single broadcast application followed by annual banded maintenance applications is the best alternative in such soils (Sanchez, 1977).

The remaining nutrients (Ca, Mg, K, Zn, Mn, Fe and Cu) are retained by most soils, with the exception of deep sands, where K and Mg leaching can result in deficiencies. Since retained nutrients are relatively available but not subject to rapid loss, the method of application for these nutrients is relatively unimportant. Generally, broadcast applications are preferred over banded applications, particularly for Zn (Hearn, 1981). Soils containing vermiculite in the clay fraction can fix significant amounts of K. Although availability of fixed K is often adequate for crops such as maize, release rates appear to be insufficient for rapidly growing, well-fruited cotton in some situations (Cassman *et al.*, 1990). Banded applications may increase uptake efficiency on such soils. Deep placement of K has been explored as a means of increasing rooting depth and improving drought tolerance (Tupper and Ebelhar, 1990). While these efforts seem promising, direct comparison between surface and deep applications have not been reported.

Historically, cotton fertilizers were uniformly distributed over the soil surface either by hand or with distribution equipment (Christidis and Harrison, 1955). The soil was then cultivated to incorporate the fertilizer into the rooting zone and then bedded prior to planting. The practice of placing fertilizer in the seed furrow was shown to be more effective than broadcast applications, but potential injury from fertilizer salts limited applications to relatively low rates. Band placement was later shown to allow higher rates without danger of seedling mortality. By placing fertilizers near the row, nutrients are near the developing roots and at a higher concentration soon after germination. The relatively high concentration of nutrients slows fixation of some nutrients and also improves uptake. Compared with broadcast applications, banding decreases the amount of soil in contact with the fertilizer, further reducing fixation potential, particularly for P. Since higher concentrations result from banding, leaching potential increases. This may explain why differences in broadcast and band decline as application rate increases (Vilela and Ritchey, 1985). Banded applications may result in less residual effect than broadcast applications (Christidis and Harrison, 1955).

Placement of fertilizers less than 2.5 cm from the seed often results in injury. Best results are usually obtained when narrow bands are placed 5–8 cm from the seed and 5–8 cm below the seed. Under-the-row placement disturbs the seed bed and can reduce germination and yields.

Several studies have shown little difference between band and broadcast applications of K where soil fertility levels are moderate to high, particularly on highly weathered or sandy soils (Luckhardt and Ensminger, 1968; Evans *et al.*, 1970; Malavolta, 1985). Improved bulk-handling and distribution equipment allows fertilizer to be broadcast much more rapidly than it can be band-applied. With increased farm size and highly mechanized farming systems, many US growers have returned to broadcast applications of fertilizers to save time and labour during the critical planting season. A relatively recent combination of the two systems uses broadcast applications to apply the bulk of the recommended fertilizer, but band-applies a starter fertilizer containing approximately 12–20 kg N and 15–25 kg P ha^{-1} at planting. The starter fertilizer is typically a liquid ammonium polyphosphate, but dry diammonium and monoammonium phosphates have been equally effective. In responsive soils, effects include increased lint yield, greater height at first square, earlier fruiting and improved stress tolerance. In many cases, these responses improve competitiveness with weeds, increase earliness and allow more timely application of herbicides (Funderburg, 1988; Guthrie 1988; Hodges and Baker, 1990). Most responses have been in rain-fed areas.

Where leaching rains are possible, split applications of N and, in some cases, K are recommended (Malavolta, 1985). In some cases, split applications not only decrease leaching losses and improve yields, but result in less weed and insect pressure early in the season (Sawhney and Sikka, 1960; Perichiappan *et al.*, 1987). In a similar fashion, application of fertilizers with irrigation water can improve uptake, decrease leaching potential and decrease application costs.

Foliar applications are useful in correcting in-season deficiencies of some nutrients. Urea and B have been foliarly applied throughout the southern USA, and use of urea is increasing in the western areas (Hake, 1989, personal communication). Micronutrient deficiencies caused by high pH are not easily corrected by soil applications. Foliar applications of Zn, Fe and Mn have been used effectively in areas where deficiencies occur (Hearn, 1981).

Cultivar Sensitivity to Nutrient Deficiencies

There are many examples of nutrient–cultivar interactions in the literature, ranging from differential uptake of nutrients to improved stress tolerance. Although a complete review of this topic will not be attempted, the following examples illustrate the importance of nutrient status for cultivar performance.

Working with improved cultivars in a year with above-average rainfall, Olson and Bledsoe (1942) reported lower nutrient requirement indices than Fraps (1919) and McBryde and Beal (1896). Recent Upland cultivars often exhibit a more determinate fruiting habit, resulting in earlier fruiting and more compact plants. Under similar soil and climatic conditions to those of Olson and Bledsoe (1942), Mullins and Burmester (1990) concluded that nutrient requirement indices and uptake rates of N, P and K by four modern cultivars were similar to those reported by Olson and Bledsoe (1942) at similar yield levels, but that a higher percentage of dry matter was allocated to fruit forms rather than vegetative growth.

Bhatt and Appukuttan (1971) compared a long-branched, bushy cultivar (MCU-1) with a short-branched, more compact cultivar (PRS-72). Although yields were similar, total nutrient uptake was greatest for the long-branched cultivar. The ratio of nutrients allocated to fruiting bodies versus vegetative growth was much greater and nutrient concentrations in the vegetative growth were much lower at maturity in the compact plants than in the bushy plants. Halevy (1976) similarly reported a higher ratio of reproductive growth to vegetative growth for Acala 4–42 than for Acala 1517-C. There was also a marked difference in K uptake, which appeared to be related to a smaller root system on the latter cultivar. In later studies, with yields of up to

2.76 t lint ha^{-1} (Halevy *et al.*, 1987), Acala SJ-2 showed even higher percentages of total nutrient uptake in the lint and seed.

Tisdale and Dick (1939, 1942) showed that K requirements were lower for Fusarium wilt-resistant cultivars than for wilt-susceptible cultivars. Recent work in the San Joaquin Valley of California (Cassman *et al.*, 1989) has shown that the cultivar Acala GC-510 is less sensitive to a widely occurring K deficiency syndrome than Acala SJ-2. In this case, partitioning of K within the plants was similar, but GC-510, with a more extensive root system, was able to extract more K from the soil after first bloom.

Selection under varying nutrient regimes or levels of stress can result in marked differences in cultivar tolerance to stress. Tolerance to high levels of Al and Mn can profoundly affect development and yield of cotton. Some cultivars show surprising tolerance to Al while others are extremely sensitive (Foy *et al.*, 1967). Foy (1969) attributed Mn tolerance to a breeding history in locations with high soil Mn levels. Tolerance to Mn or Al did not imply tolerance to both. Thus, Rex and Pima S-2 were quite tolerant of high Mn levels and moderately sensitive to Al, while Acala 4-42 was sensitive to Mn and tolerant to Al. Coker 100A was found to be sensitive to both Al and Mn. Smithson (1972) found fertilization with B increased yields of one line by 70% and of another by only 26%. These differences were also attributed to past selection under low B conditions.

Soil Nutrients and Disease Susceptibility

Over 100 years of research shows that soil nutrients play an important role in disease susceptibility of cotton. Nutrients may influence disease susceptibility in a number of ways. They may directly promote or inhibit the disease agents or antagonists of the disease agent. They may also affect the function of tissues within the cotton plant which resist or encourage pathogens. Even though genetic make-up in large measure controls the resistance to disease, resistance is expressed through complex physiological and biochemical processes that are linked to the nutritional status of the plant or of the plant pathogen. Several major cotton diseases are caused by soil-borne pathogens. All invade the root or the portion of the stem below the soil line. Thus, soil conditions may strongly affect invasion of the host and survival of the pathogen. A more complete review of this topic is given by Bell (1989).

pH

Microflora are often adapted to a narrow range of soil pH, above and below which they compete less effectively with other organisms. Races 1, 2 and 6 of *Fusarium oxysporum* f.sp. *vasinfectum* are commonly associated with root-knot nematodes in acidic, sandy soils, while races 3, 4 and 5 are more frequently found in conjunction with reniform nematodes and inhabit clay soils (Bell, 1989). In a survey of cotton fields in Texas, Taubenhaus *et al.* (1928) found Fusarium wilt (races 1 and 2) in 55% of the fields with pH 5.5–6.4, in 13% of the fields with pH 6.5–7.4 and in 2% of the fields with pH above 7.4. These races of Fusarium wilt have been observed at pH values as high as 8.1, but were associated with high populations of nematodes and saline conditions (Blank, 1962).

Verticillium wilt (*Verticillium dahliae* Kleb.) is most common in neutral to alkaline loam and clay soils in irrigated regions. Acidic soils apparently contain sufficient concentrations of Al and Mn and possibly other acid-soluble trace elements to inhibit the growth and development of *V. dahliae* (Bell, 1989). The fungus grows best at pH 6.0–7.0, and liming of acidic soils above pH 6.3 increased severity of the disease in plants inoculated by stem injection (Shaeo and Foy, 1982) and in field studies (Young *et al.*, 1959). Severity can be decreased somewhat by addition of elemental S to lower the pH of neutral and alkaline soils, but the practice is expensive and not always effective (Rudolph and Harrison, 1939; Hinkle and Staten, 1941).

Phymatotrichum root rot (*Phymatotrichum omnivorum*) occurs primarily in alkaline Vertisols of the south-western USA and northern Mexico. Little disease occurs in soils with pH less than 6.0, while the most severe infestations are found in a pH range of 7.5–8.5 (Ezekiel *et al.*, 1930).

NITROGEN

Applications of ammonia and ammonium salts can reduce populations of *Pythium ultimum*, *Fusarium* spp., *Phymatotrichum omnivorum* and *Sclerotium rolfsii* Sacc. through toxic effects and inhibited growth (Smiley *et al.*, 1970). Sensitivity to ammonia may be related to the lower incidence of Fusarium wilt (races 1, 2 and 6) in neutral and alkaline soils (Bell, 1989). Although ammonia may inhibit these seedling disease pathogens, band applications of ammonia and ammonium-containing fertilizers can cause seedling injury at high rates or when placed too close to seeds (Hood and Ensminger, 1964). In addition, N fertilization tends to decrease numbers of *Penicillium funiculosum*, an antagonist to *Fusarium* spp., *Verticillium* spp. and *Rhizoctonia solani* (Blair and Curl, 1974). Additions of N may also increase pathogenicity of *R. solani* (Zyngas, 1962; Ramasami and Shanmugam, 1976).

Increased N rates have been associated with decreased severity of bacterial leaf blight caused by *Xanthomonas campestris* pv. *malvacearum* in Upland cotton (Rolfs, 1915; Presley and Bird, 1968). Several studies reviewed by Bell (1989) indicate a similar reduction in incidence of Phymatotrichum root rot and Macrophomina root rot (*Macrophomina phaseolina*) with N fertilization. When N and P were applied, maximum yield and economic return were achieved, but root rot percentage was unaffected. Manure buried under the row and urea both reduced root rot incidence, while $Ca(NO_3)_2$ aggravated root rot in some cases.

Numerous studies (Presley and Bird, 1968; Huber, 1981; Bell, 1989) conclude that N applications increase severity of both Fusarium and Verticillium wilt, especially when K is deficient. Significant increases were noted where manures were applied (Young *et al.*, 1959; Tucker and Tucker, 1968). Although wilt incidence increases, yields and profits are usually increased in proportion to applied N.

Depending on conditions, all N sources have been reported to increase wilt severity. Ranney (1962) found that high concentrations of urea increased Verticillium wilt less than other sources in sand cultures, but recommended the use of equal rates of ammonium and nitrate N over the use of either form or urea alone for improved control of Verticillium wilt. When applied several weeks prior to planting, urea decreased wilt percentages in the field. Propagule numbers were reduced in the field following application of ammonium and urea sources (Chernyayeva *et al.*, 1984, cited in Bell, 1989). The use of properly balanced N, P and K fertilizers limits the increase in Verticillium wilt incidence while maximizing yields. Split N applications or mid-season side-dress applications may further decrease wilt severity and increase yields (Neal and Sinclair, 1960; El-Zik, 1985, 1986).

Greater vegetative growth associated with high N rates favours conditions within the plant canopy for increased incidence of boll rot (Scarsbrook *et al.*, 1959; Prentice, 1972). High N rates may also increase weed and grass pressure late in the season, restricting air flow near the base of the plant. Rank vegetative growth has been reported to decrease gossypol contents (Wadleigh, 1944). This, along with greater difficulty in obtaining good insecticide coverage, can lead to increased insect damage and allow greater entry of boll-rotting organisms. Frequent irrigation or rainfall accentuates the problem.

PHOSPHORUS

Additions of P generally have little effect on incidence of seedling disease, but may facilitate recovery from infections such as *Thielaviopsis basicola* (Zyngas, 1962). In P-deficient neutral and alkaline soils of India and Africa infested with race 3, 4 or 5 of *F. oxysporum* f.sp. *vasinfectum*, P fertilization decreased the incidence of wilt (Naim and Shaaban, 1965; Sadisivan, 1965; Ebbels, 1975). When applied with high rates of N, P may increase severity of Fusarium wilt, especially if K is deficient (Presley and Bird, 1968).

A favourable P status in the plant is apparently required for development of the Verticillium wilt fungus within the plant, and P fertilization where P is deficient may increase severity of Verticillium wilt (Davis *et al.*, 1979). Application of P with very high N rates increased severity more than N alone (Longenecker and Hefner, 1961). Combinations of P with lower N rates increased wilt when K was deficient, but had no effect when K was adequate (Presley and Dick, 1951; Baard and Pauer, 1981). Thus, the effects of P on Verticillium wilt are dependent upon whether a P deficiency exists and the availability of other elements, particularly N and K.

Phosphorus applications may decrease severity of Phymatotrichum root rot (Presley and Bird, 1968), but numerous studies indicate increased severity on a variety of soils with application of P alone or as the predominant element in a mixed fertilizer (Christidis and Harrison, 1955; Bell, 1989).

POTASSIUM

While K has tremendous effects on susceptibility to Fusarium and Verticillium wilt, its effects on other diseases are not so well defined, (Huber and Arny, 1985). Potassium may either increase (Blair and Curl, 1974) or decrease (Zyngas, 1962; Ramasami and Shanmugam, 1976) the pathogenicity of *R. solani*. Potassium has also been reported as an important factor in preventing or decreasing the severity of leaf blights caused by *Cercospora gossypina* Cooke and *Alternaria alternata* Auct. s. Wiltshire (Miller, 1969). Rolfs (1915) reported a decrease in severity of bacterial leaf spot with the application of K. Presley and Bird (1968) concluded that N additions reduced the incidence of bacterial leaf spot only when K was adequate. Although K levels in *Phymatotrichum*-infested soils are usually high, K addition decreased the severity of root rot in some cases (Tsai and Bird, 1975).

Early studies showed that K fertilization could increase resistance to Fusarium wilt (Rast, 1924), but results were inconsistent from location to location (Bell, 1989). Walker (1930) found, however, that K had little effect on Fusarium wilt when disease pressure was high and a susceptible cultivar was grown. Later studies in Arkanas, Mississippi, Alabama and Texas clearly demonstrated that K deficiency predisposes cotton to Fusarium wilt, but effective control through K fertilization depended on the level of cultivar resistance and the population of associated nematodes. In field studies, susceptible cultivars required twice the K addition of resistant cultivars to obtain maximum suppression of Fusarium wilt. No benefits of K fertilization were obtained with susceptible cultivars and very high root-knot nematode populations (Miles, 1936; Tisdale and Dick, 1939, 1942).

In controlled nutrient studies, increase of K concentration from 50 to 500 mg l^{-1} was required to suppress Fusarium wilt on a susceptible cultivar of *G. barbadense*, while a resistant cultivar required only 100 mg l^{-1}. In the presence of reniform nematodes (*Rotylenchulus reniformis*), K concentrations of 1000 and 500 mg l^{-1} were required for wilt suppression in the susceptible and resistant cultivars respectively (El-Gindi *et al.*, 1974).

Where available K is marginal, N and P additions can increase yields, resulting in greater K demand and greater incidence of Fusarium wilt. Similar results are reported in India, Africa and Brazil (Bell, 1989).

Although fertilization with N and K may affect susceptibility directly through effects on populations of *Fusarium* sp. and fusarial antagonists, the primary effects appear to be on host defence responses (Sadasivan, 1965; Bell, 1989). Equal effects of K on inoculated plants in sterile and non-sterile soil indicate that primary effects are not directly related to changes in soil micro-flora. Furthermore, decreasing K levels are required for wilt suppression with increased genetic levels of host resistance. Resistance has been correlated with several fungitoxins produced by the cotton plant, as well as sensitivity of cotton to phytotoxins produced by *Fusarium* sp., but we know little of how K and other nutrients affect these plant characteristics (Bell, 1989).

Effects of fertilization on susceptibility to Verticillium wilt are very similar to those described for Fusarium wilt. Adequate K availability is critical for maintaining resistance. Numerous studies around the world have shown that K fertilization of deficient soils reduces wilt severity (Presley and Dick, 1951; Young *et al.*, 1959; Abdel-Raheem and Bird, 1967; Hafez *et al.*, 1975; Ashworth *et al.*, 1982; Bell, 1989), but effectiveness depends on low to moderate levels of infestation and use of resistant cultivars. Where K availability is high, addition of K has little effect on Verticillium wilt severity (Presley and Dick, 1951).

Considerable interest has been directed to the K deficiency syndrome which came to prominence in California and has now been reported in Mississippi, Arkansas, Israel and Africa. Unexpectedly severe K deficiency symptoms appear on cotton in soils which produce normal maize crops. These symptoms are frequently followed by infestation with Verticillium wilt. Some workers propose that *V. dahliae* and perhaps other pathogens aggravate K deficiency in cotton by reducing plant tissue levels (Weir *et al.*, 1989). Heavy K fertilization, solarization or fumigation eliminates most of the K deficiency symptoms, reduces Verticillium wilt and increases yields. Others argue that the K deficiency syndrome results from a unique combination of K-fixing soils, K-deficient subsoils and heavy K demand by rapidly fruiting cultivars (Cassman *et al.*, 1990).

Hillocks and Chinodya (1989) report that a similar K deficiency syndrome predisposes cotton plants in Africa to epidemics of Alternaria (*Alternaria macrospora*) leaf spot. Additions of K delayed invasion in Upland varieties.

Excessive K fertilization results in greater leaf and stem tissue, but does not affect lint and seed production (Bennett *et al.*, 1965; Kerby and Adams 1985). This excess vegetative growth can lead to greater incidence of boll rot.

CALCIUM

Calcium is critical in the early development of cotton seedlings. Deficiencies result in disruption of carbohydrate movement from leaves to stems and roots, causing stunting and other symptoms typical of soil-borne fungi. (Wiles, 1959; Howard and Adams, 1965; Puente, 1965; Ashworth *et al.*, 1982; Christiansen and Rowland, 1986). When soils are low in Ca or there are other stresses which can cause limited uptake of Ca, additions of Ca increase emergence, increase seedling vigour and decrease disease incidence

(Christiansen and Rowland, 1986; Bell, 1989). Deficiencies may also affect incidence of leaf spots (Presley and Bird, 1968).

OTHER NUTRIENTS

Benefits of micronutrients in decreasing disease susceptibility apparently occur only when deficiencies are present. Soaking seeds in 0.1% $MnSO_4$ decreased the number of diseased seedlings in the field by 25% in Chinese studies (Bell, 1989). Other micronutrient additions have improved germination in greenhouse studies but have not proved effective in limiting susceptibility to seedling diseases in the field (Bell, 1989).

Additions of Zn are reported to decrease Fusarium wilt incidence in India (Sadasivan, 1965), but results were not consistent for other micronutrients. In greenhouse studies, B and Mn gave some slight Verticillium wilt reductions, but gave no benefit in the field (Desai and Wiles, 1976). Shukla and Raj (1987) tested five cultivars under conditions of Zn stress and found differences in the development and severity of visual symptoms, the ability to recover from early-season Zn stress and in response to corrective treatments. Under the conditions of this study, Zn concentrations in the plant eight weeks after sowing were not well correlated with final yields of lint and seed. Additions of Zn, Mo, B and Mn to the soil or as foliar sprays decreased Verticillium wilt severity in Russia. Foliar sprays applied at squaring and flowering and during fruit development are the most common means of adding these micronutrients (Bell, 1989).

Increased levels of Na in the plant apparently enhance resistance of cotton to Verticillium wilt. Incidence of the disease declines with increasing salt content of the soil. Where electrical conductance of saturation extracts exceeds 5 mS cm^{-1}, wilt is greatly reduced, and little wilt occurs above 10 mS cm^{-1} (Christensen et al., 1954).

Phymatotrichum root rot has been associated with soils low in available Mg and high in soluble Ca, but no causal link has been established (Bell, 1989). Drenches with Zn and Fe have resulted in slight decreases in Phymatotrichum root rot in some cases (Matocha et al., 1986; Mostaghimi et al., 1987).

Soil Nutrients and Nematode Attack

Nematodes normally cause more problems on sandy soils which are naturally deficient in several nutrients. As a result, nematodes feeding on cotton roots may aggravate nutrient deficiencies. Where soils are deficient, K fertilization tends to increase plant tolerance and yield, but also increases nematode density as root volume and nutrition of the nematodes increases. These generalizations apply to root-knot (*Meloidogyne incognita*), stunt (*Tylenchorhynchus latus*) and lesion (*Pratylenchus brachyurus*) nematodes in sandy soils (Oteifa, 1983; Oteifa and Diab, 1961; Oteifa et al., 1965; Jorgenson 1984), but not to reniform nematodes (*Rotylenchulus reniformis*), which are more commonly found in loams and clays. While K fertilization of deficient soils increases tolerance and yields in these soils, the number of reniform nematodes per unit weight of roots is decreased (Bell, 1989).

Where P is severely deficient, P fertilization increases plant tolerance to nematode attack, but increases nematode numbers as with K. At higher P levels, P additions may decrease host tolerance if P interferes with uptake of marginally available micronutrients such as Zn (Smith *et al.*, 1985). Inoculation of cotton seedlings with root-knot nematodes decreased Ca and Mg levels of cotyledons by 10–20% over the check during a six-day period. Both root-knot and reniform nematodes reduce K content of roots and leaves (Oteifa and El-Gindi, 1976). Nutrient reductions are most pronounced in leaves, and decrease with increasing availability of soil K. Root-knot nematodes reduce K content more than reniform nematodes. Nematode-induced reductions in K and perhaps other nutrients may play a role in predisposing plants to other diseases such as Fusarium wilt (Bell, 1989).

References

ABDEL-RAHEEM, A. AND BIRD, L.S. (1967) Effect of nutrition on resistance and susceptibility of cotton to *Verticillium albo-atrum* and *Fusarium oxysporum* f.sp. *vasinfectum*. *Phytopathology* 57, 451 (Abst.).

ADAMS, F. (1975) Field experiments with magnesium in Alabama – cotton, corn, soybeans, peanuts. *Alabama Agricultural Experiment Station Bulletin* 472.

ADAMS, F. (1981) Alleviating chemical toxicities: liming acid soils. In: Arkin, G.F. and Taylor, H.M. (eds), *Modifying the Root Environment to Reduce Crop Stress*. American Society of Agricultural Engineers Monograph 4, American Society of Agricultural Engineers, St Joseph, Michigan, pp. 267–301.

ADAMS, F. (1984) Crop response to lime in the southern USA. In: Adams, F. (ed.), *Soil Acidity and Liming*, 2nd edn. Agronomy Monograph 12, American Society of Agronomy, Madison, Wisconsin, pp. 211–65.

ADAMS, F. AND WEAR, J.I. (1957) Manganese toxicity and soil acidity in relation to crinkle leaf of cotton. *Soil Science Society of America Proceedings* 21, 305–8.

ADAMS, F., PEARSON, R.W. AND DOSS, B.D. (1967) Relative effects of acid subsoils on cotton yields in field experiments and on cotton roots in growth-chamber experiments. *Agronomy Journal* 59, 453–6.

ALIMOV, Kh. AND IBRAGIMOV, S. (1976) Trace elements in different cotton cultivars. In: *Field Crop Abstracts* 29, Abstr. 652.

AMIN, J.V. AND JOHAM, H.E. (1960) Growth of cotton as influenced by low substrate molybdenum. *Soil Science* 89, 101–7.

AMIN, J.K. AND JOHAM, H.E. (1968) The cations of the cotton plant in sodium substituted potassium deficiency. *Soil Science* 105, 248–54.

ANDERSON, J.T. (1894) Fertilizers required by cotton as determined by the analysis of the plant. *Alabama Agricultural Experiment Station Bulletin* 57, 16.

ANDERSON, O.E. AND HARRISON, R.M. (1970) Micronutrient variation within cotton leaf tissues as related to variety and soil location. *Communications in Soil and Plant Analysis* 1, 163–72.

ARAKERI, H.R., CHALAM, G.V., SATYANARAYANA, P. AND DONAHUE, R.L. (1959) *Soil Management in India.* Asia Publishing House, New York.

ASHWORTH, L.J., Jr, GEORGE, A.G. AND MCCUTCHEON, O.E. (1982) Disease induced potassium deficiency and Verticillium wilt in cotton. *California Agriculture* 36, 18–20.

BAARD, S.W. AND PAUER, G.D.C. (1981) Effect of alternate drying and wetting of the soil, fertilizer amendment, and pH on the survival of microsclerotia of *Verticillium dahliae. Phytophylactica* 13, 165–8.

BARBER, S.A. (1958) Relation of fertilizer placement to nutrient uptake and crop yield. I. Interaction of row phosphorus and the soil level of phosphorus. *Agronomy Journal* 50, 535–9.

BASSETT, D.M., AND MACKENZIE, A.J. (1978) Plant analysis as a guide to cotton fertilization. In: Reisenauer, H.M. (ed.), *Soil and Plant Tissue in California.* Bulletin 1879, Division of Agricultural Sciences, University of California, Berkeley, pp. 16–17.

BASSETT, D.M., ANDERSON, W.D. AND WERKHOVEN, C.H.E. (1970) Dry matter production and nutrient uptake in irrigated cotton (*Gossypium hirsutum*). *Agronomy Journal* 62, 302–3.

BAT'KAEV, ZH., ABDRAIMOV, A. AND BUDARINA, A. (1989) Fertilization of cotton on an old arable land. *Field Crop Abstracts* 43, Abstr. 4362.

BELL, A.A. (1989) Role of nutrition in diseases of cotton. In: Engelhard, A.W. (ed.), *Soilborne Plant Pathogens: Management of Diseases with Macro and Micro Elements*, American Phytopathological Society Press, St Paul, Minnesota, pp. 167–204.

BENNETT, O.L., ROUSE, R.C., ASHLEY, D.A. AND DOSS, B.D. (1965) Yield, fiber quality and potassium content of irrigated cotton plants as affected by rates of potassium. *Agronomy Journal* 57, 296–9.

BERGER, J. (1969) *The World's Major Fiber Crops, their Cultivation and Manuring.* Centre d'Etude de l'Azote, Zurich, Switzerland.

BHATT, J.G. AND APPUKUTTAN, E. (1971) Nutrient uptake in cotton in relation to plant architecture. *Plant and Soil* 35, 381–8.

BHOLE, B.D. AND VARADE, P.A. (1988) Economic response characteristics of cotton nitrogen, phosphate and potash under varying soil fertilities in Vertisols. *Field Crop Abstracts* 43, Abstr. 4363.

BLAIR, W.C. AND CURL, E.A. (1974) Influence of fertilization regimes on the rhizosphere microflora and Rhizoctonia disease of cotton seedlings. *Proceedings of the American Phytopathological Society* 1, 28–9.

BLANK, L.M. (1962) Fusarium wilt of cotton moves west. *Plant Disease Reporter* 46, 396.

BRAUD, M. (1974) The control of mineral nutrition of cotton by foliar analysis. *Coton et Fibres Tropicales* 29, 215.

BRAUD, M. (1987) Cotton. In: Martin-Prevel, P., Gagnard, J. and Gauter, P. (eds), *Plant Analysis as a Guide to the Nutrient Requirement of Temperate and Tropical Crops.* Lavoisier Abbonnements, Paris, pp. 499–512.

BUOL, S.W., HOLE, F.D. AND MCCRACKEN, R.J. (1989) *Soil Genesis and Classification*, 3rd edn. Iowa State University Press, Ames, Iowa.

BURHAN, H.O. (1969) Rotation responses of cotton in the Gezira. In: Siddig, M.A. and Hughes, L.C. (eds), *Cotton Growth in the Gezira Environment*, Agricultural Research Corporation, Wad Medani, Sudan, pp. 51–7.

BURHAN, H.O. (1971) Response of cotton to levels of fertilizer nitrogen in the Sudan Gezira. *Cotton Growing Review* 48, 116–24.

BURHAN, H.O. (1973) Level of pest control and the response of cotton to fertilizer N. *Cotton Growing Review* 50, 225–33.

BURHAN, H.O. AND BABIKIR, I.A. (1968) Investigation of nitrogen fertilization of cotton by tissue analysis. I. The relationship between nitrogen applied and the nitrate-N content of cotton petioles at different stages of growth. *Experimental Agriculture* 4, 311–23.

BURHAN, H.O. AND JACKSON, J.E. (1973) Effects of sowing date, nitrogenous fertilizer and insect pest content on cotton yield and its year-to-year variation in the Sudan Gezira. *Journal of Agricultural Science* 81, 481.

BURHAN, H.O. AND MANSI, M.G. (1970) Effects of N, P and K on yields of cotton in the Sudan Gezira. *Experimental Agriculture* 6, 279–86.

CAKMAK, I. AND MARSCHNER, H. (1986) Mechanism of phosphorus induced zinc deficiency in cotton. I. Zinc deficiency enhanced uptake of phosphorus. *Physiologia Plantarum* 68, 483–90.

CASSMAN, K.G., KERBY, T.A., ROBERTS, B.A., BRYANT, D.C. AND BROUDER, S.M. (1989) Differential response of two cotton cultivars to fertilizer and soil potassium. *Agronomy Journal* 81, 870–6.

CASSMAN, K.G., KERBY, T.A., ROBERTS, B.A. AND BROUDER, S.M. (1990) Reassessing potassium requirements of cotton for yield and fiber quality. In: Brown, J.M. (ed.), *Proceedings of the Beltwide Cotton Production Conference*. National Cotton Council of America, Memphis, Tennessee, pp. 60–4.

CHINA COOPERATIVE RESEARCH GROUP (1989) Study of the effect of multiple factors contributing to good quality and high yield of cotton. *China Cottons* 1, 27–30.

CHRISTENSEN, P.D., STITH, L.S. AND LYERLY P.J. (1954) The occurrence of Verticillium wilt in cotton as influenced by the level of salt in the soil. *Plant Disease Reporter* 38, 309–10.

CHRISTIANSEN, M.N. AND ROWLAND, R.A. (1986) Germination and stand establishment. In: Mauney, J.R. and Stewart, J.M. (eds), *Cotton Physiology*. The Cotton Foundation, Memphis, Tennessee, pp. 535–41.

CHRISTIDIS, B.G. AND HARRISON, G.J. (1955) *Cotton Growing Problems*. McGraw-Hill Book Company Inc., New York.

CLINE, M.G. (1944). Principles of soil sampling. *Soil Science* 58, 275–88.

CLINE, M.G. (1945) Methods of collecting and preparing soil samples. *Soil Science* 59, 3–5.

CONSTABLE, G.A., ROCHESTER, I.J., AND COOK, J.B. (1988) Zinc, copper, iron, manganese, and boron uptake by cotton on cracking clay soils of high pH. *Australian Journal of Experimental Agriculture* 28, 351–6.

COPE, J.T., Jr (1981) Effects of 50 years of fertilization with phosphorus and potassium on soil test levels and yields at six locations. *Soil Science Society of America Journal* 45, 342–7.

CRAWFORD, J.L. (1981) Nutrient deficiencies. In: Watkins, G.M. (ed.), *Compendium of Cotton Diseases*. American Phytopathological Society, St Paul, Minnesota, pp. 65–8.

DARGAN, K.S., SINGH, O.P. AND GUPTA, I.C. (1982) *Crop Production in Salt Affected Soils*. Oxford and IBH Publishing Company, New Delhi.

DAVIS, R.M., MENGE, J.A. AND ERWIN, D.C. (1979) Influence of *Glomus fasciculatus* and soil phosphorus on Verticillium wilt of cotton. *Phytopathology* 69, 453–6.

DESAI, D.B. AND WILES, A.B. (1976) Reaction of foliar application of microelements. In: Brown, J.M. (ed.), *Proceedings of the Beltwide Cotton Production Research Conference*. National Cotton Council Memphis, Tennessee, p. 21.

DONALD, L. (1964) Nutrient deficiencies in cotton. In: Sprague, H.B. (ed.), *Hunger Signs in Crops*, 3rd edn. McKay, New York, pp. 59–98.

EBBELS, D.L. (1975) Fusarium wilt of cotton, a review with special reference to Tanzania. *Cotton Growing Review* 52, 295–339.

EL-GINDI, A.Y., OTEIFA, B.A. AND KHADR, A.S. (1974) Interrelationships of *Rotylenchulus reniformis, Fusarium oxysporum* f.sp. *vasinfectum* and potassium nutrition of cotton, *Gossypium barbadense. Potash Review* 23, 1–5.

EL-ZIK, K.M. (1985) Integrated control of Verticillium wilt of cotton. *Plant Disease Reporter* 69, 1025–32.

EL-ZIK, K.M. (1986) Half a century dynamics and control of cotton disease, dynamics of cotton diseases and their control. In: Brown, J.M. (ed.), *Proceedings of the Beltwide Cotton Production Research Conference*. National Cotton Council Memphis, Tennessee, pp. 29–33.

ERGLE, D.R. (1953) Effect of low nitrogen and sulfur supply on their accumulation in the cotton plant. *Botanical Gazette* 114, 417–26.

ERGLE, D.R. AND EATON, F.M. (1951) Sulphur nutrition of cotton. *Plant Physiology* 26, 639–54.

EVANS, C.E., SCARSBROOK, C.E. AND ROUSE, R.D. (1970) Methods of applying nitrogen, phosphorus, and potassium for cotton. *Alabama Agricultural Experiment Station Bulletin* 403.

EZEKIEL, W.N., TAUBENHAUS, J.J. AND CARLYLE, E.C. (1930) Soil reaction effects on Phymatotrichum root rot. *Phytopathology* 20, 803–15.

FAO (1987) *Fertilizer Strategies*. FAO Land and Water Development Series, No. 10, FAO, Rome.

FOY, C.D. (1969) Differential tolerance of cotton varieties to excess Mn. *Agronomy Journal* 61, 690–4.

FOY, C.D., ARMINGER, W.H., FLEMING, A.L. AND LEWIS, C.F. (1967) Differential tolerance of cotton varieties to an acid soil high in exchangeable Al. *Agronomy Journal* 59, 415–18.

FRAPS, G.S. (1919) The chemical composition of the cotton plant. *Texas Agricultural Experiment Station Bulletin* 247.

FUNDERBURG, E.R. (1988) Effects of starter fertilizer on cotton yields in Mississippi. In: Brown, J.M. (ed.), *Proceedings of the Beltwide Cotton Production Research Conference*. Memphis, Tennessee, pp. 496–8.

GARDNER, B.R. AND TUCKER, T.C. (1967) Nitrogen effects on cotton. II. Soil and petiole analysis. *Soil Science Society of America Proceedings* 31, 785.

GUTHRIE, D.S. (1988) Cotton response to starter fertilizer applications. In: Brown, J.M. (ed.), *Proceedings of the Beltwide Cotton Production Research Conference*. National Cotton Council, Memphis, Tennessee, p. 496.

HAFEZ, A.A.R., STOUT, P.R. AND DEVAY, J.E. (1975) Potassium uptake by cotton in relation to Verticillium wilt. *Agronomy Journal* 67, 359–61

HALEVY, J. (1976) Growth rate and nutrient uptake of two cotton cultivars grown under irrigation. *Agronomy Journal* 68, 701.

HALEVY, J., MARANI, A. AND MARKOVITZ, T. (1987) Growth and NPK uptake of high-yielding cotton grown at different nitrogen levels in a permanent plot experiment. *Plant and Soil* 103, 39–44.

HEARN, A.B. (1981) Cotton nutrition. *Field Crop Abstracts* 34(1), 11–34.

HIATT, A.J. AND RAGLAND, J.L. (1963) Manganese toxicity of Burley tobacco. *Agronomy Journal* 55, 47–9.

HILLOCKS, R.J. AND CHINODYA, R. (1989) Relationship between Alternaria leaf spot and potassium deficiency causing premature defoliation of cotton. *Plant Pathology* 38, 502–8.

HINKLE, D.A. AND BROWN, A.L. (1968) Secondary nutrients and micronutrients. In: Elliot, F.C. and Porter, W.K. (eds), *Advances in Production and Utilization of Quality Cotton: Principles and Practices*, Iowa State University Press, Ames, Iowa, pp. 281–320.

HINKLE, D.A. AND STATEN, G. (1941) Fertilizer experiments with Acala cotton on irrigated soils. *New Mexico Agricultural Experiment Station Bulletin* 280.

HOCKING, P.J., REICOSKY, D.C. AND MEYER, W.S. (1987) Effects of intermittent waterlogging on the mineral nutrition of cotton. *Plant and Soil* 101, 211–21.

HODGES, S.C. AND BAKER, S. (1990) Effect of starter composition and placement on cotton in Georgia. In: Brown, J.M. (ed.), *Proceedings of the Beltiwde Cotton Production Research Conference*. National Cotton Council, Memphis, Tennessee, pp. 483–4.

HODGES, S.C., CRAWFORD, J.L. AND PLANK, C.O. (1989) Fertilizing cotton. *University of Georgia Cooperative Extension Service Bulletin* 966.

HONISCH, O. (1975) Boron deficiency of cotton in Zambia. *Cotton Growing Review* 52, 189–208.

HOOD, J.T. AND ENSMINGER, L.E. (1964) The effect of ammonium phosphate and other chemicals on the germination of cotton and wheat seeds. *Soil Science Society of America Proceedings* 28, 251–3.

HOWARD, D.D. AND ADAMS, F. (1965) Calcium requirement for penetration of subsoils by primary cotton roots. *Soil Science Society of America Proceedings* 29, 558–62.

HUBER, D.M. (1981) The use of fertilizers and organic amendments in the control of plant disease. In: Pimentel, D. (ed.), *Handbook of Pest Management* Vol. 1 CRC Press, Inc. Boca Raton, Florida. pp. 357–494.

HUBER, D.M. AND ARNY, D.C. (1985) Interactions of potassium with plant disease. In: Munson, R.C. (ed.), *Potassium in Agriculture*. American Society of Agronomy, Madison, Wisconsin.

INNES, N.L. (1971) Impressions of cotton production problems and research in India. *Cotton Growing Review* 48, 163–74.

JOHAM, H.E. (1951) The nutritional status of the cotton plant as indicated by tissue tests. *Plant Physiology* 36, 76–89.

JONES, E. (1972) Principles for using fertilizers to improve red ferrallitic soils in Uganda. *Experimental Agriculture* 8, 315–32.

JONES, E. (1976) Soil productivity. In: Arnold, M.H. (ed.), *Agricultural Research for Development*. Cambridge University Press, London, Chapter 3.

JONES, U.S. AND BARDSLEY, C.E. (1968) Phosphorus nutrition. In: Elliot, F.C., Hoover, M. and Porter, W.K., Jr (eds), *Advances in Production and Utilization of Quality Cotton: Principles and Practices*, Iowa State University Press, Ames, Iowa, pp. 213–54.

JORDAN, H.V. AND ENSMINGER, L.E. (1958) The role of sulfur in soil fertility. *Advances in Agronomy* 10, 407–34.

JORGENSON, E.C. (1984) Nematicides and nonconventional soil amendments in the management of root-knot nematode on cotton. *Journal of Nematology* 16, 154–8.

KALLINIS, T.L. AND VRETTA-KOUSKOLEKA, H. (1967) Molybdenum deficiency in cotton. *Soil Science Society of America Proceedings* 21, 507–9.

KAMPRATH, E.J. AND FOY, C.D. (1985) Lime–fertilizer–plant interactions in acid soils. In: Engelstad, O.P. (ed.), *Fertilizer Technology and Use*, 3rd edn. Soil Science Society of America, Madison, Wisconsisn, pp. 91–152.

KAMPRATH, E.J. AND WATSON, M.E. (1980) Conventional soil and tissue tests for assessing the phosphorus status of soils. In: Khasawneh, F.E., Sample, E.C. and Kamprath, E.J. (eds), *The Role of Phosphorus in Agriculture*. American Society of Agronomy, Madison, Wisconsin, pp. 433–70.

KAMPRATH, E.J. AND WELCH, C.D. (1968) Potassium nutrition. In: Elliot, F.C., Hoover, M. and Porter, W.K., Jr (eds), *Advances in Production and Utilization of Quality Cotton: Principles and Practices*. Iowa State University Press, Ames, Iowa, pp. 255–80.

KAMPRATH, E.J., NELSON, W.L. AND FITTS, J.W. (1957) Sulfur removal from soils by field crops. *Agronomy Journal* 49, 289–93.

KERBY, T.A. AND ADAMS, F. (1985) Potassium nutrition of cotton. In: Munson, R.C. (ed.), *Potassium in Agriculture*, American Society of Agronomy, Madison, Wisconsin, pp. 164–200.

KISSEL, D.E., SANDER, D.H. AND ELLIS, R., Jr (1985) Fertilizer–plant interactions in alkaline soils. In: Engelstad, O.P. (ed.), *Fertilizer Technology and Use*, 3rd edn. Soil Science Society of America, Madison, Wisconsin, pp. 153–96.

LE MARE, P.H. (1972) A long term experiment on soil fertility and cotton yield in Tanzania. *Experimental Agriculture* 8, 299–302.

LE MARE, P.H. (1977) Experiments on the effects of phosphorus on the manganese nutrition of plants. Part 3. The effect of calcium phosphorus ration on manganese in cotton grown in Bugami soil. *Plant and Soil* 47, 621.

LIU, W.E., PI, M.M. AND WANG, Y.H. (1986) Diagnosis of boron deficiency in cotton plant. *Field Crop Abstracts* 42, Abstr. 5588.

LOMBIN, G. AND MUSTAFA, S. (1981) Potassium response of cotton on some inceptisols and oxisols of Northern Nigeria. *Agronomy Journal* 73, 724–9.

LONGENECKER, D.E. AND HEFNER, J.J. (1961) Effect of fertility level, soil moisture and trace elements on the incidence of Verticillium wilt in Upland cotton. *Texas Agricultural Experiment Station Progress Report* 2175.

LUCKHARDT, R.L. AND ENSMINGER, L.E. (1968) Fertilizer use on cotton, 1953–1955. *Arkansas Agricultural Experiment Station Report Series* 78.

LUTRICK, M.C., PEACOCK, H.A. AND CORNELL, J.A. (1986) Nitrate monitoring for cotton lint production on a Typic Paleudult. *Agronomy Journal* 78, 1041–6.

MCBRYDE, J.B. (1891) A chemical study of the cotton plant. *Tennessee Agricultural Experiment Station Bulletin* 4, 120–45.

MCBRYDE, J.B. AND BEAL, H. (1896) Chemistry of cotton. In: True, A.C. (ed.) *The Cotton Plant US Department of Agriculture Bulletin* 3, 81–142.

MCCALL, E.R. AND JURGENS, J.F. (1951) Chemical composition of cotton. *Texas Research Journal* 21, 19–21.

MCCLUNG, A.C., de FREITAS, L.M.M., MIKKELSEN, D.S. AND LOTT, W.L. (1961) Cotton fertilization on Campo Cerrado soils, State of São Paulo, Brazil. *IBEC Research Institute Bulletin* 27.

MCHARGUE, J.S. (1926) Mineral constituents of the cotton plant. *Journal of the American Society of Agronomy* 18, 1076–83.

MALAVOLTA, E. (1985) Potassium status of tropical and subtropical region soils. In: Munson, R.C. (ed.), *Potassium in Agriculture*. American Society of Agronomy, Madison, Wisconsin, pp. 164–200.

MALIK, M.N.A., MAKHDOOM, M.I., SHAH, I.H. AND MALIK, U.M. (1987) Amelioration of nitrogen deficiency through foliar feeding in cotton. *Field Crop Abstracts* 42, Abstr. 4611.

MAMEDOV, M. AND ISMAILOV, T. (1989) Nitrification inhibitors and nitrogen fertilizers. In: *Field Crop Abstracts* 42, Abstr. 5587.

MAPLES, R., KEOGH, J.C. AND SABBE, W.E. (1977) Nitrate monitoring for cotton production in Loring–Calloway silt loam. *Arkansas Agricultural Experiment Station Bulletin* 825.

MARSCHNER, H. (1986) *Mineral Nutrition of Higher Plants*. Academic Press, New York.

MATOCHA, J.E., MOSTAGHIMI, S. AND HOPPER, F.L. (1986) Effect of soil and plant treatments on Phymatotrichum root rot. In: Brown, J.M. (ed.), *Proceedings of the Beltwide Cotton Production Research Conference*. National Cotton Council, Memphis, Tennessee, pp. 24–6.

MEREDOV, K.M. (1987) Efficiency of local fertilizers of Turkmenistan. In: *Field Crop Abstracts* 42, Abstr. 1135.

MIKKELSEN, D.S., de FREITAS, L.M.M. AND McCLUNG, A.C. (1963) Effects of liming and fertilizing cotton corn soybeans on Campo Cerrado soils, State of São Paulo, Brazil. *IRI Research Institute, Inc. Bulletin* 271, New York, USA.

MILES, L.E. (1936) Effect of potash fertilizers on cotton wilt. *Mississippi Agricultural Experiment Station Technical Bulletin* 23.

MILEY, W.N. (1982) Plant tissue analysis as a cotton production management guide. In: Brown, J.M. (ed.), *Proceedings of the Beltwide Cotton Production Conference*. National Cotton Council, Memphis, Tennessee, pp. 56–60.

MILEY, W.N., HARDY, G.W., STURGIS, M.B. AND SEDBERRY, J.E., Jr (1969) Influence of boron nitrogen and potassium on yield, nutrient uptake and abnormalities of cotton. *Agronomy Journal* 61, 9–13.

MILLER, J.W. (1969) The effect of soil moisture and plant nutrition on the Cercospora–Alternaria leaf blight complex of cotton in Missouri. *Phytopathology* 59, 767–9.

MOSTAGHIMI, S., MATOCHA, J.E., BRANICK, P.S., EL-ZIK, K. AND BIRD, L.S. (1987) Suppression of Phymatotrichum root rot through chemical and biological treatments of soil and plant. In: Brown, J.M. (ed.), *Proceedings of the Beltwide Cotton Production Research Conference*. National Cotton Council, Memphis, Tennessee, pp. 35–7.

MULLINS, G.L. AND BURMESTER, C.H. (1990) Dry matter, nitrogen phosphorus and potassium accumulation by four cotton varieties. *Agronomy Journal* 82, 729–36.

MUNRO, J.M. (1987) *Cotton*, 2nd edn. Longman Scientific and Technical, Harlow Essex, UK.

NAIM, M.S. AND SHAABAN, A.S. (1965) Relation of nitrogen fertilizers to growth and vigor of *Fusarium* wilted cotton. *Phytopathologia Mediterranea* 4, 145–53.

NEAL, D.C. AND SINCLAIR, J.B. (1960) Nitrogen supplement in the Louisiana–Mississippi River Delta as a possible control for Verticillium wilt of cotton. *Plant Disease Reporter* 44, 478

OLSON, L.C. AND BLEDSOE, R.P. (1942) The chemical composition of the cotton plant and the uptake of nutrients at different stages of growth. *Georgia Agricultural Experiment Station Bulletin* 222, 16.

OTEIFA, B.A. (1953) Development of the root knot nematode, *Meloidogyne incognita*, as affected by potassium nutrition of the host. *Phytopathology* 43, 171–4.

OTEIFA, B.A. AND DIAB, K.A. (1961) Significance of potassium fertilization in nematode infested cotton fields. *Plant Disease Reporter* 45, 932.

OTEIFA, B.A. AND EL-GINDI, A.Y. (1976) Potassium nutrition of cotton, *Gossypium barbadense*, in relation to nematode infection by *Meloidogyne incognita* and *Rotylenchulus reniformis*. In: *Fertilizer Use and Plant Health. Proceedings Colloquium International Potash Institute* 12, 301–6.

OTEIFA, B.A., EL-GINDI, A.Y. AND DIAB, K.A. (1965) Cotton yield and population dynamics of the stunt nematode, *Tylenchorhynchus latus* under mineral fertilization trials. *Potash Review* 23, 1–7.

PAGE, A.L. AND BINGHAM, F.T. (1965) Potassium–magnesium interrelationships in cotton. *California Agriculture* 19 (11), 6–7.

PAGE, A.L., MILLER, R.H. AND KEENEY, D.R. (eds) (1982) *Methods of Soil Analysis. Part 2*, 2nd edn. Agronomy Monograph 9, American Society of Agronomy, Madison, Wisconsin.

PERICHIAPPAN, P., GOPALASWAMY, N. AND PANCHANATHAN, P. (1987) Effect of times of fertilizer application and weed management on irrigated cotton. *Field Crop Abstracts* 42, Abstr. 2872.

PETERSON, R.G. AND CALVIN, L.D. (1965) Sampling. In: Black, C.A., Evans, D.D., White, J.L., Ensminger, L.E. and Clark, F.E. (eds), Methods of *Soil Analysis. Part 1*. Agronomy Monograph No. 9, American Society of Agronomy, Madison, Wisconsin, pp. 54–72.

PETTIET, J.V. (1988) The influence of exchangeable magnesium on potassium uptake in cotton grown on Mississippi Delta soils. In: *Proceedings of the Beltwide Cotton Production Research Conference*, National Cotton Council, Memphis, Tennessee, pp. 517–18.

PLANK, C.O. (1989) *Plant Analysis Handbook for Georgia*. University of Georgia Cooperative Extension Service, Athens, Georgia, pp. 28–30.

PLANK, C.O., HODGES, S.C. AND SEGARS, W.I. (1989) Field crops. In: Plank, C.O. (ed.), *Soil Test Handbook for Georgia*. University of Georgia Cooperative Extension Service, Athens, Georgia, pp. 85–134.

PORTER, H.G., BROWN, G.S., BYRNE, A.O., JERNIGAN, J.E., NICHOLS, G.E. AND RICHMOND, T. R. (1974) Cotton in the Soviet Union. *USDA Foreign Agricultural Service* M-254.

PRENTICE, A.N. (1972) *Cotton with Special Reference to Africa*. Longman, London.

PRESLEY, J.T. AND BIRD, L.S. (1968) Diseases and their control In: Elliot, F.C., Hoover, M., and Porter, W.K. Jr (eds), *Advances in Production and Utilization of Quality Cotton: Principles and Practices*. Iowa State University Press, Ames, Iowa, pp. 347–66.

PRESLEY, J.T. AND DICK, J.B. (1951) Fertilizer and weather affect Verticillium wilt. *Mississippi Farm Research* 14, 1–6.

PRESLEY, J.T. AND LEONARD, A.O. (1948) The effect of calcium and other ions on the early development of the radicle of cotton seedlings. *Plant Physiology* 23, 516–25.

PUENTE, F. (1965) Some effects of soil temperature and phosphorus and calcium levels on cotton seedling growth. *Dissertation Abstracts* 26, 4157.

RAMASAMI, R. AND SHANMUGAM, N. (1976) Effect of nutrients on the incidence of *Rhizoctonia* seedling disease of cotton. *Indian Phytopathology* 29, 465–6.

RANDALL, G.W., WELLS, K.L. AND HANAWAY, J.J. (1985) Modern techniques in fertilizer application. In Engelstad, O.P. (ed.), *Fertilizer Technology and Use*, 3rd edn. Soil Science Society of America, Madison, Wisconsin, pp. 521–60.

RANNEY, C.D. (1962) Effects of nitrogen source and rate on the development of Verticillium wilt of cotton. *Phytopathology* 52, 38–41.

RAST, L.E. (1924) Control of cotton wilt by the use of potash fertilizers. *Journal of the American Society of Agronomy* 14, 222–3.

RAY, H.E. AND TUCKER, T.C. (1962) Soil and petiole analysis can pinpoint cotton's nitrogen needs. *Arizona Cooperative Extension Service, Agricultural Experiment Station Folder* 97.

REISENAUER, H.M., QUICK, J., VOSS, R.E. AND BROWN, L.A. (1978) Soil test interpretation guides. In: Reisenauer, H.M. (ed.), *Soil and Plant Tissue Testing in California*. Bulletin 1879, Division of Agricultural Sciences, University of California, Berkeley.

REUTER, D.J. (1986) Temperate and sub-tropical crops. In: Reuter, D.J. and Robinson, J.B. (eds), *Plant Analysis: an Interpretation Manual*. Inkata Press, Melbourne, Sydney, pp. 48–50.

ROLFS, F.M. (1915) Angular leaf spot of cotton. *South Carolina Agricultural Experiment Station Bulletin* 184.

ROTHWELL, A., BRYDON, J.W., KNIGHT, H. AND COXE, B.J. (1967) Boron deficiency of cotton in Zambia. *Cotton Growing Review* 44, 23–8.

ROUSE, R.D. (1960). Potassium requirements of crops on Alabama soils. *Alabama Agricultural Experiment Station Bulletin* 324.

RUDOLPH, B.A. AND HARRISON, G.J. (1939) Attempts to control Verticillium wilt of cotton and beeding for resistance. *Phytopathology* 29, 752.

RUSSELL, E.W. (1973) *Soil Conditions and Plant Growth*, 10th edn. Longman, London.

SABBE, W.E. AND MACKENZIE, A.J. (1973) Plant analysis as an aid to cotton fertilization. In: Walsh, L.M. and Beaton, J. (eds), *Soil Testing and Plant Analysis*. American Society of Agronomy, Madison, Wisconsin, pp. 299–313.

SADASIVAN, T.S. (1965) Effect of mineral nutrients on soil microorganisms and plant disease. In: Baker, K.F. and Snyder, W.C. (eds), *Ecology of Soil-borne Plant Pathogens, Prelude to Biological Control*. University of California Press, Berkeley, pp. 460–70.

SANCHEZ, P.A. (1977) Advances in the management of Oxisols and Ultisols in tropical South America. In: *Proceedings Soil Environment and Fertility Management in Intensive Agriculture*. Society of the Science of Soil and Manure, Tokyo, Japan. pp. 535–68.

SANCHEZ, P.A. AND UEHARA, G. (1980) Management considerations for acid soils with high phosphorus fixation capacity. In: Khasawneh, F.E., Sample, E.C. and Kamprath, E.J. (eds), *The Role of Phosphorus in Agriculture*. American Society of Agronomy, Madison, Wisconsin, pp. 471–514.

SAWHNEY, K. AND SIKKA, S.M. (1960) Agronomy. In: *Cotton in India*. Indian Central Cotton Committee, Bombay, pp. 135–43.

SCARSBROOK, C.E., PEARSON, R.W. AND BENNETT, O.C. (1959) The interaction of N and moisture levels on cotton yields and other characteristics. *Agronomy Journal* 51, 718–21.

SHAO, F.M. AND FOY, C.D. (1982) Interaction of soil manganese and reaction of cotton to Verticillium wilt and Rhizoctonia root rot. *Communications in Soil Science and Plant Analysis* 13, 21–38.

SHUKLA, H.C. AND RAJ, H. (1987) Influence of genotypical variability on zinc response in cotton. *Plant and Soil* 104, 151–4.

SKINNER, J.J., NELSON, W.L. AND COLLINS, E.R. (1946) Potash and lime requirements of cotton grown in rotation with peanuts. *Agronomy Journal* 38, 142–51.

SMILEY, R.W., COOK, R.J. AND PAPENDICK, R.I. (1970) Anhydrous ammonia as a soil fungicide against *Fusarium* and fungicidal activity in the ammonia retention zone. *Phytopathology* 60, 1227–32.

SMITH, G.S., RONCADORI, R.W. AND HUSSEY, R.S. (1985) Development of *Meloidogyne incognita* on cotton as affected by the endomycorrhizal fungus, *Glomus intraradices*, and phosphorus. *Journal of Nematology* 17, 514

SMITHSON, J.B. (1972) Differential sensitivity to boron in cotton in the Northeastern State of Nigeria. *Cotton Growing Review* 49, 350–3.

SMITHSON, J.B. AND HEATHCOTE, R.G. (1977) Effect of rate and time of nitrogen application on yields of cotton in northern Nigeria. *Experimental Agriculture* 13, 1–8.

STANDLEY, J.P., CLARKE, E.A., OCKERBY, S.E., AND MAYER, D.G., (1988) Modification by crop rotations of the nitrogen fertiliser requirements for irrigated cotton and soil test calibration. *Queensland Journal of Agriculture and Animal Science* 45, 129–39.

STANFORD, G. AND JORDAN, H.V. (1966) Sulfur requirements of sugar, fiber, and oil crops. *Soil Science* 101, 258–66.

TAUBENHAUS, J.J., EZEKIEL, W.N. AND KILLOUGH, D.T. (1928) Relation of cotton root rot and Fusarium wilt to the acidity and alkalinity of the soil. *Texas Agricultural Experiment Station Bulletin* 389.

THENABADU, M.W. (1968) Magnesium–sodium interactions affecting the uptake and distribution of potassium and calcium by cotton. *Plant and Soil* 29, 132–43.

THOMAS, G.W. AND HANWAY, J.J. (1968) Determining fertilizer needs. In: Nelson, L.B. (ed.), *Changing Patterns of Fertilizer Use*. Soil Science Society of America, Madison, Wisconsin, pp. 39–45.

TISDALE, H.B. AND DICK, J.B. (1939) The development of wilt in a wilt-resistant and in a wilt-susceptible variety of cotton as affected by N–P–K ratio in fertilizer. *Proceedings of the Soil Science Society of America* 4, 333–4.

TISDALE, H.B. AND DICK, J.B. (1942) Cotton wilt in Alabama as affected by potash supplements and as related to varietal behavior and other important agronomic problems. *Journal of the American Society of Agronomy* 34, 405–25.

TISDALE, S.L., NELSON, W.L. AND BEATON, J.D. (1985) *Soil Fertility and Fertilizers*, 4th edn. Macmillan Publishing Company, New York.

TSAI, H.Y. AND BIRD, L.S. (1975) Microbiology of host pathogen interactions for resistance to seedling disease and multiadversity resistance in cotton. In: Brown, J.M. (ed.), *Proccedings of the Beltwide Cotton Production Research Conference*. National Cotton Council, Memphis, Tennessee, pp. 39–45.

TUCKER, T.C. AND TUCKER, B.B. (1968) Nitrogen nutrition. In: Elliot, F.C., Hoover, M. and Porter, W.K., Jr. (eds), *Advances in Production and Utilization of Quality Cotton: Principles and Practices*. Iowa State University Press, Ames, Iowa, pp. 183–212.

TUPPER, G. AND EBELHAR, M.W. (1990) Fertilizer placement: past, present and future? In: Brown, J.M. (ed.), *Proceedings of the Beltwide Cotton Production Conference*. National Cotton Council of America, Memphis, Tennessee, pp. 57–60.

VILELA, L. AND RITCHEY, K.D. (1985) Potassium in intensive cropping systems on highly weathered soils. In: Munson, R.D. (ed.), *Potassium in Agriculture*. American Society of Agronomy, Madison, Wisconsin, pp. 1155–75.

VOLK, N.J. (1946) Nutritional factors affecting cotton rust. *Journal of the American Society of Agronomy* 38, 6–12.

WADLEIGH, C.H. (1944) Growth status of the cotton plant as influenced by the supply of nitrogen. *Arkansas Agricultural Experiment Station Bulletin* 446.

WALKER, M.W. (1930) Potash in relation to cotton wilt. *Florida Agricultural Experiment Station Bulletin* 213.

WALSH, L.M. AND BEATON, J.D. (eds) (1973) *Soil Testing and Plant Analysis*. Soil Science Society of America, Madison, Wisconsin.

WEIR, W.L., GARBER, R.H., STAPELTON, J.J., GASTELUM, R.F., WAKEMAN, R.A. AND DEVAY, J.E. (1989) Control of potassium deficiency symptom in cotton by soil solarization. *California Agriculture* 43(3), 26–28.

WHITNEY, D.A., COPE, J.T. AND WELCH, L.F. (1985) Prescribing soil and crop nutrient needs. In: Engelstad, O.P. (ed.), *Fertilizer Technology and Use*, 3rd edn. Soil Science Society of America, Madison, Wisconsin, pp. 25–52.

WILES, A.B. (1959) Calcium deficiency in cotton seedlings. *Plant Disease Reporter* 43, 365–7.

YAN, R.X. AND ZHU, Y.C. (1986) Economic use of nitrogen fertilizer for cotton cultivar Yanmian 48. *Field Crop Abstracts* 42, Abstr. 380.

YOUNG, V.H., FULTON, N.D. AND WADDLE, B.A. (1959) Factors affecting the incidence and severity of Verticillium wilt disease of cotton. *Arkansas Agricultural Experiment Station Bulletin* 612.

ZYNGAS, J.P. (1962) The effect of plant nutrients and antagonistic microorganisms on the damping-off of cotton seedlings caused by *Rhizoctonia solani* Kuhn. *Dissertation Abstracts* 23, 3587.

Index